Hartelijk d

al je goede diensten

Egbert

A. J. van Dijk
Woonschip „Goedereede"
IJsbaanpad, steiger 2
AMSTERDAM-Z. II
16 dec. 1980

Technology and the Future

Egbert Schuurman

Technology and the Future

A Philosophical Challenge

Herbert Donald Morton, Translator

Wedge Publishing Foundation
Toronto
1980

This book originally appeared in The Netherlands under the title *Techniek en toekomst: Confrontatie met wijsgerige beschouwingen*, published by Van Gorcum & Comp. N.V. in Assen, 1972. ISBN 90-232-09907.

Cover design: John Burnett, Toronto
Typesetting: Compositor Associates Limited, Toronto
Printing: John Deyell Company, Lindsay, Ontario

ISBN 0-88906-111-4

Contents

Table of Abbreviations

AuM   *Automat und Mensch*. Karl Steinbuch.

C   *Cybernetics*. Norbert Wiener.

CT   "Computing Machinery and Intelligence," in: *Computers and Thought*. Alan Mathison Turing.

CWF   "Scientist and Decision Making," in: *Computers and the World of the Future*. Norbert Wiener.

DiG   *Die informierte Gesellschaft*. Karl Steinbuch.

DVS   *Die Vollkommene Schöpfung*. Friedrich Georg Jünger.

EiM   *Einführung in die Metaphysik*. Martin Heidegger.

FP   *Falsch programmiert*. Karl Steinbuch.

Gcl   *Gelassenheit*. Martin Heidegger.

GG   *God and Golem, Inc.* Norbert Wiener.

H   *The Human Use of Human Beings*. Norbert Wiener.

HdH   *Hebel der Hausfreund*. Martin Heidegger.

Hol   *Holzwege*. Martin Heidegger.

IAM   *Why I Am a Mathematician*. Norbert Wiener.

IuD   *Identität und Differenz*. Martin Heidegger.

K   *Kybernetik*. Igor Andreevic Poletaev.

KiS   *Kybernetik in philosophischer Sicht*. Georg Klaus.

KuE   *Kybernetik und Erkenntnistheorie*. Georg Klaus.

KuG   *Kybernetik und Gesellschaft*. Georg Klaus.

MuE   *Maschine und Eigentum*. Friedrich Georg Jünger.

P2   *Programm 2000*. Karl Steinbuch.

PBR   *Progress in Brain Research*, Volume 2: *Nerve, Brain and Memory Models*. Norbert Wiener and J.P. Schadé, eds.

PdT   *Die Perfektion der Technik*. Friedrich Georg Jünger.

PH   *Platons Lehre von der Wahrheit; mit einem Brief über den Humanismus*. Martin Heidegger.

PuK   *Philosophie und Kybernetik*. Karl Steinbuch and Simon Moser, eds.

SGF   *Sinn, Gesetz und Fortschritt in der Geschichte*. Georg Klaus and Hans Schulze.

SiS   *Spieltheorie in Philosophischer Sicht*. Georg Klaus.

SM   "Some Moral and Technical Consequences of Automation," in: *Science*, Volume 131 (1960). Norbert Wiener.

SP   *Selected Papers of Norbert Wiener*.

SuK   *Sprache und Kalkül*. Friedrich Georg Jünger.

SuT   *Streit um die Technik*. Friedrich Dessauer.

SuZ   *Sein und Zeit*. Martin Heidegger.

SvG   *Der Satz vom Grund*. Martin Heidegger.

TdW   *Die Technisierung der Welt*. Hermann Josef Meyer.

TiS   "Technik und Kybernetik: Die Technik im Selbstverständnis des heutigen Menschen," in: *Menschliche Existenz und moderne Welt*, Volume 1. Hermann Josef Meyer.

TO   "The Technological Order," in: *Technology and Culture*, Volume 3 (1962). Jacques Ellul.

TS   *The Technological Society*. Jacques Ellul.

TuK   *Die Technik und die Kehre*. Martin Heidegger.

TuW   *Technik und Wissenschaft als "Ideologie."* Jürgen Habermas.

UzS   *Unterwegs zur Sprache*. Martin Heidegger.

VuA   *Vorträge und Aufsätze*. Martin Heidegger.

WhD   *Was heisst Denken?* Martin Heidegger.

Foreword

The book you now hold in your hand, which is a study of technology and the future, is important first of all because it is the work of a *practicing philosopher*. Professor Schuurman has attempted to account for the incredibly diverse, strongly conflicting assessments of modern technology found in our time within the circle of western thought.

In the second place, this study is important for the no less weighty reason that the author has gone about his work as a *christian philosopher*. He has explicitly spelled out the presuppositions that prevailed in his inquiry because of his christian world-and-life convictions.

Given the tumultuous state of the current debate about modern western technology, the value of a philosopher's candor in stating his own religiously determined presuppositions can scarcely be overestimated. As this study shows, the controversy now raging discloses a whole scale of gradations precisely in this area of fundamental perspectives. We are confronted today with conflicting views of modern technology: on the one hand technology is said to be an "angel of light," that is, a reliable guide to tomorrow's world, a panacea for virtually all our social ills, while on the other hand it is branded a "demon," the sinister fruit of a western compulsion to rationalize and control everything, a conquering tyrant that enslaves all who touch it, carrying them off captive in an updated version of some ancient triumphal procession.

Technology as angel and as demon – these are the views that necessarily spring from a human religious dichotomy. They are views that arise from a civilization that chronically nourishes within its heart a bitter and incessant dialectic between bending technology to its will and bending the knee before it, between utilizing technology as an expression of human freedom and smashing technology as a threat to that same autonomous human freedom. It is exhilarating and extremely informative to follow Professor Schuurman step by step in this book and to learn with him that all the variant lines of current western philosophical reflection on the phenomenon of modern technology (excellent synopses are provided) can be connected to these two poles of western humanism and, indeed, have their origin in the interplay between these two poles. This is true for *Karl Steinbuch*, who calls upon us to place our social destiny in the hands of natural scientists and technicians. It is true for *Norbert Wiener*, the father of modern cybernetics, who is a control-optimist and freedom-pessimist in one. It is certainly true for *Martin Heidegger*, who challenges the reduction of Being to being in all the processes of modern technology. And – lest we forget – it is true for *Georg Klaus*, a representative of present-day marxism in the debate about technology and the future, a thinker who finds only in the communist societal configuration the possibility of setting the necessary goals for technological development – or perhaps the gumption required to get on with that task.

It is in the first and last chapters particularly that Professor Schuurman develops his own view of the phenomenon of modern technology. Here the inquiry concerning the *meaning* of technology assumes the central place, since it is only from the womb of this question that a liberating perspective for technological development in our society can be born. But the author does not construe that meaning in vague, personally tinted "ethical" terms to be imputed to technology as some neutral or objectively given thing; no, from the outset he regards technology itself as being comprehensible only within the framework of the grand coherence of meaning that was laid down in this world of many-colored developmental potential by the Creator of heaven and earth. It is for this reason that he can boldly and openly assert that man is called to the task of technology. He explains this task as the mandate to open up the nature side of creation with the help of tools. It is for this reason, too, that he can sound the clear warning that as long as secularized motives remain in control, technological development will assume an ever more catastrophic character. The meaning of technology, which we wanted to ignore, then becomes a judgment upon our society, as if some implacable, ineluctable doom were descending on our world.

In certain respects this study raises additional questions. For example, what *is* the connection between technological development and social structure? Had he looked more deeply into their coherence, would Professor Schuurman's important case for the liberation of technology not have been rendered more difficult on the one hand, yet invested with even more persuasive power on the other? In raising such a question as this, however, it must be remembered that this study does not pretend to answer all the questions related to the future of technology and society. Rather, it leads us to the edge of these questions and helps us to pose them in the correct way. And that is precisely what a philosophy of society can and ought to do in order to be of service.

Finally, Professor Schuurman is one of those rare individuals who are authentically equipped to address the issues of this study. In him the engineer and the philosopher meet. He studied under Professor Hendrik Van Riessen and now holds the chairs of reformational philosophy at the technical universities of Delft and Eindhoven in the Netherlands, which are endowed by the Foundation for Private Chairs of Calvinistic Philosophy founded in 1947 by Professors Herman Dooyeweerd and D. H. T. Vollenhoven.

It may be useful to note that in another, briefer study Professor Schuurman has assessed the critique of contemporary technological culture made by Roszak, Reich, and other figures of the recent so-called counterculture. Entitled "Reflections on the technological-scientific culture," this essay has been published in English in his book *Reflections on the Technological Society* (Toronto: Wedge, 1977). At present he is preparing a new book to be entitled *Responsibility in the Technological Society*, to be published early in the eighties.

I do not doubt that the commanding scientific study presented here will have an important impact on the American and Canadian public. It is lucid and comprehensible even in the more difficult passages. The case is presented in compelling terms and argued with appropriate urgency.

Tua res agitur: the business of the hour is our own.

Bob Goudzwaard

Free University
Amsterdam

Preface

Reflection on technology (the philosophy of technology) is still in its youth. Since the second world war, however, the need for such reflection has gained a special urgency: people are confronted by modern technology everywhere, by its complex development, by the treacherous problems it poses for humanity, nature, and society. It was my wish to respond to this need and to nurture reflection on technology.

In this study the main currents of recent philosophical thought about modern technology are presented and critically analyzed. My own philosophical analysis of modern technology is also presented. On the basis of this analysis and with the benefit of what others have learned from their contemplation of technology, this study concludes with the suggestion of a liberating perspective for the future, for I believe that there is a way of deliverance.

A very extensive bibliography on the philosophy of technology was published in the original Dutch edition of this study, *Techniek en toekomst: Confrontatie met wijsgerige beschouwingen* (Assen: Van Gorcum, 1972, pp. 456-534). An important complementary bibliography appears in *Technology and Culture: The International Quarterly of the Society for the History of Technology* (Vol. 14, No. 2, Part II [April 1973], University of Chicago Press). The former bibliography is perhaps weighted more heavily toward continental sources, the latter toward Anglo-Saxon ones.

I would like to express the warm appreciation I have for my translator, Herbert Donald Morton. With accuracy, enthusiasm, persistence, and in a memorable spirit of amicable collegiality, he rendered even the more difficult scientific passages into plain English. He and I would both like to thank Harry Van Dyke for reading the translation in manuscript and making many wise and helpful suggestions. We would also like to thank the editor, Theodore Plantinga, for making many excellent revisions to enhance the simplicity of the translation. Further, I am indebted to my fellow philosopher and engineer, J. C. Plooy, for preparing the indexes of persons and subjects appearing at the end of the book. Finally, I am most grateful to Josina B. Zylstra, Executive Director of Wedge Publishing Foundation, and her staff for their invaluable help and support in bringing about the English-language edition of my book. Institutionally I am greatly indebted to the Free University of Amsterdam for the financial support it provided for the translation costs.

May this study serve to promote critical and creative reflection on modern technology and its development.

Egbert Schuurman

Central Interfaculty
Free University
Amsterdam

Introduction

The Subject and Plan of this Study

Technology occupies an important place among the realities of modern, dynamic culture. That modern technology is the foundation upon which modern culture is being built is increasingly widely recognized. The future of our culture will assuredly be controlled and determined by technology to a significant degree. In fact, a survey of various futurologies indicates that their assessment of technology is inevitably the hinge upon which their expectations of the future swing open or closed.

Accordingly, reflection on the future and humanity's place in it cannot bypass technology and its development. Even less can it steer clear of the motives that force themselves upon humanity in and through technology. Therefore it is both pertinent and necessary to draw the deeper backgrounds openly into the process of reflection. We can no longer be satisfied, as we so often have been, to merely weigh the advantages and disadvantages of this or that development as we form conclusions about the direction in which technology should presumably be made to go. Such a limited approach has its value, of course, but the method is inadequate to the task of resolving the problems raised by technology as such.

The possibilities and consequences of technological development are so impressive, and everything is now so stamped by technology, and the pace of the development is so breathtaking, that right alongside its promise there appear ghastly threats to humanity and the future. Technology is becoming a planetary power. It is laying hold of the foundations of human history and imparting to that history an extremely unstable character. Reflection on the origin, unity, and coherence of technology, accordingly, is imperative at this time. The farther technology advances, the more urgent our need to understand its meaning becomes.

In this study I have undertaken to examine a great deal of the most recent reflection on the problem of technology and its future. I have done so in the belief that a critical analysis of so much reflection can serve to clarify the basic problems that have now come to be associated with technology and its development. It is my aspiration and purpose to contribute to the current debate about technology at its deepest, religious level.

The title of this study, *Technology and the Future: A Philosophical Challenge*, serves very well to indicate that the subject matter addressed here is broader than technology as such; it includes much that is affected by technology, and it embraces the questions of meaning and perspective for the future. The title also serves to bring the remarkably diverse views of the various philosophers examined here under an appropriate rubric.

The roots of the current problems of our culture lie in the past. It would therefore be improper to be silent about that past. Certainly we can say nothing meaningful about the future without making some reference to the past. Nevertheless, in discussing the history of technological development and its influence upon humanity and the whole of culture, I have been as brief as possible. Only the main lines have been presented – either as required by the exigencies of the confrontation with other thinkers or as needed for laying bare the depths of certain problems. The accent has been placed on the present state of affairs and on the future. It is today's problems that concern me in this study. The question is: What does technological development have in store for humanity and culture? Yes, what is to be expected? And what is to be done?

Chapter 1, which serves as an introduction to the challenge to diverse views of technology in the following chapters, offers a scientific analysis of technology. Far from being a summary of everything pertaining to technology, this analysis addresses the questions what technology is; what its particular meaning is; what the ground and coherence of the multiplicity of technological things and phenomena may be; and what tendencies in technological development may be significant for the future. Particular attention is paid to the relation between science and technology and to the significance and possibilities of the computer.

The discussion of the various philosophical views of technology is aimed at giving a fair overall impression of each view and of their range when considered together. They are therefore presented in appropriate detail. In the confrontation with present-day thinkers, special emphasis is placed on the problems connected with the theme of technology and the future; to the possibilities these thinkers perceive for resolving the problems raised by technology; and above all to the motives (expressed or implied) which run through their views to make them a coherent whole.

The various philosophical outlooks on technology are indeed rather divergent. Yet a *global* distinction may be made between the transcendentalists and the positivists. This distinction, the validity of which is defended at the beginning of Chapter 2, is valuable in a philosophical sense. For the transcendentalists, freedom reigns; freedom behind or above the experience at hand is the mainspring or the orientation point – or both – of their philosophies. For the positivists, the ground of philosophy *is* the experience at hand; they take their bearings from the possibilities of technology as such.

It is clear from the nature of this distinction that the transcendentalists are decidedly ill-disposed toward technology, and that the positivists rather fancy it. The transcendentalists discern a conflict between humankind and technology; they believe that technology threatens human freedom. The positivists, however, see in technology the affirmation of human power and the assurance of cultural progress.

In Chapter 2, the views of the transcendentalists Jünger, Heidegger, Ellul, and Meyer are presented in detail and examined critically. Here and there others are brought into the discussion, such as Dessauer, who, because of his favorable attitude toward technology, occupies a unique place among the transcendentalists. I have tried to point out both the areas of agreement shared by all the transcendentalists and the differences that divide them. My criticism of their outlook is tempered by an acceptance of some of their ideas.

In Chapter 3, I deal with the positivists, especially Wiener, Steinbuch, and Klaus. I also discuss Turing. Moreover, in discussing Steinbuch I deal as well with the neomarxists Marcuse and Habermas, whose views are at variance not only with the transcendentalists but also with the positivists dealt with in this study.

In contrast to the transcendentalists, the positivists regard technology as the expression of human power and greatness. The nineteenth-century belief in progress lives on in their hopeful expectations of the possibilities of cybernetics (steersmanship, the science of guidance and control), of which the computer is now the most surprising and promising result. The future, they believe, will be what it is thanks to cybernetics. It must not escape our attention that this optimistic view has its reverse side, and that some of them know it. Their conception engenders problems which rise to challenge their optimism. Their great expertise in the field of technology, and especially in the area of cybernetics as such, is accompanied by an uncertain depth of philosophical insight. The marxist Klaus, a capable philosopher, is an exception.

In the second and third chapters, each of the philosophical views to be examined is presented in synopsis. A critical analysis follows each synopsis. I am principally concerned with the question what technology is and with the implications of the various philosophical views for the future. The place of humankind is of central significance. The view of science and of its relation to technology is also of fundamental importance. I shall argue that science, served by the resources of modern technology, focuses on and transforms the countenance of praxis. In fact, the alteration has been so drastic that the transcendentalists have become terrified of technological development. They perceive in it the alienation of humanity from nature first, then from culture and one's neighbor, and finally from humanity itself. The positivists, however, believe that people, equipped with the technological-scientific method, are able every day and in every way to bring the future a little more under control. Humanity's futurology is inspired by the success of this method in modern technology – a method which includes the principles of cybernetics and which is believed by the positivists to be worthy of universal application.

On the basis of the philosophical analysis of modern technology presented in Chapter 1 and with the benefit of the confrontation with the transcendentalists and the positivists in Chapters 2 and 3, the final chapter is devoted to a thetical treatment, from a christian philosophical perspective, of the problems raised. However much the transcendentalists and the positivists may differ, they are agreed in advocating an autonomous philosophy which they eventually incorporate into a synthetic view of life. Here lies the deepest ground of their powerlessness (the transcendentalists) or their presumption (the positivists) when it comes to indicating a perspective for the future. I believe that the real problems are raised for discussion only when the ground for the crisis of meaning in our culture is authentically recognized. That ground is secularization, defined as the acceptance of a closed world. It can appear in various forms. It works, in opposite directions, through the two categories of thought that I have examined, and it results in contradictory evaluations of technological development.

With the help of a christian philosophy, I have undertaken to analyze technological development philosophically and to address in the broader sense the problems of humanity and technology and of technology and the future. I have drawn the motives of people in technology into the discussion in order to examine the extent to which these motives can distort technological development. I have contended that a liberating perspective for technological development is possible whenever philosophical contemplation proceeds from the meaning of creation and its future.

The similarity of my own thinking to that of the thinkers examined is rooted in the fact that we are all confronted by the same technological realities. The great disparity appears in our respective ways of posing the problems, in our approaches to the investigation of their causes, and in our suggestions of how deliverance from the difficulties is possible. The disparity derives ultimately from a difference in our deepest convictions about life.

I advocate an integration of christian belief and philosophical (or scientific) thought. Such an integration allows a fresh light to fall on the problems posed by the transcendentalists and positivists and on their suggested solutions to these problems. I noted earlier that their views are mutually contradictory. The one group is oriented to human freedom, the other to technological power. Freedom and power exist for them in an eternally unbridgeable dichotomy. Furthermore, these secularized motives have clearly distorted technological development. Only when freedom and power are brought into harmony – and this is only possible through an acknowledgment that created reality, including humanity, is not self-sufficient – does a meaningful, liberating perspective for technological development open up.

1

A Philosophical Analysis of Modern Technology

1.1   Introduction

A philosophical analysis of modern technology is an indispensable prerequisite for any philosophical clarification of the problems connected with the theme of technology and the future.

When *philosophy of technology* first appeared at the beginning of the twentieth century, its practitioners did not devote much effort to the structural analysis of modern technology. Their aim at that time was primarily to defend technology as an independent segment of culture. They wanted to break the domination of science and economics over technology, and they rejected the idea that technology is simply neutral. Furthermore, because of the stormy development of modern technology, it was not until later that the philosophy of technology began to concentrate on the significance of technology for culture as a whole.

Engineers share the responsibility for the absence of any structural analysis of modern technology during the period in which philosophy of technology was emerging. The engineers, deeply involved in the practice of technology, had so little interest in philosophy that they made almost no positive contribution to such an analysis. Besides, the lack of philosophical knowledge among engineers led all too easily to an overestimation of the role of technology on their part. More often than not, technological progress filled people with grand expectations for the future development of culture.

1

We must not forget that *general philosophy*, in turn, paid very little attention to technology. It underestimated the significance of technology at first, and for the sake of convenience reduced it to a science or regarded it as a neutral tool in people's hands. Philosophers were unfamiliar with technology; they lacked a basic knowledge of it.

Eventually, because of its enormous influence on all of culture, technology could no longer be disregarded. Only then did general philosophy begin to take notice of it. Nevertheless, because of the current widespread lack of a thorough knowledge of technology in our society, technology is still being disparaged as a dangerous power threatening human welfare. Technology is all too quickly blamed for the cultural crisis. Inherent in this negative appraisal is an idea which is also shared by the optimists who view progress in positive terms – the idea that technology is all-embracing and all-dominating. The optimistic and pessimistic views both lack an adequate perspective on technology. While the one view overestimates the cultural significance of technology, the other fails to appreciate the possibilities it offers.

A knowledge of technology and philosophy in their mutual interaction is essential to any effort to arrive at a structural analysis of modern technology. In the past, unfortunately, there was a great *lack of communication* between engineers and philosophers. "Their respective terminologies and the origins and the orientations of their thought are so disparate that a consensus on problems of mutual interest becomes possible only through extraordinary effort and with much good will."[1]

When weighing the beginnings of the philosophy of technology, then, we must take into account the serious difficulties it originally faced. General philosophy virtually ignored the impressive phenomenon of technology and scarcely reserved a place for it. Philosophy of technology lacked a general framework. As a result, it developed almost completely independently of general philosophy. Engineers who had a philosophical interest in technology and its development but had an uneven background in philosophy (or none at all) arrived at philosophies of technology, most of which can only be regarded as pseudophilosophical.

The proper task of general philosophy is to express the unity-in-diversity of total reality. By neglecting technology, general philosophy failed to do justice to the diversity of reality and thereby damaged even its own insight into the whole of reality. Philosophy of technology, in turn, must take full account of the fact that it cannot be developed correctly outside an appropriate general framework. Philosophy of technology deals with only one segment of culture. Its field of investigation is limited. Therefore it must be able to call upon general philosophy to account for the coherence of technology and reality as a whole. General philosophy, for example, should account for the significance of technology for culture. The fact that dynamic development has made technology a prominent cultural force renders it all the more important for the philosophy of technology to be able to appeal to general philosophy.

A philosophy of technology set within the framework of a general philosophy should provide insight into the rich diversity of technological objects, of technological products and means of production, and of technological activities, all of which are increasing in variety.

A structural analysis of technology ought to define clearly both the potentialities and the limitations of technology, thereby obviating and guarding against misconceptions and false expectations concerning it. All technological problems and all norms for technological development should be analyzed. In particular, a clear view is needed of both the *relation between technology and science* and the many cultural changes which their modern alliance has brought about.

Only with the aid of such a structural analysis does it become possible to deal philosophically with the cultural influence of technology. By giving prior attention to such an analysis, we can greatly reduce the danger of becoming too speculative about both the positive and the negative influences of technology on culture.

An analysis of modern technology is indispensable, then, for understanding technology in its development, its influence on culture, and its bearing on the future. However, such an analysis is not sufficient in itself. Nor is the availability of a framework of general philosophy sufficient. What is decisive above all else is the religious conviction out of which both general philosophy and the structural analysis of technology are approached. The sources of disagreement in philosophy are generally differences in basic beliefs.

It is my conviction that the christian faith is the proper starting point for truthful philosophizing. Faith in Christ should direct all philosophical activity. Such an approach assigns philosophy its proper place and rules out any absolutization of philosophy, any elevation of philosophy that seeks to put it in the place of religion.

The Amsterdam school of reformational philosophy, which has also been called the philosophy of the cosmonomic idea,[2] is conscious of its nonphilosophical presuppositions expressive of the christian faith. This philosophy has developed into a general philosophy within which technology may be dealt with. Because it provides a suitable place for the philosophy of technology, it can be used as a framework for analyzing the diversity of technology as well as exploring the significance of technology within the whole of reality – especially the

3

significance of technology for other sectors of culture, such as the social, the economic, the juridical, and the ethical. Furthermore, under the guidance of christian belief we are compelled to respect the limitations imposed upon our expectations for the future, limitations which are of the greatest importance for a philosophical approach to the subject of this book.

In my analysis of technology, I shall draw heavily on the philosophy of technology which Hendrik Van Riessen began within the framework of the reformational philosophy. In the process I hope to carry this line of thought further, building upon his conception and enlarging it.

Needless to say, neither the reformational philosophy nor the philosophy of technology that has been worked out within its framework is a closed system. With due allowance for the basic point of departure, the reformational approach to philosophy of technology is open to continual renewal and wishes to take account of new scientific discoveries and the most recent developments in technology. Although I shall continue in Van Riessen's line, offering occasional criticisms on less important points, I hope to benefit from what others have said about technology and (especially in this first chapter) from what they have noted about the *structure* of technology. As much as possible, the philosophy of technology offered in this book will incorporate the latest developments.

An initial comparison of classical and modern technology (1.2) will lead up to a philosophical elucidation of the following topics: technological objects (1.3), technological form-giving (1.4), and technological designing (1.5).

Although I will have to sharply limit the number of examples from practical technology, these can be found in abundance in the literature cited. Moreover, I can hardly refer constantly to the general systematics of the reformational philosophy. Therefore the reader will have to be content with a sketch of the essentials, along with references to the primary sources.[3]

1.2 Modern Technology Compared with Classical Technology

In order to bring the distinguishing characteristics of modern technology into focus as clearly as possible, I will begin by comparing it with classical technology. The differences can be accentuated by bringing out the extremes in the features of both. To accomplish this, I will have to ignore momentarily the development of classical technology into modern technology. The comparison we seek will yield modern technology's most obvious characteristics. These can then provide the occasion for an analysis of modern technology, or serve as an introduction to it.

Before proceeding to a point-by-point comparison, however, I must comment on the term *technology* and offer a specification of the sense in which I wish it to be understood.

From its etymology it appears that the word *technology* is rooted in the Greek *techné*, which had many meanings. *Techné* was used to express all human ability as distinct from nature or *phusis* – in particular, the quality of skilfulness. Further, *techné* appears to have been related both to the word *epistémé*, which meant science (knowledge) and artisanship, and to the word *poiésis*, which denoted creating, writing poetry, and (again) artisanship. We see, then, that one cannot speak of a differentiation of technology, science, and art where the Greeks are concerned. Hence the etymology of the word *technology* cannot help us, it would seem, when we ask what technology really is.[4]

The occurrence of the root *techné* in modern English does manifest some differentiation. We speak of "technology" or "technique" when we wish to denote a specific human activity and its result. We also use these terms for human skill in general and for methods employed in work or other activities.

All of this, however, is too ambiguous to serve as a basis for a scientific treatise. Therefore I will define *technology* initially as the activity by which people give form to nature for human ends, with the aid of tools.[5] For the moment, this definition will suffice. Yet, to avoid misunderstanding, it is important to note immediately that it is far from exhaustive. Its parameters make it appear too anthropocentric and too pragmatic. It will be adequate for an initial comparison of classical and modern technology. In the sections that follow, however, and chiefly in the last chapter, I will state in greater detail what is to be included under technology – especially what is to be understood by the *meaning* of technology.

Let us now proceed to a brief discussion of the differences between classical and modern technology. These differences involve environment, materials, energy, skill, tools, the steps in technical execution, cooperation in technology, working procedures, the role of people in the formative process, and the nature of technological development.

1. In the technology of classical times, people were largely surrounded by a natural *environment*. They were closely bound to the supplies and possibilities afforded by nature. Only in a few instances could they take distance from nature. In modern times, by contrast, the environment itself bears the stamp of technology. Moreover, highly

developed technological objects are at our disposal; in a certain sense, our observations have become very instrumentalized, enabling us to separate ourselves from nature.

2. People used to have at their disposal only the *materials* at hand, but in modern technology we strive to become as independent as possible from materials in the forms in which nature offers them. Since natural materials differ in properties even when they are of the same kind and since they are sometimes mixed with other, undesired materials, they are selected before use and purified whenever possible. Increasingly, the aim is to reduce matter as it is found in nature to the smallest elements possible and then to build it up again in structures deemed more desirable by people, as we can see from the development of modern synthetics.

3. In early technology, the *energy* required for technological form-giving was produced by animals and people as tractive power and muscle power. In modern technology, energy is derived from nature – either directly, as in the case of wind and water power, or indirectly, as in the burning of coal and oil and the splitting of the atom. Moreover, the direct use of solar energy will be widespread before long.

4. In earlier days, technological *form-giving* was determined by human skill in handling tools. Tools were generally uncrafted objects found in nature. Today human skill has been objectified in or projected into technological operators, which we call machines. This has resulted in a definite refinement of skills. Moreover, it has become possible to introduce altogether new skills which human beings do not possess on their own. Electrotechnology, of which the computer is an example, is based almost completely on such skills. "In modern technology man's labor is externalized both as an 'objectification' of power exertion and as an 'objectification' of formalizable mental effort. Thus man is no longer in immediate touch with the world."[6]

5. In earlier times it was human intervention that determined the *succession of steps* in the execution of technological form-giving. People merely followed certain rules in handling the materials; explanatory insight was lacking. Throughout the formative process, therefore, the accent fell on the human being. Today the succession of steps is determined by automatic coupling and controls. Form-giving is concentrated in modern tools – technological operators that work independently.

6. In an earlier technology it was usually one person that carried out the formation of a technological product. Initially the producer was also the consumer, but even when this was no longer the case, the producer still had a close relation to the consumer. Modern technology, however, is unthinkable without *cooperation* between engineers, contractors, and laborers. Production has been set apart by itself, in an

industrial enterprise. Personal contact with the consumer has been reduced to a minimum and may even be entirely absent.

7. Premodern technology was interested in the *here and now*. Its problems and solutions remained highly dependent upon the demands of practical life. Hence they were singular, if not nonrepeatable. Intentional form-giving (the plan) and technological form-giving (the execution) constantly interacted. Today's engineer, by contrast, dissociates himself from life around him; that is to say, he takes distance from the incidental problem, from the available material, and from the tools. From a distance he renders the particular problem a general one, an abstract question formulated according to the canons of science and solved according to the scientific method. Through its dissociation from practice and by virtue of its scientific character, technological form-giving has thus become independent, and technological work in the preparatory stage has been raised to a higher level.

Intentional form-giving and designing, on the one hand, and actual production, on the other, are separated in modern technology. By way of contrast with early technology, designing is new. Its function is to give a picture of the results of abstract imagination, thereby bridging the distance between intentional form-giving and actual production. "Modern technology first of all draws up a plan of action. . . . This plan can be mentally (theoretically) determined down to the last detail. Only in the second place does modern technology consist in the execution of this plan. The separation of these two activities forms the distinguishing characteristic of modern industrialized technology."[7]

The intentional form-giving of the engineer is based on mathematics, physics, and technological science. Through technological literature, through measuring techniques, and through the actual processing of the results of theory, the engineer gains information about concrete reality, information that enables him to correct the abstractions of science. The possibility of new constructions or inventions is opened up in the designing stage.

8. In contrast with the older technology, modern tools (technological operators) have tended to eliminate *people* as much as possible from the process of technological form-giving. With the aid of energy transformation processes directed by the technological operators, it is possible to manufacture objects in accordance with a theoretical plan. This is called *automation*. The realization of the plan takes place on the basis of a methodical division of functions. Where technological operators are not yet employed exclusively, workers are accorded subfunctions in the production process. This method of manufacturing usually implies mass production.

9. In short, early technology remained restricted by man's natural potentialities, that is, by the force and capacity given directly with the

human body; it remained entirely within the range of the human hand and the human senses. Hence the naturalness of premodern technology. Because of its limitations, however, there was almost no technological development. The routinelike experience was passed on from father to son, from mother to daughter, from master to apprentice. Accordingly, *early technology was undifferentiated and static*.

Modern technology, owing to its scientific foundations, has seen a tremendous advancement. It is *highly differentiated and dynamic*. It has put its stamp on the whole of culture, sweeping it right along with the swift pace of development.

Later on we shall have occasion to refer to the matter of development more than once. At this juncture, however, I would note that even in modern times, technological activity is and remains a human activity. This fact is decisive for proper insight into the limitations and possibilities of technology. One must never lose sight of the intimate connection between people and technology, especially when evaluating such a phenomenon as automation, which may sometimes appear to be rather autonomous with regard to people (i.e. in the independent operation of technological objects). In fact, the intimate connection between people and technology is crucial if we are to resist arguments to the effect that technology as a whole is autonomous.[8]

This comparative analysis yields a number of topics which will be examined more closely in the next sections. They are: technological objects (1.3), technological form-giving (1.4), and designing (1.5).

I will conclude each of these sections by suggesting the trends that will characterize the future development of technology to a large extent.

Chapter 4 will deal at some length with characteristics of technological development that have been omitted from the present discussion of the differences between modern and premodern technology.[9]

1.3 Technological Objects – Modern Tools – The Computer

1.3.1 Introductory remarks

We need a philosophical analysis of modern technology to make it clear what technology is, what its limitations are, and what its possibilities may be.

The first step is to gain insight into the structure of technological objects. These are objects that are fashioned by people for the express purpose of being put to use in technology. It is important to point out the feature that these objects have in common while showing at the same time how their technological uses may differ. I shall therefore try to trace the characteristic structure of technological objects; in other words, I shall attempt to formulate the *law* for all technological objects.

An important distinction is to be found in the fact that objects formed in technology are either technological objects or objects of another kind. The latter do not have a technological destination or purpose or end function; cars, houses, and washing machines, for example, do not (500, 502).[10]

Technological objects are not just formed technologically; they also have a technological destination. Let us call the relation people enter into with an object in order to form it an *objectifying relation*. I will speak of an *actualizing relation* when people enter into a relation with a cultural object and the object in this relation reaches its destination.[11] Without the actualization of the cultural object, the end function is latent; in the actualizing, this function is made patent by people. The objectifying and actualizing relations are closely connected: technological form-giving, after all, is guided by the desired end, while the realization of the end is founded in the realized technological form (503).

Finally, we should take note of a certain distinction between technological objects, a distinction that will later on prove to be essential to an understanding of modern technology. Technological objects may be things (more or less durable) or they may be facts or processes (subject to constant change). Friedrich Dessauer takes account of this distinction when he speaks of *Raumform* and *Zeitform* respectively.[12]

1.3.2   Facets of technological objects

In order to avoid straying too far from the subject at hand by presenting an extended treatise on the distinctive characteristics of technological objects, I will limit myself to a simple point-by-point analysis.

1. The philosophy of the cosmonomic idea (*wijsbegeerte der wetsidee*) distinguishes in every "thing" a number of *subject functions* and *object functions*. A stone, for example, has arithmetical, spatial, physical, and kinematic subject functions. All the other functions are object functions, namely, the biotic, the sensitive, the analytical, the technical or cultural-historical,[13] the linguistic, the social, the economic, the aesthetic, the juridical, the ethical, and the pistical functions. An animal, by contrast, functions subjectively in the first six modalities and objectively in the others. A person functions subjectively in all modalities of being. Nevertheless, the stone, the animal, and the person are

not exhausted by their functions; they are more than the sum of their functions. Their individuality expresses itself in the various functions, which we also call modalities or modes of being.

2. We attain insight into the specific meaning and character of a technological object by considering its *start* and *destination*. The technological object functions in coherence with people, materials and other technological objects. Yet, as a unit it always remains more than the sum of all its subject functions and object functions: as a unit it manifests itself in the diverse functions. The last subject function of the technological object is the physical one. The remaining functions are object functions. These are latent and can be actualized only in connection with human activity (503).

3. The *foundational function* of technological objects is the cultural-historical or technical function. It is in this function that the stage of technological development first finds expression. We call this function, which is an object function, the technological *form function*. The technological form function manifests the result of the free human forming, or shaping, of the things formed. In the first phase of activity people give form to physically qualified entities, and in the later phases to technologically qualified entities. The latter already have a technological form function – just think of semimanufactures. It is on the foundation of the technological form function that the function of destination can be actualized (508).

4. The *destinational* or qualifying *function* of a technological object is also an object function: it is the cultural or technological *forming function*, which must be disclosed or activated by a person in a subject/object relation. As the qualifying modality, this destinational function leads the unfolding and disclosing of all the "earlier" functions. The destinational function is likewise definitive for the eventual disclosure of the "later" object functions. Thus we can say that the destination is founded in the realized technological form, while this destinational function, in turn, plays the qualifying role in the realization of the technological form function as foundational function. The foundational and destinational functions are mutually attuned (514).

5. Technological things are to be divided into two sorts: (a) those that must be worked up, which can be said to have a *passive* forming function, and (b) those which serve people as tools in their work, which have an *active* forming function. In the subject/object relation aimed at the forming of technological objects, tools stand on the side of the human subject. This subject/object relation is an objectifying relation preceded, where active technological objects are concerned, by an actualizing relation (509).

6. For all technological objects, and for them alone, both the foundational function and the destinational function are cultural or techno-

logical functions. This typification admits of a diversity of individual destinational functions; technological objects can be classified accordingly (509; see 1.3.6).

All other cultural objects also have a cultural or technological foundational function, but their destinational function is a different one. For example, the destinational functions of the automobile and money are the social function and the economic function respectively, and so forth.

7. As we shall see later, a *technological fact* is to be understood as the independent operation of tools in concert with other objects. Subject and object functions are also to be discerned with respect to technological facts (see 1.3.4).

1.3.3   The basic structure of modern technology

The grand quest in modern technology has been to develop technological objects that can operate independently. To this end, human proficiency in forming is projected into and transferred to the technological object. A transfer of the decision-making capacity pertaining to the sequence of the activities of forming also occurs. By means of automatic switches, people make provision for the technological forming process to undergo discontinuous changes with the passage of time.

Taken together, the projection of proficiency, the transfer of "decisions," and the use of formed energy constitute the foundation of the independent operation of the modern panoply of tools and instruments. This panoply is composed of what we call *technological operators*. Apart from design and installation, people need only give a command to set the technological operator going.

The "proficiency" of the technological operator surpasses that of human beings in speed, reliability, and accuracy. Even mechanical "decisions" are realized more quickly and faultlessly than "decisions" made and implemented by people. And the power of formed energy far exceeds human power. Now, what all this means is that people equipped with technological operators can accomplish a great deal more than people without them. Moreover, there are more recently created technological operators that work in ways that bear little or no resemblance to human activity. Electrotechnology and chemical technology offer examples of such operators.

The development of technological objects as sketched here, namely, the realization of independent operation, has been made possible only by the scientific foundation and scientific method of modern technology.

Through the scientific approach of modern technology, a distinction has arisen between preparation (designing) and execution (technological forming). Human responsibility and decision-making have been transferred to the phase of preparation, and the human activity of designing has thereby come to occupy a higher place. There has in fact been an intellectualization of technological labor. Preparation along the lines of the technological-scientific method[14] leads to the accomplishment of a design for the execution; this design is complemented by the independent operation of the technological panoply. As a result, human power in the phase of execution is enormously increased.

Because it will be necessary to refer rather frequently to the significance of science and of the technological-scientific method for technology, I will summarize the basic structure of modern technology as follows: *the basic structure of modern technology is characterized by the technological operator, the scientific foundation, and the technological-scientific method.*

There are cases in which the control of technological operators from a distance is possible but still unrealized, for reasons still to be mentioned. There are also cases in which such control from a distance will continue to be impossible – at least for the time being. In these cases technology is restricted by particular or singular circumstances. In the future, with the help of cybernetics, it will be possible to alter this situation; with the feedback principle, even the singular can be controlled. This kind of problem arises in highway and water systems engineering, in mining, and in architecture – in short, in all areas where work is carried on in direct contact with concrete, singular nature rather than with semimanufactures, as is otherwise the case.

1.3.4 Technological events, facts, things, and energy processes

We speak of a *technological event* whenever people play the decisive role in the objectifying and actualizing relations. The technological event therefore functions subjectively in all modalities. Whenever people cease to play a role in the stated relations, we speak of a *technological fact* or an automatic forming process. A technological fact, accordingly, is always introduced by a technological event. If we disregard the event, we cannot understand the fact, for then we are ignoring the initiating role of people. Hence there might well be a number of occasions when disregarding the initiating role of people in a technological fact would cause problems and misunderstandings. Such misunderstandings can only be dispelled if it is realized that the fact is actually introduced by the event. Thus the technological fact is autonomous in a relative sense only – not in a complete sense. A technological fact does not have to embrace an entire forming process, of course; it may constitute only a part of it, as in those situations where people (still) play some role in the forming process.[15]

A technological fact is also a technological object. Technological things and materials are subject to differentiation in this respect, and so are energy processes. The former (i.e. technological things and materials) have a more or less permanent form function as their foundational function, which is recognizable from the existence of a possibly dormant condition in these things and materials. In the technological forming process, these things and materials manifest themselves first according to their character. The technological operator is an operator working for forming and accordingly has an active destinational function; the product is the entity in production, and it has a passive destinational function. In the forming process, the latter undergoes a lasting alteration of form in the sense of technological individualization. The operator, which always brings about an energy transformation process, maintains its form function throughout the actualizing and forming – unlike the energy, which does not.

Van Riessen makes somewhat different distinctions (514-19). He does not speak, for example, of a distinction between technological events and technological facts. He calls a technological fact an energy transformation process that implies an actualized technological operator, and he says that the fact and the thing are mutually complementary. I prefer to speak of a technological fact with respect to which humans can be abstracted, since they play no role in it, and in which the technological operator and the energy process are mutually complementary.[16]

As we shall see, this complementary relation affords points of reference for the general classification of technological objects.[17]

Before this complementary relation is dealt with more closely in a separate section, there are two remarks that must be made.

1. Not every fact arising from actualizing by a technological operator and directed to the realization of an energy transformation process has a technological destination. A fact might find its destination in communication instead, as is the case with a tape recorder, or it might have a traffic destination, as is the case with traffic lights. These facts might well contain technological facts as parts (511-14).

2. In contrast to the situation that obtains for many technological facts, the technological operator cannot be regarded as the guiding structure in the energy process in chemical processes. Here the technological operator only protects and conditions the energy process and the alteration of the form of the materials, which together are called the chemical reaction process. The energy process is the actualizing of

13

the coherence of the materials; the energy process and the alteration of the form of these materials constitute an indivisible whole (518).

1.3.5 The complementary relation between the technological operator and the energy process

We have already seen that in facts, whether technological or not, the technological operator and the energy process are distinguishable. Yet they are inseparably joined, for the actualized technological operator takes care of both the conveyance and conversion of energy, while the energy process actualizes the technological operator. Since they constitute the fact together, their meaning is contained in that of the fact.

In order to analyze the complementary relation between the technological operator and the energy process more closely, we must look again at the basic structure of modern technology. As has already been noted, the severing of the preparation (designing) from the execution is made possible by the technological-scientific method, which is founded upon mathematics, physics, chemistry, and technological science. The basic tendency of technological execution is toward the independent working of technological operators.

Now, to explain the relation between a technological operator and an energy transformation process, it is necessary to investigate the influence of the technological-scientific method on a technological fact. From the great diversity of technological facts, let us select the so-called individualization process. This is the technological fact in which things are made. The complementary relation between the technological operator and the energy transformation process also holds, of course, for other facts.[18]

During the preparation, the technological fact is analyzed in the light of the technological-scientific method. It is subdivided into fact components, each of which is realized by a technological operator and an energy process. The latter is nevertheless dependent upon the energy transformation processes of antecedent fact components, and may in addition be joined to a complex of many other energy processes. The whole, as the integration of these processes, is a coherence of transformation stages and energy transmission in which many processes are separate right from the beginning. From the beginning, nonetheless, the main process must furnish energy from a structure adapted to every particular case. The destinational function of this energy, and consequently its form function as well, therefore ought to be as neutral as possible; that is, it ought to exhibit as little individuality as possible. On the basis of this neutral energy, the technological fact component is influenced; with that, limitations are set with respect to the form function of the technological operator in the fact component (523, 527).

Although the energy transformation process is indeed the objective of the technological operator, it alone does not determine the form-giving of the latter. In the first place, the technological operator must be able to be formed. The limitations inherent in the materials, semimanufactures and manufacturing potentials guide the form-giving throughout. In the second place, given the technological-scientific method in the preparation, people have an influence on the formation of technological operators similar to that which they have on the integration of the energy processes. This method of analysis and abstraction from the technological problem is the reason why the technological operator to be made is assembled from components. In general, the components are really the parts of a whole, just as in the case of a nature-thing. They are dependent, to be sure, and limited by the particular destination of the whole, but in the technological operator some parts may nonetheless have a certain artificial independence. They are not completely limited by the singular destinational function of the whole. They are undetermined, as it were, in their destinational function. Such is the case, for example, with normed components like the screw, the rivet, the cogwheel, and so forth. Such relatively independent components have consequences for the technological operator as a whole because they do not have the nature of the individuality of the whole.

The technological operator is the result of the kind of compromise in which the envisioned destinational function and the limiting factors in the form-giving arising from a number of "neutral" components are weighed against each other. This point is of exceedingly great importance in modern technology, for the independence of the parts implies a leveling of the individuality of the whole, which further implies the neutralization of the destinational function of technological things.[19] *Technological dividing is a neutralizing dividing.* Neutralizing concerns the parts in the first instance, but it exercises an influence on the technological operator as a whole, and consequently on the energy that is to be transformed by the operator (522, 524-6).[20]

In summary, the complementary relation between a technological operator and an energy process in the fact can be described as follows. On the one hand, the desired fact directs the meaning of both the technological operator and the energy process, but on the other hand, both of these, given their more neutral structure, condition the desired fact. On the one hand, of course, the fact ought to be set within the framework of a great coherence of energy transformation processes, from which a neutralizing influence is exerted on the fact. Yet the technological operator, on the other hand, ought to be seen as a thing totality, assembled from "neutral" parts, from which a neutralizing influence is exerted upon the thing totality and thus upon the desired fact. Both the tendency toward the integration of energy processes and the tendency toward dividing up technological operators into component units are hallmarks of the unique character of modern technology. Where the technological operator and the energy process imply a lessening and a neutralizing of technological individuality, the destination of both – namely, the technological fact – is liable to leveling and neutralizing. This effect is reinforced, in addition, by the requirement that the "neutral" form functions of the technological operator and of the energy process must be attuned to each other (529).

1.3.6   The classification of technological objects

The most suitable point of reference for a classification of technological objects is the perspective afforded by technological facts, for it encompasses not only technological operators and energy transformation processes but also passive technological objects.

When the classification is undertaken from the perspective of the place of technological operators in technological facts, the role and significance of energy transformation processes and of passive technological things and materials enter quite naturally into the discussion. Conversely, a classification of technological objects undertaken from the perspective of the role and significance of energy transformation processes in technological facts readily brings out the meaning of technological operators and of passive technological objects.

Because the technological operator and the energy transformation process point to each other, a classification of the role of technological operators in the technological fact will always agree in one way or another with a classification of the role of energy transformation processes in that fact.

It will be meaningful to provide both systems of classification. The classification one prefers will depend largely on the place a person occupies in modern technology. A mechanical engineer may prefer a classification based on technological operators, while an electrical engineer may lean toward a classification based on energy transformation processes. In general, the concomitant use of both systems of classification has the advantage of providing a good view of the limitations and potentialities of technology; at the same time, the unbreakable unity of technological operators and energy transformation processes gains important and appropriate emphasis.

Here, then, is an abbreviated presentation of the two systems of classification.

1. An analysis of the *role of technological operators* in facts yields the following results.

1-a. The purpose of technological operators, in relation to energy transformation, is the fashioning of things and materials. With further analysis it is possible to gain insight into the destinational function of this form-giving.

1-a-1. Operators for assembling things. Such operators aim at technologically objectifying the arithmetical subject functions of the things (for example, operators for assembling: from a diversity of things, a unit is fashioned).

1-a-2. Operators for giving spatial form to things and materials. These aim at technologically objectifying the spatial subject functions of things and materials (for example, machine tools).[21]

1-a-3. Operators for giving physical-chemical form to materials, such as the analysis and purification of compounds, and chemical transforming and mixing. These aim at technologically objectifying the physical subject functions of the materials (for example, blast furnaces and concrete mixers).

The things and materials that are to be worked up in 1-a-1, 1-a-2, and 1-a-3 may already be cultural products; some of them may even have a technological meaning – for example, passive technological objects. The results of the working up may have an active technological, a passive technological, or some other destinational function – for example, a symbolical or societal destinational function.

A more elaborate classification of 1-a-1, 1-a-2, and 1-a-3 is possible if attention is paid to the various sorts of energy, such as mechanical, thermal, electrical, and the combinations of these.

1-b. Technological operators for the generation, transmission, and transformation of energy. These are to be differentiated according to the energy source and according to the kind of transformed energy – mechanical, electrical, and so forth.

1-b-1. Operators for starting and for connecting energy processes.

1-b-2. Operators which transform energy for the purpose of linking technological forming objects. The technological meaning here is to arrive at a temporary linking in order to make a certain form-giving possible. This is broken in turn by a new linking fact – link automation (546-8).

1-c. In connection with energy processes, the technological operators meant under 1-a and 1-b constitute facts. These facts have a meaning which is technological.

Technological operators also appear in nontechnological facts. This makes it possible, depending on the meaning of the fact involved, to draw the following distinctions.

1-c-1. Operators for reflecting a symbolical fact, such as information and measuring symbols, reflect respectively a qualitative and a quantitative aspect of a fact that is or is not technological.

1-c-2. Operators for a societal function, such as that of transportation and traffic.

The technological operators meant in 1-c-1 and 1-c-2 are encapsulated in or enkaptically interwoven into a symbolical thing totality and a societal thing totality respectively (549-51).

1-d. Computers form a separate category. In the following section it will be shown that in combination with 1-c-1 and 1-b-2 they are a control apparatus, and that as such they fulfil a technological function. When used as a thought apparatus, they have an information function.

2. A classification of the destination of a fact based on the *function of energy transformation processes* is also possible. To the extent that technological operators appear in these processes, the classification which now follows should agree in one way or another with the prior classification – a correspondence which can be explicitly indicated, as we shall see.

2-a. *Correspondence processes*. The meaning of the fact at the end corresponds in one way or another with the meaning of the fact at the beginning, which is to say that both the fact at the beginning and the fact at the end are individualized. In these processes "neutral" operators are used, serving to ensure that the great diversity of individual facts assigned to energy processes is preserved to the maximum extent possible. These processes involve *information processes* such as those of measuring and informing (e.g. flow meters), regulating or switching (e.g. regulators), and the combination of these two (so-called automatic regulating and switching). The correspondence processes also include the *communication processes*, such as those which appear in radio, television, and telephone,[22] and the *choice transformation processes* of computers, which are also known as *data-processing processes*.[23] A combination of choice transformation processes and automatic regulating and switching affords the possibility of complete automation (555). (The correspondence processes appear in the prior system of classification under 1-b-2, 1-c-1 and 1-d.)

2-b. *Individualization processes*. In the prestages of the process, energy has an indeterminate, "neutral" structure, like the electrical energy of a generator. This is allowable insofar as the various transformations bring about the required individualizing for the end fact, as for lighting, heating, and machine tools. These processes can be integrated into a branching coherence. In contrast to technological operators in correspondence processes, technological operators in individualization processes must individualize "neutral energy." At the beginning of the process, it is appropriate to speak of *starting operators*; examples of these are the gasoline engine, the steam engine, the water turbine, etc. *Linking operators* like a dynamo, a motor, etc. provide for both the transmission and transformation of energy (see 1-b-1). *End operators* individualize energy – see 1-a-1, 1-a-2, 1-a-3, and 1-c-2 (555).

2-c. *Material reaction processes*. In contrast to the two preceding processes, the transformation consists in this instance of only one step. The technological operator is not actualized in the same manner as in the preceding processes. Specifically, the technological operator is not the leading structure in the energy transformation process (see comment 2 at the end of 1.3.4).

1.3.7 The computer

At the end of our analysis of technological objects, we must pause before the computer since it will be the preeminent technological object of the future. The computer is one of the most fundamental technological achievements of humankind. It originated through the cooperation of people who were very highly qualified as scientists. Disciplined thought, inventiveness, and teamwork led to this powerful and surprising instrument of modern technology.

The unmistakable significance of the computer appears from the fact that while technological development is greatly advanced by the use of this machine, the computer is applicable to many other cultural activities as well. More than anything before it, this newest technological operator has accelerated the development of modern technology; our entire culture has now become more dynamic than ever, owing to the direct or indirect influence of the computer. One of the ways in which the increasing significance of technology for culture is revealed is by the many views being expressed about the computer and its potentialities, applications, and limitations.

What the computer produces is unforeseeable, and this distinguishes it from other technological operators. That the computer's results may contain surprises – albeit within certain limits – does not, however, mean that the computer, like the human being, is free. The computer works in a set way; its rapidity, precision, and accomplishments are so enormous that it is easy to forget that it is a tool or implement of human beings.

The potentialities of the computer furnish fertile ground for all manner of optimistic or pessimistic speculation. On the one hand, the computer excites great expectations for the future. Beautiful promises appear capable of realization at last; the accomplishments of computers will increase rapidly, bringing undreamed-of benefits. On the other hand, the surprising results arouse fear and anxiety. Could it be that man, with his computer technology, has set in motion a terrible development which he will eventually be unable to oversee and to control? Will man himself become the victim of this development sooner or later? Will the machine become its maker's master?

These optimistic and pessimistic attitudes toward the computer and the developments that accompany it constitute a distinct reason for giving some special attention to the computer. Focusing on the computer should open up the possibility of participating responsibly in the *philosophical* debate about computers. And it is self-evident that it is neither possible nor necessary to go into detail. The technological details are so manifold that even a general survey would demand disproportionate consideration. Accordingly, only some main lines related to the basic structure of the computer will be presented here.

We have already seen[24] that so-called correspondence processes occur in computers. I spoke of these somewhat more accurately as choice transformation processes.

In connection with computers, it is customary to refer to information-processing or data-processing processes. The danger in this is that one might be led to ignore the fact that information is lingual at bottom. Language indicates and signifies something. And indicating and signifying are human activities, expressions of human freedom and creativity that cannot be tied up in set rules. The language which results from human activity can certainly be formalized, and formalized language can be objectified in the computer. Yet the significance of the objectified signs and symbols is human in origin; the computer's results receive their meaning through people.

Whenever these considerations are not respected, that is, whenever people speak simply of data-processing processes without any further refinement of terms, confusion lies at the door. The question of the machine's capacity to add new information can only arise – and is in fact frequently raised – because it is forgotten that computers only work technologically. The technological processes of the computer can only process data that has been analyzed and abstracted and then stored in an energy form – the signal. Although it would be much better, in the interests of avoiding confusion, to speak of signal-transmission and choice-transformation processes, the term *data processing* has already become so well established that I shall have to use it often. An improper use of such terminology can easily lead to misplaced notions and expectations concerning the computer.[25]

In short, the basic structure of the computer is this, that a signal in the computer can find its way along either of two alternative, mutually exclusive routes. Thus a "choice" consists of two possibilities, which on theoretical grounds must have an equal chance of occurring. An equal choice of two possibilities with a chance of 50 percent (comparable to heads or tails) is said to represent a unit of information of one "bit" (a word derived from "binary digit"). One bit thus affords two mutually exclusive possibilities – yes or no, open or closed.

In a strict sense, then, what is processed is not information but the *analytical substratum* of information, and then by choosing. The number of possible choices in the computer depends on the number of independent switching elements with which the machine is furnished.

Such mathematical processes as adding, subtracting, dividing, and multiplying can be reduced to a combination of these elementary choice possibilities. Through electronic switching, they can be processed one after the other at great speed. With the help of the controlled switching elements, millions of processings can be accomplished in one second.[26]

Where the fundamental processing consists of a choice of two possibilities, numbers must be transposed into the symbols of the binary system – 0 and 1. Analyzed and quantified information must also be translated into these symbols. When this has been done, processing on the basis of a great number of choices becomes possible.

The commands which the machine is to carry out successively (the program) are fed into its memory as codified instructions in the form of a "machine language." This language varies from machine to machine, and for a person (the programmer) it is relatively difficult to master. A number of standard program languages have therefore been defined in which many of the more common routines and procedures have been given simple names. These languages are easier for the programmer to use. With the help of a translation program – called a compiler – that belongs with the machine, these programs can be "translated" into the machine language by the computer itself.

From its memory bank, the computing part of the computer receives commands to carry out processing in a certain way. In addition to the program, it contains stored information which can be used for the commanded processing. This stored information is also presented in the symbols 0 and 1. At the present time, a memory bank can store up to 10^8 bits of information.

Information theory and technological actualization are perfectly attuned to each other. The theoretical unit of information is mirrored, as it were, in the switching elements, which are either "open" or "closed" and which can thus assume two theoretically equal positions independent of each other.

The technological coherence and the mutual interaction of the switching elements, which together constitute the so-called logic system of the machine, can be constructed with the help of Boolean algebra. The "open" or "closed" positions of these switches function in this algebra as variables.

The application of the computer is principally twofold: it can be interwoven into a thought apparatus or into a control mechanism. In the former case it delivers information to people; in the latter it is interpolated between people and a fact or process, the destination of which need not be technological. In full automation, the computer is wholly integrated into the technological fact or process.

The computer as a control mechanism has great significance not only for automation and the guidance of chemical processes but also for space flight and all sorts of service-oriented activities. Here the data processing in the computer is based on information gotten from technological facts, processes, and space flight, and from the computer program.

A special case of computer application is its use as an aid in simulating a variety of technological and physiological processes. This simulation technique provides us with knowledge of these processes which would otherwise be difficult, if not impossible, to obtain.[27]

The structure of the computer is dependent on the destination of its application; a processing of analyzed and quantified information occurs.[28] From these two most important applications, it is clear that science and technology make each other productive. The computer, taken as a technological result, is at the service of science, which is in turn a durable stimulus to technological development. But the computer is important to more than technological science; because of its command of logic and mathematics as analyzing and qualifying aids for many sciences, the use of the computer for scientific research has become possible.[29] Furthermore, the computer as a thought apparatus has a large role to play in the areas of operations research, decisiology, and praxeology, as we shall see later.[30]

Although the computer processes an enormous amount of information rapidly and faultlessly, executing complicated mathematical procedures that will never be within the capacity of the human brain and giving answers to questions that people alone (unaided by the computer) would never be able to answer as fully or as well, it must not be forgotten that the computer is simply man's *instrument*. The great quantities of information involved are fed into the memory bank by people. The processing of this information on the basis of a great number of "choices" is the fruit of human analytical powers, and the program for processing the information is also a product of human powers. In all these regards, the computer is subordinate to people. The last subject-function of the computer is the physical function. The technological function is an object-function. While the computer's results may be in the nature of a surprise, and while people may never be able to achieve these results by themselves, we are not to jump to the conclusion that the computer is independent of people. The computer is at humanity's service. It helps people to analyze, find, remember, count, and sort.[31]

The computer we have been discussing up to this point is the *digital* computer. The accuracy with which such computers work is practically unlimited. Their working is discontinuous; through analysis of each continuous process it is possible, with the help of the discontinuous working process of the computer, to approximate the continuity of a process very exactly.

The working principle of the *analog* computer is continuity. Via an image of a continuous process (the analogon), it furnishes answers to the questions posed. The programming is simple and quickly realized. The precision and rigidity of its processing is inferior to that of the digital machine. Accuracy is circumscribed in proportion to the scale of the analogon. The measurements involved concern physical quantities for the most part.[32] While the digital computer can be used for many purposes, the *analog* machine can resolve only one fixed class of problems. It is used primarily where the phenomena and magnitudes involved can be defined mathematically in a system of differential comparisons.

Since the applications of the digital machine far surpass those of the analog, the former is more frequently used. The analog machine is often employed to locate the real problem quickly.

The analog computer may be used in an investigation of the strength of a certain structure for the purpose of locating the points of maximum or critical stress. The precise magnitude of these stresses is then calculated with the digital computer.

A machine composed of both digital and analog units is called a *hybrid*. Unless otherwise noted, the computer I will be referring to throughout this study is the digital computer.

1.4 Technological Forming: The Execution

1.4.1 Introduction

A comparative study of classical and modern technology reveals great differences in the area of technological forming. In the earlier technology, people performed and people decided the sequence with their own power and skill. In modern technology, human skill is projected into technological operators which have been actualized by means of formed, natural energy; the sequence in forming is delivered by automatic linkage. For people, the method of working is entirely altered in modern technology.

The independently working technological operator is made possible by the introduction of the technological-scientific method in the preparation. Technological science, mathematical physics and, frequently, chemistry together form the basis of the scientific approach of technology, while the industrial enterprises provide the opportunities for realization. For this reason, modern technological forming is characterized by the quest for automation, by worker redundancy, and by mass production (560).

It has already been noted that the most characteristic feature of modern technology is the division that has arisen between preparation and execution. The objective design furnishes the connection between these two stages. It is primarily in the preparation that people (engineers) bear responsibility in modern technology. Wherever possible, people are excluded nowadays from the postpreparatory, forming phase, and even where they are not excluded from the execution, the role of the worker in mass production has been analyzed in the preparation and immutably predetermined in the design.

Reserving for the following section (1.5) a treatment of the problems of designing, I will now discuss the influence of the technological-scientific method on *technological forming*. This is the method that governs designing, so its results will also be apparent in both technological forming and the results of technological forming. By way of introduction, then, consider the following.

In designing, the engineer has in mind both the design of the product and a plan for its production. The former, the design of the product, involves the law for a thing. That the destinational function of the product plays a decisive role in the designing of this "thing-law" goes almost without saying. The latter, the production plan, establishes as precisely as possible the law for the way of production, that is, the law for a fact. This latter design concerns the "thing-production process" taken as the coherence of actualized operators, complemented eventually, where necessary, by human labor. The transport of the thing to be made, the so-called thing-transformation process, must also be designed (566).

As far as designing is concerned, there has been a certain differentiation in modern technology, that is, a differentiation between thing-laws for products and fact-laws for production. Yet, this differentiation is accompanied by integration. Neither of the designs can be made independently of the other; rather, each must be attuned to the other. Together with other thing and fact processes, these designs have to be joined to one integrated production process.

The quest for production according to a theoretical – and thus predetermined – production plan in which the product is fashioned in conformity with a theoretical design implies the exclusion of the (free) laborer from production to the extent that this is possible, so that the chances of deviation may be reduced as much as possible. The technological-scientific method accordingly requires production by technological operators. Technological operators, however, because of the limited forming possibilities afforded by their structural inflexibility, require a division of the production process. The production process is a linkage of technological operators.

In any given production process, the technological operators actually available for the renewal or alteration of the product or of the production process will fix the limits of the possibilities of the technological-scientific method. From this it is apparent that the *technological-scientific method* and the *mechanical operators* point to each other and set the direction of each other's development. Furthermore, it is also apparent that this *interchange* is what determines the meaning of modern technological forming (568).[33]

1.4.2 The technological-scientific method

The method of science[34] is that of analysis and abstraction. This means that the field of investigation is dissected: the particular, the concrete, the here and now, and the milieu are set aside in order to facilitate penetration to the laws and to the order of the reality concerned. This scientific method leads to general or universal, lasting, and coherent knowledge of the law for a particular area of the reality being investigated. Now, the *technological-scientific method* is not to be equated with the scientific method, for while the latter seeks knowledge of the laws which hold for reality, the former is concerned with the formulation of the laws for both production and the product, that is, the laws for fashioning and for what is to be fashioned. The meaning of the technological-scientific method is technological in character, even though the method could not exist without a scientific foundation. The significance of this foundation is that the hallmarks of scientific method and knowledge are "reflected" in technological forming and its results.

The engineer is confronted with technological problems for which he seeks solutions. To this end, he analyzes the problem and breaks it down into separate parts, each of which is made into a universal (part-of-a-)problem. The engineer then tries to find a solution for every part. This holds for both the components of the thing to be produced and the parts of the production process. This function-analysis can go on until atomized, isolated, autonomous functions are attained.

Technological component functions, then, are attained through analysis. Abstraction sets each function off by itself, apart from others, so that the technological solution for a component function may be found in this way. The result consists of elemental building blocks which are neutral in their destinations and therefore universally applicable – for example, the screw, the rivet, and the welding. In this universal applicability, universal scientific knowledge is projected into technology as the result of analysis and abstraction. In other words, the universal utility of the solution to a component function is an analogy of scientific knowledge in technology.

Along with the differentiating and neutralizing separation of functions, the technological-scientific method seeks control over technological forming with respect to both place and time. To this end, technological operators ought to be of durable structure; that is, they must not change with the passage of time. This control of technological forming *from a distance* and *in time* is an analogy of science as the durable *knowledge* of reality from a distance.[35]

This control at a distance is subsequently possible with respect to the whole, by integrating the component solutions. Depending on the integrated component solutions, this function-integration can represent any one of a number of forms of individuality. *What the technological-scientific method seeks is on the one hand a neutralizing separation of function components, and on the other hand a variegated or individualizing function-integration.*[36]

The production process thus achieved, accordingly, is an analogy of the *coherence* of scientific knowledge. In the result of such a process (i.e. the mass product), something of the *universality* of scientific knowledge becomes visible, again by analogy.

An analogy of scientific abstraction is to be observed, furthermore, in the integration of component solutions. By means of abstractions, the components of the problem are isolated first, and then a general component solution is found. Consequently, the integration of these component solutions makes it necessary for the technological forming to occur in isolation from its surroundings. Appropriate measures are taken to this end. Protected as much as possible from such environ-

26

mental influences as temperature and humidity, technological forming acquires a durable character.

It should be clear from the above that the technological-scientific method displays the same features as the scientific method. "From reduction via disposability to combination"[37] is true for both. The destinations, however, are different. The one method seeks scientific *knowledge*, while the other is directed toward technological *forming* based on scientific knowledge.

It can be said in summary that the technological-scientific method of modern technology seeks the control of technological forming at a distance and in time. This control is made possible when such features of scientific knowledge as its general universal validity and its coherence and durability are projected, via the method of analyzing and abstracting technological problems, into technological forming and its products.

This projection is never entirely successful because it evokes the resistance of reality, into which projection must occur. Reality offers resistance specifically to the realization of the technological-scientific method. The application of the technological-scientific method has its limits in practice. Its potentialities are limited in principle. More about this in the following section.

1.4.3   The limits of the technological-scientific method

From the preceding section it is apparent that the scientific unfolding of modern technology in the stage of preparation leads to designs for the products and for the production processes. The result of the design is then projected into reality. This projection creates tensions as an attempt is made to reenter reality with universal, durable, unchanging thing units and production processes. Irreversible change must be eliminated as much as possible. Individuality must be neutralized as much as possible, for in a certain sense, parts with a neutral destinational function also neutralize the whole into which they are taken up. People try to make objects function isolatedly in the continuous coherence of reality (572-3).

While the individuality, coherence, and mutability belonging to reality are eliminated in the laws that are designed during preparation for technological forming, the attempt to achieve realization in conformity to these laws in technological forming itself calls forth resistance from the nature aspects of reality. The fundamental traits of this reality are, specifically, the individuality or particularity, the coherence, and the mutability of everything. Because of these fundamental traits, it is impossible to carry out the technological-scientific plan completely. There are limits for this method set by the reality in which technological forming is carried on. There are, for example, no fully universal products to be made. Neither is it possible to completely isolate the technological process from its surroundings, nor can absolute permanence be realized in a technological thing.[38] Thus the resistance of the nature aspects prevents any complete realization of control over technological forming. In one way or another, there will always be a scientifically uncontrollable remainder of reality that will make its demands and pose special problems for all attempts to achieve control from a distance (575).[39]

We turn now to the various factors that limit technological forming and offer resistance to it. Insofar as it is feasible I will indicate the means by which this resistance can be minimized, since theoretical control from a distance *is* realizable to some extent.

1. The natural *materials and energies* employed manifest differences in individuality. This is due to the fact that the milieus from which they are won can vary very strongly in character and also to the fact that changes in individuality can result from such meteorological influences as summer, winter, rain, and drought. In order to realize theoretical control from a distance to the maximum extent possible, people in modern technology strive to break natural materials down into their most fundamental building blocks and then build up entirely new materials from these elements (synthetics). Materials of a more homogeneous character are also achieved through the purification and mixing of natural materials, to prevent difficulties arising for control from a distance (semimanufactures). The artificial generation of energy also prevents irregularities – as much as possible – in the delivery of energy. Differences and irregularities in natural energy, such as water and wind power, must likewise be evened out.

2. By way of such factors as temperature, humidity, pressure, and so forth, the *milieu* in which technological forming occurs causes uncontrollable, singular, mutable influences on the control of technological forming from a distance. It is true that people strive to isolate technological forming in the milieu, but they are never entirely successful. In the "techniques" directed at the boundary between nature and culture, such as highway and hydraulic engineering, and architectural and mining engineering, isolation from the milieu is precluded in advance. Here there can be no question of *complete* control from a distance.

3. The *technological operators* employed are certainly designed with a universal structure, but each retains its singular character, however slight the differences may be. Above all, these operators undergo environmental influences that vary from place to place. Meanwhile,

they are somewhat less than immune to alteration through wear and tear, aging, breakdown, strain, temperature changes, and so forth. The *energy transformation processes* are also subject to meaning-disruptive alterations stemming from differences in origin and milieu. Moreover, the uncontrollable differences between technological operators affect these processes equally contrarily. Conversely, the energy-transformation processes may affect the technological operators disadvantageously – for example, in cases where rising temperatures occasioned by the energy processes result in alterations in the structures of the operators, which then in turn have a meaning-disruptive influence on the transformation of energy. Dippel is correct in stating that wear and tear, corrosion, and dissipation are among the most serious problems facing technology. Extending durability is one of the most difficult problems in the struggle of technology against "enemy nature."[40] Some of these problems hampering theoretical control from a distance can be partially alleviated – for example, by artificially aging the materials from which a technological operator is made. Thereby the limits of the technological-scientific method are merely shifted a bit, however, and not eliminated. The limits continue to exist.

4. To the extent that it is necessary to employ *laborers* in the forming process, whether because full automation of the production process is impossible or because it remains unrealizable for the time being,[41] the individuality and freedom of the laborers endanger the determined control from a distance.

In mining and architectural engineering and in highway and hydraulic engineering, a different situation exists in principle, for in such fields there is no question of control from a distance. The work is not scientifically determined, and the free laborer is accordingly not disruptive of its meaning.

5. The *consumers* of the products to be made all have different desires. Individual tastes and preferences with respect to the products in the market place, accordingly, offer resistance to the technological-scientific method and limit the possibilities of mass production. To make control from a distance possible in spite of this situation, it is necessary to reach a compromise. Any one uniform mass product is eliminated from consideration, and a diversity of variations of the mass product are offered instead. These can indeed be manufactured through a production process that is controlled from a distance, by introducing a variety of production processes which from time to time can be altered in such a way that the final product exhibits the greatest possible variation in the course of time. While it remains impossible to satisfy the consumer's individual desires perfectly, such an approach does make it possible for him to purchase a good product at a reasonable price. In the case of the singular final product – which may be a

variant of some mass-fashioned product – control from a distance in the last phases of manufacture is often impossible; a large increase in price is then unavoidable. Such large price increases are also found in certain instances where control from a distance is possible by means of so-called numerical control abetted by the computer, but where unit production is nevertheless very complicated and very labor-intensive. Such is the case with machine tools.

The factors just mentioned above fix boundaries for the technological-scientific method of modern technology. Perfect theoretical control of the forming process from a distance is precluded. This is due first of all to the individuality, coherence, mutual influencing, and mutability of all the materials, thing units and people involved in the technological process, and secondly to the milieu in which forming occurs. It is often necessary for people to tamper with the technological forming process in order to prevent or to correct meaning-disruptive influences. In some cases such necessary interference can itself be controlled from a distance, although no theoretical formula can be provided for this. This is the situation, for example, when automatic switches are employed to correct unacceptable deviations. It is not known beforehand when and how frequently the maximum acceptable deviation will occur. This "trick" is possible with the help of the cybernetic principle of feedback. More about this in Chapter 3.

The limits of the technological-scientific method are not, as we saw, absolutely immovable. They can be pushed back a bit by human management and interference. Yet, they can never be removed completely. Rather, these limits are sometimes even intentionally tolerated or strengthened. Such is the case when people decide against full theoretical control and thus prevent any fixation of the production process – perhaps because they anticipate technological developments that will lead to discoveries offering entirely new perspectives, discoveries by which the limits in question may be pushed back even further than would once have been thought possible. Economic reasons, too, may be the source of some products' incapacity to withstand wear and tear: the manufacturer's goal may be a more active market. People are able to adjust and correct the limits in question, but never to abolish them.

The setting of the parameters for control from a distance, insofar as this lies within the realm of the possible and the capacity of people, occurs through consultation between the designing, production and commercial departments, each of which is concerned in some specific way with one or more of the limiting factors in question. Since their respective interests and responsibilities are usually not entirely compatible, a mutually acceptable compromise must be reached.

1.4.4 Mass production and individual production

We have seen that the technological-scientific method of modern technology makes possible the utilization of technological operators. The production process is characterized by a neutralizing partitioning of things and of facts, by the construction of a whole from universally employable parts, and by the procurement of an energy process from neutralized processes. The result obtained is that in various stages of production, use may be made of mass products. For the greater part, this applies to semimanufactures. Yet, end products, too, may sometimes be mass-produced. In general, mass production is striven for wherever possible (586-7).

In brief summary, the following factors lead to mass production.

1. After the fashion of science, the engineer abstracts from individuality and analyzes component problems for which he seeks universal solutions. Under the influence of the desires of the consumer, differentiation in the solution of the whole becomes necessary. Using this method, it is possible to take components with a universal destination (which have of course been mass-produced) and to integrate them into a differentiated mass end product (589).

2. The technological operator, thanks to technological-scientific control, has a fixed facility, which implies mass products. It admits of slight variations in fashioning through the exchanging and adjustment of parts. As a result, the products which people want to fabricate with the help of a technological operator must be fitted to the discontinuity of the production potential of technological operators. In setting the production process, the design department must take into account the capacities of the technological operators at its disposal (589).

3. A further motive is that it becomes a case of internal technological economy to use mass production to compensate for the tremendous effort required to design a technological operator (589).

4. More often than not, the needs, wishes, and desires of consumers are similar. Mass production can prove sufficient, given a broad market and a reasonable price.

5. Along with these motives for mass production goes that of the *entrepreneur*, whose goal is to produce a reliable, durable end product at the lowest possible cost. In the case of mass production, the entrepreneur must see whether a sufficient market exists to justify undertaking it. He may enlarge his market through regular modification or renewal of his product. On the one hand mass production is thereby made more expensive, but on the other hand the greater variety may stimulate sales so much that costs are saved as a result. Advertising can also stimulate the market and thereby facilitate mass production.

As we have seen, these factors that lead to the necessity of mass production do not have unlimited force or validity. If a customized end product is to be made, little reason remains to develop and adapt a technological operator for it. In some phase of the individualizing function-integration, laborers will be called upon for the technological forming. The engineer, then, is a person not only of science or theory but also of construction and invention. To complete the production process and the product theoretically is at the same time to determine them: thereafter, for weighty economic reasons, they are extremely difficult to alter. Even though the engineer joins in the exercise of technological-scientific control, the fact that he is the inventor will lead him to oppose any unwarranted rigidification of the plan of production. The more thoroughly established the technology, the more important the inventions that are required to achieve a new breakthrough. This is the reason why the development of a fully automated production process is sometimes postponed. This need not mean repudiation of mass production, for it always remains possible to utilize people in the production process (590). I will have more to say about this in the following section.

Compromise between the engineers and the entrepreneurs is always required in deciding whether and to what extent mass-production methods will be employed. This pertains to both the means of production and the product, and it covers the degree of customizing to be incorporated once mass production has gained the nod (591).

For the *end product* the question of mass production is of great importance. In the case of the mass product, one touches on the ultimate destination of integrated production in the face of individuality in the marketplace. And the decisive question, then, is whether it will be possible to maintain mass production in the final stage or whether it will be necessary to go over to customized production.

If complete mass production should indeed prove feasible, the complex of individual desires of the consumer must be analyzed and the many individual facets must be placed beyond consideration in order to arrive finally at the fixation of several types. The consumer will then have to be satisfied with a leveling of variety in the market structure. Thus the individuality of the automobile with respect to the destinational function is neutralized as much as possible in the uniform product. The different types of automobiles, in all their variety, display such a rich diversity that leveling with respect to final mass production is scarcely a drawback.[42] In general, the mass production of an end product that is intended for a service relation to people is more difficult to accept than the mass production of an end product that is to serve only for technological forming or for cultivation, as is the case in agriculture with artificial fertilizers (586, 596).

A. G. M. Van Melsen points out in this regard that a mass product as end product has an unmistakable leveling effect on the user. The (theoretically) average user as a profile of the many users with their various choices and desires is ultimately the true user. Every user, says Van Melsen, is in the long run a representative of the average user. People adapt to what technology requires: standardized goods for standardized users, for the convenience of a mass.[43]

The following remains to be said in conclusion. In all mass production, the factors discussed above (1.4.3) which offer resistance to the practice of the technological-scientific method dictate that differences will appear in the products, whether semiproducts or end products. These differences derive primarily from the imprecise functioning of technological operators and from variations in raw materials. Depending on the destination of the mass product, these divergences must be held within set limits. Norms for acceptable deviation are established even for semimanufactures which, after processing, are utilized in the integration of a thing. If only the slightest deviation is tolerable, extremely costly production may be required. Then a solution may perhaps be found in a rather less refined, normal production, with the product of the desired high quality being sorted out and selected from among the general run. The remainder would be reserved for some other purpose. In this manner, inexpensive production may result in a small number of high quality products (597).

In mining engineering, architectural engineering, highway engineering, and hydraulic engineering, mass end products seldom occur. Here, as we saw in the previous section, full control of technological forming through the technological-scientific method is not possible, since the object must function in a singular milieu. Nevertheless, when it comes to constructions that are common in these "techniques," every effort is made to produce parts that are universally applicable.

In architectural engineering, people use prefabricated elements. The result is a leveling in the domestic sector. The same phenomenon occurs in bridge building, where the assembling of neutral elements results in a marked leveling of individuality in the end product. The technological-scientific method thus makes itself felt in these "techniques" too, although insuperable difficulties will always remain to block general control of technological forming from a distance.[44] In electrotechnology, by contrast, general control from a distance is not only possible but also necessary, since it constitutes the very heart and *kernel* of this technology. Electrotechnology could not exist apart from the technological-scientific method and control from a distance (598).

1.4.5   The role of people in modern technological forming

It is important to inquire about the extent to which technological-scientific control of the production process may have altered the role of people in the execution. In general, the situation would appear to be as follows.

1. The utilization of technological operators results in the exclusion of people from the forming process. In comparison with earlier times, great advantages are thereby gained. With respect to both skill and energy, the technological operator is far more capable than the laborer. Above all, much time is saved, and the quality of the products is far superior. The latter is possible because stricter limits for fashioning result in a more regular piece of work.

2. While full automation is feasible in mass production, *people* must always establish the original conditions in which the technological operator is to function. These conditions concern the milieu in which technological forming will occur, the energy systems of the technological operator itself, and the destination of the process. After establishing the original conditions, people give the command to start operation. To the extent that a laborer continues to participate in the *form-giving*, he has (or shares) a *controlling* task in which the element of freedom is still real.

Free formation by people is certainly essential to the manufacture of the individualized end product. The engineer may set the law for such an end product, so that in this respect there is indeed a restraint on the laborer. Where nonautomated assembly is the conclusion of technological objectification, however, human freedom is manifest in people's ability to participate in deciding how and when assembly will occur, even though the result is established beforehand by the pertinent specifications.

Finally, there is room for handwork in the *servicing* of technological operators. Admittedly, this will be reduced as more extensive use is made of the possibilities of cybernetics. Such servicing addresses nuances of the process that may be beyond the reach of science or uncontrollable from a distance. The various disruptive influences of wear and tear, aging, and so forth are beyond theoretical control. As we have seen, they set the limits of the technological-scientific method. To overcome these meaning-disruptive factors, the laborer has at his disposal adjustment objects and switching, regulating, and control objects (530-8, 603, 612-19).

Technological labor of a controlling or servicing type is carried on in modern technology by laborers who have attained a very high degree of technological expertise. Some even command exclusive capabilities.

34

In general, the essence of the matter is to be found in *manual dexterity* where *controlling* work is concerned, and in *technological insight* where *servicing* work is concerned (574, 601, 605).

3. Handwork is in a grave position in the case of mass production, where all that is desired of the laborer is a subfunction already fixed by theoretical control, one which must be repeated time and again. When this subfunction has been made as elementary as possible, one may even speak of a *standard procedure*. This procedure may still be too complicated for a technological operator; or there may be so many steps involved that, even when they are quite simple, automation remains economically not feasible. This occurred, and still occurs, in the assembling of radios and television sets. Thus the reasons why people in a given situation have not yet converted to unlimited application of automation may be either technological or economic at bottom.

The situation of these handworkers is far from satisfactory: to perfect and maintain theoretical control, the individuality and freedom of the laborer are eliminated as meaning-disruptive. The laborer is expected to perform just like a technological operator. Through repetition of the most simple acts, he is expected to exclude all mistakes of thought or skill. He is permitted neither to think consciously nor to act freely; rather, he must allow his instinct for rhythm and routine to dominate. "Insofar as ... man is a laborer, direction stems from the results of his passion for work or from the mechanical propensity of his machines."[45] The consequence of this standardization of labor is a leveling influence on the personality; such labor deadens the mind. In the specialization of his work, the laborer is deprived of an overview of the whole. Hence he lacks insight into the coherence of the production process.

The result of this devaluation of human labor is that those who are involved lose their initiative; given such standardization of labor, responsibility and freedom are things of the past. The reduction of human personality to a predetermined function leads all too easily to the loss of any sense of calling. The factory has taught the laborer that his individuality and free personality are not useful but harmful – a situation that is aggravated by the technological and economic integration of the enterprise (601-7).[46]

Such difficulties may be alleviated somewhat by providing for the laborer to switch his post on the production line from time to time. While it is possible to let him move along the line to attain an overview of the whole production process, his work is still predetermined. It can be carried out just as well – and even better – by a technological operator.

4. In practice there are all sorts of transitional types between the

free laborer (point 2) and the standardized laborer (point 3). Furthermore, even the freedom of the former, while its possibilities are greater, is relative; he does not enjoy the freedom of the artisan. *The modern laborer, whatever the responsibilities he may bear, is subject to severe limitations* (601).

5. It is important to note that with the progressive development of technology, ever fewer people will be able to find work in technological forming. Full automation liberates people from work on the production line.

Yet the problem for these people is not thereby solved. Full automation requires very highly qualified workers. Those on the production line often fail to achieve the level demanded. They are doomed to unemployment or to other jobs that have not yet been automated.

The same development explains the shortage of natural scientists and of personnel who are qualified in technological-scientific respects. Van Melsen states synoptically: "For any type of labor which is objectified in the machine, whether this labor formerly was intellective or manual, operative or directive, can remain fruitful only if there are human beings who intellectively and materially control this objectivation and thus make it human again."[47]

The retraining of redundant laborers for new functions will, as a rule, be difficult. In many instances, the production line will have reduced them to mere parts, as it were. Moreover, progressive rationalization will gradually eliminate inhuman work of the type to which they have become accustomed. To realize and maintain full automation, a level of academic achievement is required that relatively few are able to attain. Strangely enough, work is now demanding the kind of responsibility that was systematically exorcised in the first place. Education – forward-looking education at that – will be required to meet the challenge of the future.[48]

6. It is likely that entirely new crafts will arise to parallel this development. In these crafts the modern panoply of tools and instruments will be utilized, as will the new materials. The disadvantages of modern mass production and of the work required in modern technological forming will thereby be moderated in some measure. The leveling of products and the threat of unemployment can be broken through and prevented by means of such new, future crafts based upon modern technology. The attendant rise in costs will not be an impediment, given increasing prosperity based on technological development.

In concluding this section, it can be stated that while many workers were once reduced to inhuman labor, the tendency of modern technological development is such that they will now be entirely eliminated from the production process. Reeducation will be feasible only for a very few. A solution for others will have to be found in the direction of modern artisanship, which could result in a reawakening of the so-called independent trades. Let us hope that the new demand for technologically informed people who can think creatively is a challenge that will be satisfied by the coming generation. Education – self-renewing education at that – must prepare people for new scientific tasks and new technological-scientific tasks. The central burden of human labor in technology will come to be found in the preparation, in the designing. In short, ongoing technological development now demands a total elevation of sensibilities and of response from all who are involved in technology.

1.5 Technological Designing: The Preparation

1.5.1 Introduction

Earlier I remarked that in modern technology, designing has been uncoupled from execution with regard to both place and time. The connection between these two sectors of modern technology is found in the objective design. The objective design is a deposit of the productive fantasy. Its destination is to be the blueprint for technological forming.

In my explanation of technological objects (1.3) and technological forming (1.4), I had occasion to mention the designing of these objects and the blueprinting of their execution. I will now return to this matter to concentrate explicitly – if briefly – on the various aspects of technological designing. Here the role of the engineer merits special attention.

The modern engineer is not satisfied with intuition or with merely practical experience of uniform reality. He wants to penetrate to the very order that rules the things and the facts that he seeks to fashion. Thus he does not concentrate on a particular concrete problem; rather, he steps back from it. He takes distance from the milieu, from the here and now. He addresses himself to the law for the technological and tries to formulate it as an abstraction from the specific technological problem. Familiarity with the law accordingly becomes the basis upon which the engineer attempts to make designs for technological things and facts. While it is true, in view of all this, that designing is scientifically founded and characterized by the scientific method, it must not be forgotten that designing in itself is never scientific but always *technological* in nature. Designing transcends science, as it were. Consider how the following tripartite delineation of designing indicates this.

1. Initially the technological problem is described, analyzed, and scientifically processed and formulated.

2. Such established knowledge (of the laws) of things and facts as may be expected to contribute to a solution of the problem is gathered. When this has been done, the productive fantasy can begin to function.

3. The engineer imaginatively conceives technological solutions to details, or even to the problem as a whole, and then proceeds to formulate them scientifically; that is, he positivizes the product of his fantasy in a design of the law for the technological object.

Demarcation of these three phases of designing is seldom clear in practice. Only in the sudden springing into consciousness of something entirely new – in invention – is the distinction clear. For the rest, there is a constant interplay of the three phases. Theoretical and experimental operations are employed to ensure that the objectification in the design will provide a possibility for future realization in the execution.

1.5.2 Engineering

The initial work of the engineer is to formulate the technological problem scientifically and to ask which technological laws may offer a solution to it. A problem usually arises in an environment already marked and conditioned by technological objects, technological production, products, and basic materials. In designing, however, the milieu is temporarily ignored.

Designing is always characterized by its distance from technological forming in the stricter sense. The problem is always separated from the particular concrete coherence via abstraction and analysis; it is dissected into component problems and then posed in its general sense. The designer attempts to grasp the law for the technological component problem in his imagination in such a way that he can represent it in symbolic form. A specific problem is solved by taking the solutions to component problems and adjusting them to each other and so achieving integration or, in other words, a design. This design, in the form of a technological specification or drawing, serves as the instructions for eventual technological forming (643-4).[49]

Designing is the formulating of a complicated technological law. It will always remain to be seen if a given formulation is correct. Accordingly, as the engineer is busy working on his design, he seeks contact from time to time with the milieu of the technological problem he is trying to solve. There is a constant interplay between the design and the antecedent environment, which is technologically conditioned. The engineer experiments. He so arranges the environment that he gains an answer to a question about technological order. He then sets the answer back out on the theoretical plane and utilizes it in the design. Such interplay is particularly important in complicated designing, where progressive specialization of the design is increasingly necessary. In such instances, established technological knowledge and the productive fantasy are inadequate to the achievement of a finished design. An engineer must engage in experimentation in order to probe the uniformity of the milieu of the technological problem (625).

The theoretical plane is the *basis* of technological designing. It is from there that the technological imagination takes its departure. This imagination or fantasy is the key to designing; its aim is an objectified design. Designing is accordingly consummated as intercourse between fantasy and theory. Its fruit is the *intentional* technological forming of a design that can be deposited in an objectified design. The object fashioned in keeping with this design comes into existence in correlation with this design as the law that governs this object (624, 626).

It is a chronic problem of the engineers of our time that the technological knowledge of any given engineer does not increase in proportion to the pace of technological development and is accordingly deficient with respect to the whole body of relevant technological knowledge. *Specialization* and *teamwork* are required to alleviate the unfortunate consequences of this situation.

As a result of this increasing, unavoidable specialization, the engineer loses contact with the great number of possibilities and variations in technology. He is no longer in a position to perceive coherence in its manifold diversity and is no longer sufficiently familiar with problems of the most general import or with problems of the most striking particularity. This entails a certain impoverishment in specialization which can engender difficulties for technological development. Working in teams is intended to prevent such difficulties. Yet, the task of the modern engineer is the heavier as a result. The dimension of necessary cooperation is upon him. The engineer must be prepared, in view of the inadequacy of his personal knowledge, to place his confidence in others. Nonetheless, from such association derives the striking benefit of specialization: more can be achieved in technology (643, 646).

In conclusion, it may be stated that a particular problem is rendered suitable for scientific formulation, subjected to designing (now often a team activity), and brought to a systematically constructed solution. The design is subsequently actualized in the milieu in which the problem arose. To the extent that conditions may vary from those envisioned when the problem was originally posed, the milieu must adapt itself – or be adapted – to the engineer's solution.

There is a final special feature of the work of the engineer to be noted: the fundamental technological facts that concern him may be applied in diverse technological situations. Consider the following examples: the transformation of energy from one kind to another; controlling, measuring, safeguarding, testing, tuning, switching, regulating; the purifying of natural products. It must not be forgotten, however, that inventions may afford entirely new solutions for the parts of a technological design as well as for the design as a whole. This means new possibilities and, of course, new problems.[50]

1.5.3 Technological science

Technological science (which, etymologically speaking, should really be the exclusive meaning of the term *technology*),[51] is a cultural science because it *attempts to formulate knowledge concerning human activity that is inherently technological in its significance*.

Such knowledge concerns first of all the order of technology – the laws for technological things and facts. This knowledge derives from the regularity of established production processes and objects and is constantly enlarged by the fresh projection of intentional forming. Intentional forming is itself technological in nature, as we have seen, even if it is founded upon and stimulated by the scientific method. Yet the fruits of intentional forming belong properly to the technological *knowledge* of the engineer. As we noted, technological science is always the basis from which the productive fantasy takes flight; intentional forming always returns, via the projection of the design, to this same basis. Accordingly, intentional forming increases knowledge pertaining to the technological. The *meaning* of technological science is in the first place the conservation of knowledge recorded in theoretical formulations. Secondly, technological science clarifies the intuitively grasped insight as it pertains to technological forming. Finally, technological science is the basis from which technological forming can be scientifically managed and controlled. The engineer is thus the builder and the elaborator of technological science (637).

Technological science is a distinctly cultural science because free human beings play a decisive role in its field of inquiry – the whole terrain of technology. This is true especially in designing and inventing, but it is also true in the execution to the extent that the technological-scientific method does not – or does not yet – make control from a distance with the help of automation possible. To the extent that automation is realized, the limits of technological-scientific control become visible as the limits of a certain area of technological science.[52] But beyond these fixed boundaries, technological science has a larger field of inquiry: within its domain is the *free person* in technological forming, in designing, and in inventing. That technological science is accordingly a cultural science is all too easily forgotten, especially in engineering circles (639, 648-51).

In summary, technological science, as the long way around for designing in modern technology, affords the following *advantages*. Technological problems are systematically dissected. Nature is analyzed with an eye to its possible technological functions. From the standpoint of method, solutions for a concrete technological problem can best be sought in elementary solutions to component problems and in possible integrations of component solutions. Through the scientific way of working, many problems are resolved at a stroke, with such universally applicable solutions as solder and the rivet. Solutions are recorded and thereby preserved. Control of technological forming from a distance affords possibilities for automation and mass production. The industrial context in which production occurs, accordingly, is greatly enhanced, and new technological possibilities are more easily realized as a result.

However, there are also disadvantages tied to the technological-scientific way of working. In brief they are the following. The professionalization and depersonalization of technology are unavoidable. There is a danger that technological development will seize up or petrify because of the theoretification of technology. Mass production entails leveling. Human responsibility in technological forming is all too easily diminished. (The best example, of course, is mind-killing work on the production line.) In a technological setting, the person can become alienated from the meaningful technological unity of the end result simply because all his attention in scientific analysis and abstraction and in finding solutions is focused on a *component problem* (which, of course, is by no means unimportant). The final product and its consumer can all too easily disappear from view.

1.5.4   The relation of physics to technological science

Physics endeavors to formulate the laws which control the physical aspect of reality. In technology, physical subjects are the points of departure for form-giving. This being the case, people in technology require both technological science and physics as the basis for designing. The difference is that physics addresses general knowledge of the physical, while technological science addresses the laws for the order of technology and the laws for technological things and facts. More specifically, technological science strives for knowledge pertinent to technological forming – the forming accomplished with and from the physical subject, which is thus disclosed in its technological significance. To the extent that people are absent in technological forming, physics and technological science are closely akin. The terminologies of these sciences are similar more often than not; the differences in the symbols used derive from their different bases. The symbols utilized in technological science are often borrowed from images afforded the designer by previously accomplished technological forming (641-2).

Because technological forming involves physical subjects and because people are being increasingly eliminated from technological forming, it is possible to make wider use of mathematics in the technological sciences.

There is an interplay between physics, technological science and technology. On the one hand physics is the basis of the other two, but on the other hand, both technology and the insights earned in technological science afford opportunities for the advancement of physics. Noteworthy in this regard are observations and experiments made with the help of the technological apparatus; they furnish answers to questions concerning the physical order. In such cases the technological apparatus functions in the service of physical theory formation. Such assistance of physics by technology can lead to new physical knowledge, which can in turn abet further technological development. A discovery – such as a sudden insight into a law for the physical – can be a condition for or an introduction to an invention; with every discovery, the framework within which inventions can occur is enlarged. The invention can then prove useful in the attainment of new physical knowledge, and so forth.[53]

As we shall have occasion to observe, the scientific foundation encompassing mathematics, physics, and often chemistry and technological science as well is of tremendous significance for the cultural power of modern technology and for the consequences of this power for the whole of culture.

In view of the numerous misconceptions afoot concerning technology, it is not superfluous to point out again that modern technology as such is not a science. As W. H. Raby remarks, it is "more than mathematics and physics because, even today, physics is theoretical by nature. Physics lacks, technologically speaking, the creative element."[54]

1.5.5 Inventions

Continuity of development is effected in the ongoing interplay between the productive fantasy and theoretical reflection and formulation. In

42

the intentional forming of an invention, however, the invention forsakes such interplay at the critical moment and in an inimitable, unanticipated manner gets on the track of a new technological entity within the productive fantasy. An invention entails a discontinuous development process.

It is difficult to pinpoint the beginning of an invention. Various factors such as need, chance, demand, venturesomeness, and imagination, whether alternately or in combination, usually play a role. Nevertheless, invention is more than the sum of such factors, for they do not serve to explain adequately the element of the unexpected in invention.

Invention occurs on a higher plane in modern technology than it did in classical technology, for invention in modern technology has at its foundation mathematics, physics, and technological science. This means that whenever the first trace of a new technological entity appears in the productive fantasy, the inventor attempts to work out his invention theoretically and to size it up quantitatively. Thus he completes it. In this way an invention becomes a fitting design for the objectification of concrete technological structures. In the decisive moment of invention, theory plays a negligible role. Subsequently, the new design for a universally applicable unit of the technological order is added to the terrain of technological science, which provided the basis for the invention in the first place.

Delivered to the theoretical level, an invention becomes subject to neutralizing function-articulation and individualizing function-integration. It is not itself a neutralized building block of technology, but it forms the frame within which it becomes possible to make neutral building blocks in different varieties. The neutral building blocks which derive from an invention can be utilized in a great variety of ways in widely differing situations.

Inventions may be divided into the basic invention, which pertains to a total structural entity, and other inventions, which improve only the elements of such an entity. Each kind of invention may involve either things and materials, in the first place, or facts in the second.

As we noted earlier, technological science is very important as the theoretical basis for invention. Yet it is at the same time a danger, since theoretical petrification is not inconceivable at this niveau. The increase of theoretical devices and their use threatens to sterilize the productive fantasy. On the other hand, invention may offer the possibility of breaking through an unpromising or rigidifying development, and so of achieving fresh perspectives. Computer development has served to make this quite clear.

Modern physics and chemistry are also, as noted, both a basis and a framework for the possibilities of technological inventions. In intentional forming, invention transcends technological science, physics, and chemistry. For the goal of invention, after all, is the solution of a technological problem. Invention implies a cultural task (671). Owing to its theoretical basis, invention is more likely to occur in connection with teamwork than in connection with isolated individual effort.

W. H. Raby is correct in pointing out[55] that technological invention and ensuing development are dependent not only on mathematics, physics, chemistry, and technological science but also on the material situation of technology itself, which is prerequisite to actualization. Babbage had worked out a calculating machine in theory – on paper, that is to say – as early as 1823, but its actualization was frustrated by many difficulties. Economic limitations aside, the technological possibilities were not yet present. Remarkably enough, the situation that exists in computer technology today is just the reverse: the capabilities of the computer exceed man's capacity to put them to use. It can be said that in this case actualization is really far in advance of new technological ideas. Human fantasy is inadequate to the discovery of all the computer's potential uses.[56]

Every invention in modern technology entails two kinds of newness: the newness of the invention itself as a technological law-entity, and the newness of every object fashioned along the lines of this new law-entity. With respect to the latter, it should be noted for the sake of completeness that at the design niveau an invention can be endowed with any of many possible variations in technology – for two reasons. First, there is still freedom for individual form-giving outside the objectified design. Second, every discovery becomes an object itself, suitable for further technological development.

It is not always easy to say when we are confronted with a discontinuous rather than with a continuous development in technology, and therefore with an invention. There are borderline cases. This is most often so in instances where the invention is incorporated into a continuous technological development or where it serves to introduce new developments.

Dessauer, in whose philosophy of technology invention occupies a central place, distinguishes inventions in continuous and discontinuous technological development as "developmental inventions" and "pioneering inventions" respectively.[57] With this distinction he suggests that the productive fantasy always leads technological development and that it sometimes breaks through it in a surprising way.

From a legal viewpoint (in connection with patenting) it is important to define what is required of an invention. It must involve a new, qualified, general, and integral entity of the technological order of things or facts. The inventor must orient himself in two directions. In the direction of individualization he must investigate precisely what is

44

and is not comprehended in his invention; in the opposite direction, the inventor must answer the question whether his invention is perhaps a particularization of some still more general facet of the order of technology, in which case more new inventions might be possible (669-72).

Invention is really an unanalyzable activity founded upon human freedom. Psychology indicates that creativity, as the expression of the freedom of persons, can be promoted through special education, through "brainstorming," and through the learned behavior of tolerating proffered criticism. Totally consistent with this are the following marks of the inventor: a flexible style of thinking, playfulness, youth, an inquiring mind, the capacity to view freshly what is already known and thereby to deviate from traditional solutions, a high grade of intuition, a tendency towards perfectionism, introversion, great intellectual ability coupled with courage, and a weak social posture.[58] All these stipulations are but conditions for invention. At bottom, invention is a mystery.

1.5.6 Theoretical and experimental operations

We have seen that the engineer, whether he is initiating intentional forming or concluding it, avails himself of technological science and its possibilities in designing the laws for the technological. In the former instance (i.e. initiating intentional forming), he does so in order to achieve a theoretical formulation of a technological problem; in the latter instance, he does so in order to positivize a discovery in an objective design. In technological fantasizing and the creative advancement of technology, people are able to operate with symbolical formulations of the technological order. Technological science makes it possible for intentional forming to be accompanied by scientific operations. These scientific operations give technological creativity an instrument for designing (673).

These scientific operations are set up according to the scientific method, and their meaning is the *symbolical* objectifying of the technological order. These operations may be divided into two groups: (1) *theoretical operations* and (2) *experimental operations*. Each will be discussed briefly.

Re 1. Theoretical operations serve in the first instance to make it possible to deal with technological questions and problems. They may also aim at deriving the desired, concrete solution from previously established formulations of the law. The symbols employed in these operations are a rendering of the technological law. They are aimed at stimulating insight into this law; they incite the imagination. Above all, they facilitate the utilization of theoretical operations in the support and guidance of intentional forming.

The structure of modern technology allows two types of symbols to be distinguished. Symbols for things indicate the relative durability of technological operators, while symbols for facts reflect the variable structure of energy transformation processes. These symbols can also be differentiated for the constant and variable magnitudes of technological objects; that is, they can represent various technologically disclosed aspects of these objects – the physical, the spatial, or the arithmetical aspect (674, 679).

Further distinctions divide theoretical operations into *analysis*, *drafting*, and *calculation*. These operations are mutually complementary, and the uses made of them depend on the simplicity or complexity of the technological problems involved. In one case, calculation seems called for; in another, analysis. Given the pretermissions in the delineation of a problem, analysis may pass over into calculation. Pretermissions are consistent with the tension between theory and individuality, with the changeableness and coherence of reality itself, which we discussed earlier. Accordingly they are permissible, for the law pointed to *in the theory* never corresponds perfectly to the particular and unique in reality as such. This matter is an instructive one to keep in view, for pretermissions in the theory may necessitate corrections in the pertinent individual design. When that is indeed the case, fresh approaches to reality must be made for the sake of securing additional interaction for the theoretical operations. A very common "approach," which makes calculating meaningful after pretermissions, is that of establishing the safety coefficient of stress-bearing technological constructions. In the matter of *pretermissions* and fresh *approaches*, the connection between calculating and the technological is constantly maintained. Calculating, then, is not to be construed as an exclusively mathematical affair (680, 685, 690, 691).

Important types of deductive theoretical treatment are: *calculation*; the graphic formulation and presentation of the order of technological things in *schemes* (which indicate functions) and *drawings*; the graphic formulation and presentation of the order of facts in *diagrams* (which show discontinuity) and graphics (which show continuity); and the *technological specification*. Theoretical operations, then, are just as applicable in technological science as they are in physics, for two reasons. First, technological forming involves structures which, as we saw earlier, have the physical function as the last subject-function. Second, in control from a distance, at least, man does not appear in technological forming (677).

Calculation involves relations between arithmetical symbols of technologically characteristic magnitudes – relations which man subjects to mathematical operations to advance knowledge of the law. Analysis lets us down at this juncture (685).

The graphic treatment of thing-laws in *schemes* and *drawings* involves the symbolical representation of the laws only for things as such, that is to say, for things in abstraction from the energy transformation processes which are intended to be actualized with these things. (To provide a household example, the energy transformation process of vacuuming is not actualized until the vacuum cleaner, a thing, is switched on.) The design is objectified in the form of *geometrical symbols* and is complemented with *arithmetical symbols* and *short descriptions* on the drawings. The geometrical operation is completed with the drafting and often goes hand in hand with calculating. The drawing may be more than the reiteration of a schematic structure; it may perhaps be intended to furnish a guideline for a particular production process (686).

A *graph* or a *diagram*, in contrast to a drawing or a scheme, always provides a representation of a fact characteristic; it involves the energy-transformation process (689).

Finally, a *specification* is a description of a fact – and of a fact of production, at that. Like the construction drawing or blueprint, it serves as a guideline for fabrication (689).

Re 2. The designer-engineer will switch to *experimental operations* if the theoretical operation (especially calculation) is no longer feasible because he lacks an adequate point of departure or perceives that ultimate failure is inevitable. Irresolvable differential equations are a case in point. *Experimentation* must be utilized to furnish solutions. Chemical technology existed for a long time simply by the grace of experimentation. We are reminded in this regard of qualitative chemical analysis, of classical organic chemistry, and so forth.[59] Here calculating is strictly a secondary matter. More common, however, is that a designer seeks *to establish or correct a theory* through testing.

It is characteristic of experimentation that a designer so arranges reality that he can gain a clear answer to a specific question. The answer is attained, via measuring instruments, in the form of arithmetical or geometrical symbols which are subsequently processed into knowledge of the law for technological reality (673, 692).

There is interplay between theoretical operations and experimentation. The two complement each other. On the one hand, the designer particularizes generally formulated laws in his theoretical operations, while on the other hand, the experimental method necessitates his taking the particular result of some test – even if the setup of a test already abstracts many particular characteristics – and using it as a point of departure from which to arrive ultimately at laws of general validity. Over against the deductive method of theoretical operations stands the inductive method of experimental operations. What this means, of course, is that apart from certain central matters, the final results of a calculation do not present us with many of the characteristics that will in fact appear in the functioning of the object in question. It also means that there are numerous matters implicit in the results of an experiment that were not comprehended in the original question. This state of affairs is traceable to the singular, coherent and changeable portion of technological reality that cannot be brought under theoretical control. In calculation, the designer strives to approach singular, coherent, changeable reality. In experimentation, he attempts to eliminate such particular influences through better concentration on the question – for example, the influences of milieu and cohesion on a measuring circuit (693, 695).

Along with the other advantages of technological science,[60] the operations discussed above offer the designer the possibility of working with symbols that can guide and stimulate his productive fantasy and which can further technological development at a high niveau.

1.5.7 The utilization of computers in designing

It is a most interesting fact that the engineer is able to utilize the computer in designing – especially in theoretical and experimental operations. Hence he is able to accomplish a great deal more in much less time. Theoretical data can be so complicated and comprehensive that no one, whether working alone or in cooperation with others, can grasp them. Here the computer offers a way out. The same is true, more or less, of experimental operations. These can also be of such a nature that they are beyond execution, whether because their preparation would require too much time or because the number of factors involved in an experiment would be too large to allow reasonable insight into the results. With the computer it is sometimes even possible to imitate (or simulate) an experiment. In place of the experiment, it is then the computer that answers the questions posed. This is especially important if the experiment itself would be too dangerous to carry out, or even impossible.

Theoretical and experimental operations converge in the utilization of the computer. Thanks to the computer, the engineer's designing capabilities have now been enormously enhanced. The fruits may be observed in the development of space flight, the development of the modern chemical industry, the development of communications technology, and the development of the automation of both the technological and the nontechnological. Think of the automation of service-furnishing processes, for example.

Although the inventor, the builder, and the user of a science were once united in the person of the artisan, modern technology based on science has led to a remarkable differentiation of activities. Now the latest development in technology is that not only science but even the technological panoply of tools and instruments has become basic to technological development. The computer and the developments tied to it have led to the full automation of the execution. Furthermore, everything relating to designing that is predetermined can be taken over by the computer. In fact, in all sectors of modern technology, including the preparation, the machine is increasingly taking over predetermined work. As a consequence, it is solely in designing that emphasis is now being placed on the typically human; that is, on freedom in responsibility and on the creativity or productive fantasy that serves to express freedom in responsibility. The trend in modern technology is to limit the human role exclusively to creating and controlling.

This tendency of modern technology is anything but problem-free. While emphasis is indeed being placed on the creative as the expression of human freedom, technological creativity is becoming steadily more dependent on independently functioning technological operators. Creativity is thus becoming encapsulated, as it were. And as time goes by, it will become increasingly difficult to alter and renew technological development at all: technology, owing primarily to the influence of science, has acquired a universal or planetary character. Individuals are powerless in the face of this development. Will designing not be increasingly limited to the alteration of details? Or will that competence, too, disappear as machines are equipped to improve themselves and to design and bring forth other machines?[61]

The question whether it is possible to break through the dilemma thus sketched, in which the possibilities for human creativity are being enhanced at the same time as they are being throttled by the growing complexity of technology, is an important one. Or is this perhaps a false dilemma?

Moreover, if human creativity should actually attain a higher plane through this same technology and so come into possession of even greater possibilities – my own thinking tends in this direction – would the question whether a later generation could attain a similarly high level of technology and creativity not become most urgent? Even more important, would a later generation accept and embrace this high level and respond to the challenge issuing from it?

In Chapter 2 we will encounter some philosophical views of technology in which technological development is regarded as an autonomous development, as a suprapersonal power, and as a threatening fate.[62] The *transcendentalists* propose to resist technological development. In Chapter 3 we will confront some variations on the alternate theme: people are just now coming authentically into their own since they are just now becoming able, through technology, to hold their future in their own hands. According to the *positivists*, the successful technological-scientific method should be applied universally.

In Chapter 4 I will present a more or less thetical exposition of the many questions that will have been raised by then with regard to the theme of technology and the future.

2

The Transcendentalists

2.0 The Main Division: Transcendentalists and Positivists

At the beginning of this second chapter I shall briefly explain my way of classifying some of the thinkers who have dealt with technology.

In Chapters 2 and 3, the views of a number of contemporary philosophers who have taken an interest in modern technology will be set forth and critically examined. Chapter 2 will deal with the so-called transcendentalists, and Chapter 3 will take up the positivists.

In a philosophical sense, the distinction between transcendentalists and positivists has been of value since the time of the philosophy of Descartes. The transcendentalists find the starting point of their philosophy within man, and their philosophical thought focuses, for the most part, on the inner aspect of the human subject. In other words, the transcendentalists, in one way or another, honor the transcendental directedness of the experience of the human subject. Their archimedean point, that is, the origin, stability, certainty, and assurance of their philosophical activity and its results, resides within the human subject. While in earlier centuries the archimedean point was self-sufficient reason, for many transcendentalists in our century it is autonomous reason as freedom *behind* or *above* experience – often with some participation of the *ratio*. This freedom idea is the "transcendental" presupposition for the philosophizing of the transcendentalists to be examined in this chapter.

The positivists, by contrast, choose their archimedean point in the given reality outside man. They take as their base the facts that are *at hand*. In their philosophy they demand consideration above all for the determined nature of that experience. Observable experience is their condition for philosophizing.

The distinction between the transcendentalists and the positivists is of the greatest possible importance for any contemporary study of views of modern technology or any consideration of man's future based upon this technology.

In general the transcendentalists perceive a threat to the human subject in the power of scientific technology. They represent the notion that man's freedom is threatened by technology. They see a conflict between man and technology. Consequently, most transcendentalists assume a hostile stance toward modern technology, characterizing it as an autonomous power.

With the positivists the matter is reversed. They regard modern technology with favor. They see in technological development a confirmation of human power and an advancement of culture.

Accordingly, the transcendentalists devote a great deal of attention to the past. They do so in order to probe the question why technology has become a threat to man and why it appears likely to become an even greater threat in the future. For them the future of the technological world is decidedly not a meaningful one. Thus their "solution" to the problem of modern technology is largely inspired by the notion that the past was better. Nostalgia for the past or a desire to return to nature leads them either to resign themselves to technological development or to flee from it. Given their assumption, they can find no meaningful perspective for technological development.

The positivists call attention to the past only long enough to make it clear that the present is to be preferred to it, given the possibilities of modern technology. According to them it is primarily the progress of technology that has contributed to the elimination of human suffering. It is precisely from the development of technology that they gain their hopeful expectations for the future. In fact, technological control provides them with their model for the control of society, of human history, and of man himself. Just as technology in the sense of control of inanimate material has brought forth progress, so technological control of man, society and the future will usher in unheard-of well-being and prosperity. In short, the positivists are generally inspired by the notion that the future can be assured through technology. Their optimism with respect to technological development stands in sharp contrast to the pessimism of the transcendentalists.

The transcendentalists and the positivists are united in their advocacy of an autonomous philosophy. Over against technology's threat to freedom (in the philosophy of the transcendentalists) stands the confirmation of human power through technology (in the philosophy of the positivists).

The distinction between the transcendentalists and the positivists, however, does not imply an authentic division or separation. As we shall see later, the two views are in fact related. The precise character of their affinity merits thorough exploration.

In Chapter 4 the various problems which arise in Chapters 2 and 3 will be examined from a reformational philosophical perspective. This philosophical perspective, which rejects any notion of an *autonomous* philosophy, can cast a different and better light on the problems raised by transcendentalists and positivists.

2.1 Introduction to the Transcendentalists

In this chapter we will examine a selection of contemporary philosophical viewpoints on modern technology and its relation to culture.

In the first place, the transcendentalists all share a conscious concern that proper consideration be given to the transcendental directedness of all experience, including the experience of technological factuality. They strive to understand the origin, development, and future of technology. In their critical view of technology, they appear to set themselves against any worldview that is both natural-scientific and technological. They downgrade science and technology, and some of them even fail to distinguish between the two. To them science itself is really a technique, a theoretical technique. Thus they repudiate the positivists (see the following chapter), who borrow the norms for technological development from technology itself and more often than not utilize technology as a model for the control of interpersonal relations, of society, and of man himself. In short, the transcendentalists oppose philosophical technicism, that is, the view that absolutizes technological-scientific thought and results in practice in a technocracy bent on subjecting everything to the power of technology.

In the second place, the transcendentalists all share an appreciation of the relative autonomy of technology and of its tendency toward absoluteness. They construe technological power – however much they may otherwise differ in their interpretation of it – as an *autonomous power*. Intrinsically implied here is the consequence that they must eventually give up the struggle. In their alarm they seek a solution that is inherent in their presupposition! They do seek to transcend technological development, but in the sense of dodging it and yielding it the right of way. Such fatalistic resignation to technology or adaptation to it clears the field, more often than not, for a flight from technology, for a speculative hope of deliverance, for the safeguarding of some kind of transcending freedom, or for a return to nature. Some of them also make efforts to restrict the autonomy of technology.

Among the transcendentalists to be examined in this chapter there is a pronounced reluctance to found the assessment of technology upon such singular phenomena as environmental pollution or the atomic bomb. They all regard the problems of technology as involving more than merely incidental difficulties. For them the problem is universal and all-embracing; it is the problem of technology's growth toward technological collectivism, toward an omnicompetent technocracy. They are concerned, in other words, with what they perceive as a structural problem. Although their own starting point and the view of technology issuing from it may deprive them of the possibility of a meaningful perspective for the future, they are nevertheless justified in demanding that we consider the consequences of absolutizing technology and its possibilities.

According to the transcendentalists, modern technology eventuates inevitably in a total technocracy in which the individual is robbed of his freedom and transmuted into a manipulated mass person. The views reported in this chapter reflect the same "modern consciousness" of human loss that has been expressed by so many of the protest movements of our time. The views reported shed light on the causes and composition of the current feelings of discontent. It would appear that these feelings are closely connected to the development of modern technology.

I have made a selection from the vast quantity of available literature. The most important considerations have been that the views presented should be of a philosophical nature and that they should so complement each other as to afford a fairly comprehensive picture of current reflection on the development and future of technology. My selection was also influenced in part by my desire to clarify the differences between the transcendentalists and the positivists as much as possible.

For the sake of fairness, the main features of the thought of each philosopher will be reported more or less fully. Where one publication has followed another, I will point out any shifts of viewpoint which may have occurred and probe the possible causes and significance of such shifts. The purpose of my exposition is to shed light on the basic standpoint from which the transcendentalist view is presented and to clarify the deepest motive of these thinkers, that is, the motive that pervades and governs not only their view of the origin, development, and future of technology but also their potential solutions to the problems sketched. It is this motive that makes their view a coherent whole.

I shall deal with each thinker in two sections. After presenting a synopsis, in which only details will be criticized, I will go on to a critical examination of the main issues raised. The questions to be considered are: the nature of technology; its origin and development; its philosophical background; the relation between science and technology; the characteristics of modern technology; the relations respectively between man and nature, man and technology, and technology and the future; dangers and portents; the demonic element in technology; technology as earth-exhausting predatory cultivation; the autonomy of technology and technocracy; possible perspectives for the future; hope of deliverance from the process of technologization; and so forth. Thus my critique will touch on the entire area between the origin and future of technology. It is the future, however, that will receive the emphasis. The question of the meaning of technology is of central importance in this study.

I shall deal in turn with Friedrich Georg Jünger, Martin Heidegger, Jacques Ellul, and Hermann Meyer. Repetition will be avoided as much as possible. These authors are in agreement on many points; therefore we must be careful not to let important individual nuances escape our attention.

In connection with the examination of Heidegger, we will also take a look at Friedrich Dessauer's philosophy of technology. He, too, is a transcendentalist. His reflections on technology and its history proceed from man's original mandate to be a former of culture. There is not a breath of cultural pessimism about him. As an idealist he is really a cultural *optimist*. Because Dessauer advocates a synthesis between autonomous philosophy and christian faith, the remarks made in Section 2.0 about the transcendentalists as well as the remarks made just above apply to him only in a limited respect. In a certain sense this is true also for Ellul, who writes about technology as a Christian.

2.2 Friedrich Georg Jünger: On Perfecting Technology

2.2.1 Introduction

Jünger[1] is a novelist, a poet, and an essayist. Therefore we need not expect a systematic treatment of technology from him. Nevertheless, his view of technology is clearly philosophical in its intent. He concerns himself with technology, with what he considers its suprapersonal power and also with possible deliverance from its demonic development. He tries to understand technology as a cultural force in its origin and unity, and he finds words to express what many people experience in the face of technological development.

Jünger reminds one of Spengler, whose view of technology[2] was based on his fundamental thesis that man is a predator. For man as a predator, technology is a tactic; it is a means to power. But his will to power is at the same time his destiny, his destruction. Man literally expends himself in the will to (technological) power.

In Nietzsche's vein, Spengler's philosophy is marked by a heroic element, despite the final outcome (the demise of the West). Technology is man's crowning achievement, but it leads to his total destruction. To Spengler, optimism is folly; bravely man ought to pursue to the end the course that is set for him. Only dreamers believe in ways of escape.[3]

In Jünger, Spengler's heroic element has given way to anxiety and fear. Spengler would probably have regarded Jünger as a dreamer believing in escape routes. Jünger concurs in Spengler's judgment that the will to power is the great stimulus for technological development. They are also agreed that modern technology means no lessening of labor, and that technology and organization persistently threaten life with death and eventually destroy it.[4]

The considerable influence of contemporary technology stems, in Jünger's opinion, from man's current propensity to turn to technology in orienting himself for the future. Utopias were previously grafted upon the power and greatness of the state. (Francis Bacon is clearly an exception.) Technology has now largely usurped that role of the state. "A utopia requires a scheme that is susceptible to rational development, and technology is the most useful scheme of that sort that is presently available" (PdT 10).[5] Initially, expectations of the future based on technological development were fairly optimistic. Such was the case with Comte, for example. Later, skepticism appeared: Jünger notes Wells and Huxley as examples of it.[6] He himself belongs to the category of those who regard technological development as a threatening, all-destroying development. Through technology, the very opposite of what is generally proclaimed will actually be achieved. This is apparent from the fact that as technology progresses, free labor declines and forced labor increases. Moreover, man disposes over less and less authentically free time (*Musse*), since even his so-called leisure is governed by technology. "We must regard technological organization as a coherent whole, and we shall then see that there can be no thought of a lessening of the amount of labor, but rather that it is constantly increased by the progress of technology..." (PdT 16).[7] The machine does indeed take over much of man's work, Jünger agrees, but this does not imply that the number of workers – by which he means more than just those who work with their hands – decreases. Human bodies, he argues, are more necessary than ever for all the new collateral tasks. "We all become workers to the degree that we become dependent upon apparatuses and organization" (MuE 233).[8]

The deepest ground of Jünger's pessimism is his conviction that technology leads ultimately to unfreedom and thus to "nonexistence." Through technology, death penetrates life. This is the main substance of what Jünger has to say about technology and its development.

We shall now examine various subordinate themes: technology and property; technology and organization; technology as earth-exhausting, predatory cultivation; technological thinking; man and technology; technology and technocracy; technology as diabolism; technology and the future; and the basis for the hope of deliverance from the threat of destruction posed by technology.

2.2.2 Technology and property

According to Jünger, it is widely believed that technology confers wealth and freedom upon humanity. This belief is incorrect, for our wealth and freedom are in fact threatened by technology. Wealth and freedom are inextricably joined – wealth not in the sense of "having" but of "being." All talk of a technological wealth is seriously mistaken. In technology people are always striving to make something, but what they achieve does not match their attendant losses. Technology reduces them from "being" to "not-being" – to penury, to unfreedom, and to uncertainty. Property traditionally conferred upon its owner a sense of permanence, of security. This sense is worn away by technology. Technology ushers in a propertyless, demonic world.[9]

Property, which Jünger contends is continuous with time and space, is threatened especially by the machine. "It [the machine] does this in that it pushes aside the personalism inseparably associated with property. It separates the owner from his property in that it interjects mechanical specifications into property and makes it accessible in a mechanical manner, a manner which really cannot be identified with the notion of property" (MuE 251). There is enmity, as it were, between technology and property – enmity that intensifies rather than abates. It may be that technology was originally engendered by property, but there is no room for property in the technological collective. "While the technological collective flourishes within the capitalist economy, it at the same time undermines the structures of this capitalism, which wants to be an owner's capitalism. The movement has a force of law that brooks no resistance" (MuE 290-1).[10]

Technology does not consume private property only; it would be equally mistaken to think that other, nonpersonal property is increased through technology. "The distinction lies between property and the technological collective, not between private and public property" (MuE 267). The promises of communism, therefore, are also not attainable. The principles of both capitalism and communism are in conflict with technological development. Despite all appearances of success and increasing prosperity, the human condition in modern technology is that of pauperism, Jünger says.[11] He accordingly rejects the social and economic theories of the last century as a matter of course.[12] Marx failed to understand the machine and technological development because he approached technological development from an economic perspective. "Marx did not understand anything about the machine. He knew nothing about it, for if he had understood it, he would not have attempted to comprehend it as a tool and a component of a world founded upon economic laws – which it is not. It is anything but" (MuE 227). Marx gave assurances that the machine would usher in economic certitude and prosperity, but just the opposite has been brought about through the *consumption* of property: "...it has become senseless even to speak of economy, for it is only the process of predatory cultivation by the technological collective that materializes" (MuE 240).[13]

According to Jünger, the conception of Marx and Engels that the machine is a means of production is incorrect. Technology's power to consume is always greater than its capacity to produce; technology is not governed by the laws of economics.[14] Marx failed to perceive that since the machine generates more and more forced labor it cannot deliver freedom, Jünger says. Machines can be made without limit, to the utter destruction of whatever free labor may in some restricted measure be present.

Jünger maintains that neither human freedom nor the advancement of man's dominion is to be achieved through technological development. Technological development will erase the distinction between capitalism and communism.[15] That the true nature of technology so rarely comes to light is attributable to the stormy development of technology, which has ensured that only speed and change should be striking, and also to the belief in the progress of technology, which has obstructed a correct appraisal. "Man's belief that something good will accrue to him from the hand of the machine is sheer obstinacy. The advocates of the collective can go and push all happiness, faith, righteousness, and peace out in front of themselves like a wheelbarrow. It is a sour task, for the bearer of the burden never reaches the delectable freight; it remains always before him, suspended in a different dimension of space and time" (MuE 244).

If we are to understand the true nature of technology, we must understand its dynamism and also the belief in progress. Technology is found to be an all-consuming power. Jünger finds the laws of thermodynamics at work here: "...the laws of thermodynamics obtain just as well for the technological collective as a whole. It is a loss-collective which, since it destroys its substrata, must extend itself by force" (MuE 330).

From the above it is particularly clear that Jünger laments the loss of property on the ground that this particular loss entails the forfeiture of humanity's security and its confidence as well. Implicit in the *dynamism* of technology is an exchange of possessions, such as they are, for technological means.[16]

Jünger's concern is not really the loss of property as such; rather, he laments the plight of people who seek their own *secure* place in and through property. The new mobility of property puts humanity in jeopardy. The tenor of Jünger's view of technology and property involves what he perceives as modern technology's disquieting, dynamic proclivity toward one mammoth collective. And this speaks volumes, since it is in fact not correct to say that modern technology empties property of meaning; Jünger deceives himself in this regard. Certainly, technology lends mobility to property, since it must be made productive, but this is by no means the same thing as the loss of property.[17] It is precisely through technological development that many have gained possession of things which they could scarcely have dreamed of just a decade or so ago. For the rest, however, Jünger's view that people ought to have property in order to better bear responsibility is not incorrect. Many accordingly applaud technological development as a good thing that can lead to *property growth*.

From what follows it will become apparent that Jünger's greatest concerns are modern technology's *dynamic* character, which leaves nothing unaffected, and its impersonality. Thus it is understandable that he does not advocate a return to a feudal state of affairs; he favors rather a return to nature, the fixed, eternal, and unchangeable.[18]

2.2.3 Technology as predatory cultivation

In view of the above, it is hardly surprising that Jünger should characterize technology as predatory cultivation. Technology fully exhausts all natural resources. "All-affecting, ever-spreading predatory cultivation is the hallmark of our technology. It is predatory cultivation alone that makes it possible and enables it to unfold" (PdT 29).[19]

Technology as predatory cultivation everywhere leaves behind itself the ineffaceable traces of destruction. Increasingly, technology renders the earth a dead earth. The devastation is the mightiest in the case of radioactivity. "Wherever there is radioactive waste, the earth has been made unfit for human habitation" (PdT 30).[20]

Indeed, the destruction perpetrated by technology is not confined to the human environment and nature. Man himself is consumed for technology's sake. This is clear from the exploitation of workers in industry.[21]

59

Technology as depradation, as ruination, threatens even itself with annihilation. It is this danger, in fact, that is the stimulus for technological development. "The power of technology is universally threatened by decay and squandered in senseless consumption; the more precipitously it seeks to escape destruction, the more stubbornly and rapidly it is hounded by it" (PdT 32).

However correct Jünger may be when he claims that technology often threatens nature, man, and technological development itself, he exaggerates to such an extent that his credibility is impaired. This is a consequence of his conviction that technology is a suprapersonal, menacing power.[22] His view of technology ignores the true relation between man and technology. Man, after all, *does* bear responsibility for the mandate given him to unfold nature; in this sense technology *is* a striving against death and for culture.[23] It is true that man often goes about his task in a prodigal, reckless, myopic, egoistic manner. Nature is violated, and other creatures are exterminated. People foul their environment with rubbish and waste.

Yet, all this is assuredly not necessary. In keeping with their calling, people ought to develop technology in harmony with nature. From a purely *technological* standpoint, environmental pollution is preventable. The problems involved are ones of economy for both industrial concerns and consumers, all of whom should simply refuse to abandon their milieu to pollution, by willingly accepting the financial burdens necessary to prevent it.

Since Jünger deems technology an autonomous power, he has no room for any norms with which technological development ought to comply. On the basis of norms, that warped technological development to which Jünger is properly alert might well be given a new direction, a meaningful direction.[24]

2.2.4 The "perfection of technology"

Jünger believes that with the latest developments, technology as predatory cultivation is nearing its consummation.[25] By *Perfektion* Jünger understands *in the first place* that all possible stimuli for technological development are made subject to technological development itself. However often technology may have been regarded in the past as subordinate to economy, Jünger ventures a contrary opinion.[26] "Technology does not serve economic laws; economics is subject to an increasing degree of technologization" (PdT 35). He continues: "Technology, taken as a whole, has no capacity to earn, nor is it possible for it to have such a capacity. It unfolds itself at the expense of economy and aggravates economic distress. It fosters an economy of waste which becomes the more apparent as the pursuit of technological perfection is the more successful" (PdT 35). Jünger again points to the influence of technology on economy when he writes later: "Turnover and circulation become the hallmarks of a money economy dominated by automatic technology because here all money must be mechanically mobile and available. Everything in such a system must be available, for all the instruments of economic power stand at the service of the technological total process. Its dynamics determine the nature of the monetary system and the nature of the capital system as it determines the system of credit and finance, which has become just as dynamic" (PdT 44-5).[27] Neither does self-perfecting technology leave any room for religious, political and social factors. "There is a power struggle going on here which, because of its poverty, is successful and fatal" (PdT 35).

In the second place, Jünger uses the term *Perfektion* for the automation and automatism of technology. "Only with this automation did our technology acquire its typical stamp, which distinguishes it from the technology of all other times. Only hereby does it attain the consummation which we begin to discern in it" (PdT 39).[28] In this completed technology, both the worker and the capitalist are victim or captive. "Automation now grasps him, allowing him no leeway" (PdT 39).[29]

Jünger sees the *clock* as the harbinger of perfected technology. By means of the clock, "dead time" had already penetrated life. Now, given a technology that respects nothing, life is threatened with total death. "Dead time thrusts forward. Life enters the service of an all-pervasive automatism by which it is regulated" (PdT 174).[30]

2.2.5 Technology and organization

It is noteworthy and praiseworthy that Jünger, in contrast to others,[31] distinguishes between technology and organization. He asserts, correctly, that technology and organization are unbreakably connected and that they depend upon each other.[32] They stimulate each other. "All organizing of a technological nature enlarges the mechanism, and all mechanization enlarges the rational organization" (PdT 89). The aim and significance of the two cannot be distinguished. It is the hallmark of organization, too, that it does not enhance wealth so much as it apportions poverty. "As poverty is distributed, something happens that cannot be impeded: it spreads" (PdT 24).[33] This goes on until what remained to be divided has vanished. As examples Jünger mentions whaling and oil production. No one desires to regulate and apportion anything of which there is an overabundance, Jünger says. Only shortage and need, which are caused by technology, call for organizational measures.

Jünger pretty well summarizes his own view of the interplay between technology and organization and the consequences of that interplay when he says: "Technological progress is, by definition, tied to the aggrandizement of organization, to a constantly expanding bureaucracy which demands remarkable numbers of employees – employees who produce nothing and whose numbers increase in proportion to declining production. It is not the farmers, craftsmen and workers who increase in number with the progress of organization, but officials, functionaries and bureaucrats" (PdT 26). Elsewhere Jünger says that these people, too, are workers.[34]

Jünger's assessment of both technology and organization is clearly one-sided. He sees only the gloomy side of things. Certainly, people have been drawn into closer contact and made more dependent upon each other by the interplay of technology and organization. Yet, instead of being a loss, as Jünger considers it, this new proximity may be qualified as a gain: people are now in a position to accomplish much more than they once could. People working together can undertake tasks which an individual could never aspire to. Moreover, while technological development does have a tendency toward integration, it also has a parallel, individualizing tendency. No matter how severely technological development presses people upon each other, specialization offers possibilities for every person to develop his own gifts and powers. The tendencies toward integration and specialization can both prove pernicious if people are not acknowledged in their freedom and their responsibility. Jünger has seen clearly that integration often does advance at the expense of individual freedom. In addition he has seen that the individual can become isolated in his work, to the great detriment of cooperation.

The problems involved here will continue to engage our attention.[35]

2.2.6 Man and technology

Clock time is dead. As such it is the precondition for the natural sciences and technology, says Jünger. Given knowledge of the forces and movements in nature, it is possible for people to intervene; they can fashion and guide nature. Once this process is started, it cannot be restrained. "It is everywhere observable that with the appearance of mechanical instruments, which turn up wherever dead time awaits them, dead time penetrates living time" (PdT 54).[36]

In Jünger's estimation, the wheel is the preeminent example of "dead time." The wheel leads technological progress everywhere. Finally, humanity itself becomes a wheel in and of technological development.[37] Technology robs people of their freedom and puts pressure on them. It subjects people to mechanical exigency.[38]

This is clearest when we consider the relation between people and their work. In Jünger's opinion, no liberation from labor is to be expected through technology; rather the opposite, since monotony and the amount of work must increase. "Whoever would insist that all work that *can* be mechanized *must* be mechanized cannot base his case on the notion that mechanization will lead to the liberation of the worker. It increases not only the mechanical movement and the consumption tied to this movement but also the quantity of work" (PdT 66).

Just as a machine can be laid out in component parts, so the work done by and with a machine can be divided.[39] At the same time, work is uncoupled from the worker "...it [work] is disjoined from his person; it becomes independent" (PdT 68). In other words, just as those parts of a machine that from a technological viewpoint are entirely alike are interchangeable, the workers are interchangeable.[40] Through this technological development, the worker has become by far the most fruitful object of exploitation.[41] Man is fully encapsulated by the machine and the organization: "He disappears as a person and is still visible only as a functional working component" (MuE 309). Originally people had a certain resilience under this assault, but this decreased as the character of technological development became increasingly brutal. Thus viewed, says Jünger, man is not to be defined as an economic product, as Marx suggests, but as a technological product. "The assumption that in a classless industrial society exploitation could disappear is pure utopianism and messianism" (DVS 260).[42]

Human freedom has been driven from the field by this development. The exploitive character of technology has turned people into beasts of burden and slaughter. Their number is increased immensely by technology.[43] Human freedom makes way for anxiety. "In this anxiety there is a kind of hidden expectation, an expectation that seems like the premonition of an impending great explosion" (MuE 330).

The initial resistance of the workers to the all-embracing, destroying process of technology gained concrete expression in labor organizations, which aimed at protecting workers against entrepreneurs. Far from being successful, they actually increased the workers' dependence on technological development, putting them in shackles. "For the labor organizations arise in connection with the broadening apparatus" (PdT 30).[44]

Even socialism, which originated in the discontent of the workers, accomplishes the opposite of what was, and is, intended. Precisely by means of their own organizations, both organized labor and socialism have unmistakably advanced technology, played into the mechanical grip of depredation, and frustrated their own goals. Social justice has become adaptation to the force of mechanical laws.[45] In this regard Jünger says of socialism: "Socialism is, then, the mentality which comes willingly, without reservation, and steadfastly to the aid of technological thought as it aims at exploitation and depredation, and which encourages and impels this thought on every terrain" (PdT 78).

Jünger's view of the relation between people and technology is certainly one-sided. Of course there are examples enough at hand when he seeks to illustrate inhuman activities in modern technology. The employee is all too frequently the standardized worker executing standardized procedures. Yet the other side is also true, and it is this other side that holds promise for the future. Jünger fails to see that people will be phased out of the execution and that the crux of human work will eventually be found in the preparation, in designing. He fails to realize that, as that happens, emphasis will be placed increasingly upon creativity as the expression of human freedom.[46]

Jünger concentrates too heavily on the (undeniably present) unhealthy situations in technology. On the whole he shows no appreciation of the positive fruits and possibilities. We will see that this posture is determined by his presuppositions.

2.2.7 Technology and technocracy

It is only after he has said all this that Jünger finally raises the question what technology is. He is of the opinion that technology is the "mimicking" or "mirroring" of nature – but then of nature as the great machine which technological thought or natural-scientific thought construes it to be.[47] "Only thought that interprets the world as a machine could manage to make smaller machines in which the course of mechanical power is imitated" (PdT 94).[48]

In the beginning, technology – and the thought inseparably connected with it – rules only inanimate nature; eventually life, too, is subjugated, no less rigorously, to "mechanica." Via its control of people, technology becomes all-encompassing. "Mechanical theories of man characterize advanced stages of technology" (PdT 143).[49] Moreover, as technology advances, its unbroken consuming power becomes so great that even the territorial and political organization of the state must undergo fundamental alteration. "In the struggle for power, the technician aims to subjugate the state and to replace its organization with a technological one" (PdT 97).[50] The result of this is that the state as a technological organization is also brought to completion; it "...obeys a completed automatism." Yet, it is "precisely this destination which abolishes the state as state" (PdT 179). The all-encompassing tendency of the state, in Jünger's view, is toward this growing *technocracy*.

We have already seen that in Jünger's opinion modern technology has been made possible by natural science.[51] It is an expression of the technocracy, according to Jünger, that science – and he means not only natural science but also sociology, statistics, psychology, and medicine, for example – has become subordinate to the technological collective. "It is an expression of this displacement of power that we should now find the scientist as a functionary in the institutes and laboratories of industry, where his knowledge is utilized technologically. The scientific disciplines are made helping disciplines of technology . . ." (PdT 100).[52] Technology has penetrated the university: " . . . the university becomes an institute of technology and thereby serves technological progress . . ." (PdT 107). Through the decline of the university into a technological instrument, technology has acquired unprecedented possibilities; its planetary power is enormously enlarged; it sets its stamp on everything. Jünger perceives the spreading technocracy in uniform technological products, in technological dynamics as mechanical movement *(Mobilmachung)*, in sports,[53] in music, in fun time, and in free time.[54] Moreover, the technocracy is fertilized by its own results. It is strengthened by the increase of mass forming and powerfully stimulated by the ideologies, which are in Jünger's view the religions of the technocracy.[55] Originally Jünger believed that the technocracy would countenance a variety of opinions, conceptions and ideologies. Later he claimed that the planetary power of technology would bring about a single world ideology. The basis for this unification, he argued, is that technology "works the same way everywhere, in church, factory, and home, with the Christian and the non-Christian. Whether it is someone in America or Australia who serves an automaton makes no difference" (DVS 10).

For Jünger, that technology becomes technocracy is a fact given with technology itself. He regards this development as unavoidable and autonomous, characterizing it as a demonic development.[56]

2.2.8 Technological thought

To this point I have confined myself primarily to one line in Jünger's reflections on technology: his sketching it as a suprapersonal power in its frightening development. The other line – its relation to the first will require attention later – is Jünger's showing that technological thought precedes technology itself. This thought is the origin of technology, and it leads technological development. Technology is coming to perfection through technological thought's coming to its perfection. "What does it mean to say that technology attains perfection?" "Nothing other than that the thought which created it and made it grow reaches a conclusion and encounters limits that are dictated by the methods" (PdT 128).

The origin of technological thought and thereby also of modern technological development is to be found, Jünger believes, in Descartes. *Res extensa* is inanimate and can be fully described and fixed, that is to say, explained in a mechanical way.[57] Since it is inanimate, man need not be timid about intervening. "That the *res cogitans* rises up as the sole lord and master of the world process, that it shall do this sharply, irresistibly, and without regard to objections, lies decided in the thought of Descartes" (PdT 41). The shrewd intellect can have its way indefinitely acquiring food, booty, or loot.

Jünger ranks Francis Bacon next to Descartes. Bacon furthered technological development as an empiricist. "The rationalism of Descartes and the empiricism of Bacon are aimed at an enveloping causality which must increase with every expansion of mechanics" (PdT 43).

Thus technological development depends on technological thought and would be inconceivable without it.[58] If technology makes a "voracious" impression, it is because of the thought that lies behind the technology. And because that thought is totalitarian, says Jünger, technology has no limits such as those which economics and politics may be said to have. Rational technology demands an independent place for itself, a place not subject to religious, political, economic, or social considerations. All such considerations "are excluded by this thought, and indeed can be excluded by it since they have no compelling coherence with it" (PdT 35).

Newton's mechanical view of time also contributed greatly to technological development, according to Jünger. Given this conception of time, it was possible to achieve certain discoveries in the exact natural sciences. Those discoveries stimulated technology enormously. "How stringently ordered by the measure of time everything is here, and how inexorably technology advances to the subjugation of totality to chronological order – work, sleep, rest, and recreation!" (PdT 51).

Technological development becomes a tyrant, for " . . . the thought of the clockmaker gains predominance" (PdT 52). It steadily enlarges the areas of its sway. It becomes a question of importance whether there are indeed limits or not, and if so, just where they lie. "If one thinks of the earth as a great clock and of every conceivable movement belonging to it as mechanically measurable and calculable, then knowledge of this central mechanism may be the goal of scientific-technological thought; yet the application of this knowledge will be nothing other than the comprehensive mechanization of man" (PdT 52).[59]

The death precipitated by technology as predatory cultivation, which threatens man as well as nature, is thus called forth *ultimately* by technological thought: it is " . . . a death which springs from causal thinking and its mechanical conception of time" (PdT 194).

2.2.9 Technology as diabolism

On the one hand, as we saw, technology is an all-encompassing and all-governing power. Self-perfecting technology possesses unprecedented power. It tracks across the face of the whole earth, touching everything and leaving ineffaceable traces. Because of technology, the whole of life takes on a different rhythm and appearance. In short, modern technology represents an unparalleled power.

On the other hand, this technology is brought forth by the technological thought of *people*. In the final analysis, Jünger cannot get around the fact that technology has a human origin: "The automaton always presupposes man; if that were not so, it would be no lifeless apparatus but a demon with a will of its own" (PdT 123). Nonetheless, he still wishes to characterize technology as demonic. "Experience conveys the insight that the machine has a *force peculiar to itself* and that man must be careful not to come into conflict with it" (PdT 125 – italics added). The "being careful" would seem to indicate that man might somehow be able to escape the threat of technology. Yet this notion is only illusory, since technological coercion and power turn against man preeminently. "The extremely rapid movement which arises through technology clutches man, too, who perceives technological progress as his own progress. Technology is the making mobile of everything that is not mobile. And even man has become mobile; he follows without resistance the automatic movement; yes, he would like to see it go even faster" (PdT 144).[60]

Not only would a person from another time, lacking all understanding of our technology, perceive that diabolical powers are at work in technology; even the person of *today* is recognizing his collapse under the burden of technology, "...and that perfected technology leads to an overtaxation of his powers, which he cannot long endure" (PdT 158).

Jünger argues that the basis of the diabolical development goes back, as Nietzsche and Spengler[61] proved, to the *will to power*. And here Jünger has returned to man: "Will to power, to the subjugation and harnessing of natural law" (PdT 97).[62] This will to power constantly loses ground as technology exhausts nature. Accordingly, this will can never be satisfied. And it stops at nothing to strengthen itself. "The will to power aims at the procurement of power for itself; it wants power because it is poor in power, because it is hungry after power" (PdT 171). Finally, technology as it has been called forth turns against the will to power itself.

In short, the passion for power which rules technology releases forces by which humankind itself is annihilated. These forces are what Jünger christens the demonic element in technology. "Ordinarily, it is said, the demons slumber; they must first be awakened; man must intrude upon their sphere before they can be activated. There can at present be no further doubt that they are fully awake" (PdT 172).

2.2.10 Technology and the future

Jünger does point to the will to power as a source of modern technology. He is nevertheless impressed, when looking at the character of technological-scientific thought, by the automatic nature of the technological development that leads to destruction. It is thus a question of importance whether humanity might not be able, somehow, to make a fresh start and in so doing thwart the destructive development. Jünger is of the opinion that technological development must run its course and work itself out – even to the point of passing right through the destructive development. Does there remain, then, a possibility of deliverance? Is Jünger perhaps less pessimistic when he poses this question than we believed at first? We shall consider these questions in the next two sections.

Given technology, people generally have great expectations for the future. Jünger in no way shares this popular view, however. "The machine is not some god bestowing happiness, and the epoch of technology does not end in a peaceable and charming idyll" (PdT 157). Technology as predatory cultivation and diabolism yields the opposite of happiness. It may seem as though technology disposes over a horn of plenty, but this appearance is deceiving. "The demon uses man to create the appearance, but this appearance is always a distorted one. Man deludes himself into thinking that he can live in freedom, peace and prosperity on the very earth he so relentlessly plunders and devastates" (PdT 185).

It is understandable that on this basis Jünger designates completed technology as total war, both literally and figuratively. This total war leaves no reserve untouched; it implies total destruction.[63] Yet the catastrophe is not conspicuous: "The destroying force of technological thought remains intact in the face of the ravagings, which witnesses to the progress of technology and indicates that it is approaching a state of perfection" (PdT 195). There can be no thought of a victory through technology: "The consumption is so enormous that it swallows up the victory" (PdT 196). "Man never again controls the mechanistic laws that he himself set in motion. Rather, these laws control him" (PdT 197).

Jünger offers constructive criticism of this development. Yet the question arises as to whether this criticism can be meaningful if technology does indeed overwhelm us ultimately as a demonic fate.

Jünger's reply to this question is ambivalent. On the one hand, in keeping with his view of technology, he holds that criticism of this catastrophic development of technology is itself embraced by this development and accordingly offers no perspective at all. "The question arises as to what extent such a plan [for a technological collective] which aims at embracing everything remains possible, since what it aims at embracing includes whatever criticism might be directed against it" (MuE 352). On the other hand, it has already been shown that Jünger locates the origin of technology in technological thought since Descartes and in the will to power which is at work behind it. There might be a possibility of deliverance in the repudiation of this origin. According to Jünger, however, it is necessary that technology reach completion first. "The plan succeeds because it neglects nothing. And it fails because it neglects nothing. *This is the point upon which everything turns, the point where man is cast back upon a new sort of thinking*" (MuE 352 – italics added).

What Jünger means by this statement remains shrouded in mystery, especially in his first publications. It is all the more mysterious because the outcome of technology is said to be total death. In his very first work, *Perfektion der Technik*, Jünger expresses a desire to dissociate himself from technological thought and to make room for the nontechnological *ratio*.[64] Just what that might be is not clear. In contrast to the technicians, who have an unconditional confidence in their capacity to conquer the future mechanically, Jünger has no pat answers for the future.[65] "However, if he [the technician first of all] perceives that these enormous ravages are formed beforehand in the thought of the technician, and that this thought evokes them and even breeds them, and that the world of corpses and ruins and the prodigious desolation which encircles man is a corollary, a product of this thought, then he has gained a great deal" (PdT 194).

What Jünger has in mind becomes clearer later when he advocates cherishing nature and property in place of using and abusing them. "Utilization without maintenance is theft" (MuE 271). We have all become such awful exploiters that we can scarcely imagine what our new relation with nature must be, Jünger declares.

In *Maschine und Eigentum* a few sentences appear on the subject, though they are anything but clear. Jünger urges us to resist the coming *Automatismus*, lest we be devoured by it. Our resistance must, of course, have a different source than technology. Man, who has forgotten and even "unlearned" how to treat the earth as a mother, is no longer a son of the earth. He finds himself rather on the cartesian globe, which is dead and which can be used at will like an inanimate ball. Jünger fails to suggest a new direction here. Of this he is well aware, for at the end of his book he declares that his only intention has been "to sketch the consuming power of this *automatism*" (MuE 363 – italics added), and that the reader has read it with discernment if he has understood that. "This knowledge will not be our salvation. The earth needs man as a caretaker and keeper. We must learn again how to treat it as a mother. Then we will prosper upon it" (MuE 363).

The question remains whether there is a possibility of deliverance on the far side of destruction.

2.2.11   A basis for hope?

In summary we can state that technology, in Jünger's opinion, is a deadly power evoked by cartesian thought and inspired by the will to power. Technological power is a suprapersonal power bearing the character of necessity. In *Perfektion der Technik* Jünger claims that a turn for the better can only be achieved on the basis of another kind of thought. From the conclusion of *Maschine und Eigentum* it can be learned what Jünger means by this. The caring for and keeping or tending of nature, of mother earth, must come to occupy the place now given to violating and abusing nature. Jünger seeks perspective for the threatening future of technology by going back, as it were, to nature. Thus he seeks perspective by denying the meaning of techno-logy – and by implication the meaning of science as well, which he scarcely distinguishes from technology. Jünger prefers to forget, if indeed he perceives it at all, that complying with nature can bring no freedom either, since such compliance will only expose humanity once again to nature's caprices, to its recalcitrance and its unanticipated assaults. Or if he does perceive it, he would prefer the caprices of mother earth to the madcap chase after total death.

In *Sprache und Kalkül*,[66] a later publication, Jünger returns to the question whether there is any possibility of deliverance from the threatening development of technology. From this publication it can be seen how heavily Jünger was influenced by the "later" Heidegger; we could perhaps say that thanks to Heidegger, he is now a bit less at a loss about pointing to a basis for the hope of deliverance from the catastrophic development. Yet the mystery remains intact.[67]

Jünger now construes *Sprache* (speech, language) as the origin and precondition of science and technology. Since Descartes, however, *Sprache* has become *Gegenstand*, the "object" of technology, of science, and especially of technological thought. This, however, does not detract from the fact that science and technology always remain bound to the "mother tongue."[68] "That *Sprache* cannot be controlled, that it is not arbitrarily at our disposal, means at the same time that it cannot be regarded as an instrument. Yet every conceivable *Kalkül* aims at this" (SuK 8). *Sprache* is the basis of *Kalkül*, even if *Kalkül* strives to make itself the master of *Sprache*. Yet, because this usurpation is never successful, Jünger perceives a possibility here for future deliverance from the all-affecting process of technology.

It is "technological thought" itself that Jünger labels "Kalkül." "Thought which occupies itself with *Kalkül* is a *calcul*ating, mis*calcul*ating thought, and *Kalkül* must be understood here as the model of such thought" (SuK 7). This *Kalkül, which coincides with logical-symbolic calculating*, has attained its greatest strength in modern logical calculating, in statistics, in cybernetics, and in electronic calculating machines and thought machines, Jünger says. He is persuaded that people fail to discern the preponderance of "calculation" thus conceived when they assert that cybernetics is the bridge between the sciences. He believes that when the principles of cybernetics are applied to sciences, they often come to be dominated by *Kalkül*.[69] And this occurs the very moment that *Kalkül* has placed the greatest distance between itself and *Sprache*. Nevertheless, these "machineries, too, presuppose *Sprache* and are inconceivable without it" (SuK 11).

If *Sprache* is indeed the basis for science and technology, then it must also be the case that science and technology are of good report, that they are possible and justifiable on this basis, and that we are accordingly under no compulsion to reject them. It is noteworthy that in this last publication Jünger no longer speaks of technology as predatory cultivation and diabolism. He speaks rather of science and technology as having definite boundaries; for example, they may not make people objects. "What is objectionable about thought machines is to be found not in their scientific-technological aspect but in their application in politics, bureaucracy and police work. As people devise them for calculation and notation, so shall they be calculated and registered! New apparatuses are aimed at that, but should we not also consider the extent to which they might be a means of help for helpless humanity? Who really considers that?" (SuK 11). In the face even of the impermissible transgressions, however, there remains comfort, for *Sprache* can never be encompassed by technology and science: rather, *Sprache* itself encompasses even the unplanned. "*Sprache* exists for everyone and is therefore our comfort and our comforter; it is our own innermost self (*Innigkeit*)" (SuK 12).

This does not mean that Jünger is now prepared to deny the existence of the tremendous perils that humanity faces as a result of technology. He even warns once more against taking an instrumental view of technology; people cannot just lay technology aside as instrument and so be rid of it. "This laying technology aside is beyond the capabilities of man and is presently something impossible" (SuK 13). Man is more likely to be taken up as an instrument in the technological plan himself and rendered subservient to it. This can be seen quite clearly in the relation between the organization of human work and the apparatus which that organization makes necessary. Via the scientific object, man is reduced to a machine.[70] Because of this development, there is no difference in place between man and the machine throughout the whole of the technological plan. Just as all of the equipment and component parts are numbered and normed lest technology get out of hand, so man becomes a number. Man is thereby taken up into the functioning or busyness (*Betrieb*) and is lost: "...for nowadays man is marked in excessive measure by the functioning, and by the apparatus, and by organization, and by the organization of labor. Accompanying this development are far-reaching alienation, uncertainty, and anxiety, for man has no established place and no certitude between the tremendous prostheses of the labor plan, which is a mechanical plan. He lies between these prostheses as between the arms of tongs, which clench him" (SuK 26). Estranged from themselves, people are also alienated from one another.[71] If he is to be delivered from this great danger of technology, man must be vigilant. He must not restrict himself to only one sector of consciousness, to technological thought, allowing it to rule other sectors. Rather, he must make ample room for "lingual awareness" (*sprachliches Bewusstsein*). In this room is to be found the basis for a hope of deliverance: "Reflection, yes, but not the reflection of a person pursuing an obsession, hedged about narrowly by darkness. *In Sprache there is sanctuary.*" "*We* ourselves are *Sprache*; and where we are not *Sprache* we can no longer meet one another" (SuK 27 – italics added).

Looking briefly at this development, we see that after perfecting the idea of a technology that runs automatically, which he characterized as predatory cultivation and diabolism and which he regarded as resulting finally in total destruction, Jünger returned in the end to *man*, who was also the starting point of technological development. Man must resist the devouring power of technology by tending and keeping nature. It remains curious that even now, the new way of thought can be entered upon only via the crucible of destruction. The new way of thought means seeking comfort and drawing power from *Sprache* – speech or language – which is man himself.[72] Via an unavoidable, diabolical development evoked by man, Jünger returns, as by a great detour, to man again. It is man who must supply the means of deliverance.

2.2.12 Critical analysis

2.2.12.1 Introduction

Jünger is important because he has expressed explicitly something that has been experienced by many who have had contact with modern technology, namely, that humanity has been made subordinate to technological development and groans under its burden. Jünger has correctly called attention to the wastage and pollution caused by technology. It is more important than ever that questions of conservation and a clean environment be taken seriously. This does not detract, of course, from the fact that Jünger has lost all sense of proportion in his assessment of the whole of technology and of its development. He does not take into account the fact that technology has brought a lightening of the load for many and has enormously increased material prosperity.

Because he gives natural science an important place in his view of technology, Jünger is able to perceive a number of real characteristics of modern technology. He correctly notes, for example, that the development of modern technology is a development in the direction of independently working machines.[73] It is also true that modern technology brings about a division of labor by means of planning and setting norms.[74] Moreover science sees to it that technology acquires a planetary or – as I would prefer to put it – universal character.[75] And massification, which is one of the most notorious consequences of modern technology, does not escape Jünger's sharp eye.[76]

Jünger's themes recur repeatedly, but not always in context. Dessauer has shown that he is inconsistent at times.[77]

My critique will be confined to the main points of Jünger's argument. I will keep it brief, since various subjects will occur again in more elaborate form in the other transcendentalists to be examined. In the pages below we will look critically at Jünger's views on the origin of technology and technology as a suprapersonal power. We will also take up his answer to the question what technology is, and we will examine his anthropology. In the final section I will endeavour to demonstrate that Jünger's vista-less dialectic of humankind versus technology prevents him from pointing out a meaningful direction for technological development.

2.2.12.2 The origin of technology

It is clear that by *technology*, Jünger means always *modern* technology. He traces its origin to modern thought, which arises in western culture with Descartes, according to him. He concurs with Nietzsche and Spengler in affirming that the will to power is at work behind this technological thought. "That the *res cogitans* rises up as the sole lord and master of the world process, that it shall do this sharply, irresistibly, and without regard to objections, lies decided in the thought of Descartes" (PdT 41).[78]

In addition to the thought of Descartes, Jünger points to distress or need as a cause of modern technology, even though exigency is generally increased rather than alleviated by technological development.[79] This, too, is an intensifying stimulus to technological development.

It is certainly true that science attained great importance for technology after it was addressed to practice. Jünger perceives this quite nicely. The great question is whether he perhaps forces the history of technology into a straitjacket by neglecting to investigate the influences which such factors as religion and warfare have had upon technological development. His notion that technological-scientific thought is *the* cause of technology is one-sided in the extreme. In the next subsection we will see that it results in his having to interpret the phenomenon of modern technology as an inexorable development. This, in turn, makes it extremely difficult, if not impossible, for Jünger to point out a way of escape before what he regards as the sinister development of technology.

A related objection is that Jünger does not respect the continuity in the history of technology. For him technology is *modern* technology and is to be rejected as such. He has nothing to say about the relation between earlier technology and modern technology. Reflection on classical technology could have opened Jünger's eyes to the fact that people and technology really belong together. A great question arises here, a question that cannot be evaded: How, given this relation, is modern technology to avoid becoming an autonomous power or an automatism to which humanity falls victim?

2.2.12.3   The suprapersonal power of technology

I should like to remark first of all that Jünger points out a variety of traits in technological development which deserve attention. He notes that technology is dynamic. This dynamism leaves nothing untouched; all that is old passes away. And it penetrates the whole of culture, creating the impression that humanity does not have technology in hand but has become the victim of technological development. Moreover, technology threatens to become a technological collective seemingly without room for human personality. In and through technology, humanity becomes a function.

Whereas Spengler, whose view of technology is based on his thesis that man is a beast of prey, regards technological development as characterized by the *misuse of power*, Jünger sees man's position in technological development as one of *powerlessness*. When Jünger elucidates this proposition by pointing to the independent working of modern machines, he forgets that this represents only a *relative* independence. Not a single machine can be understood apart from humankind, for it is man who made it and lets it work.

Yet Jünger's assertion that technology, as predatory cultivation and diabolism, has become a suprapersonal power and a law unto itself is not put forth without certain additional considerations. "The movement has reached a point at which it begins to run self-sufficiently, automatically, and with mechanical necessity; whereupon it can no longer simply be shut down, since it contains a *uniformity (Gesetzlichkeit)* and a *historicity (Geschichtlichkeit)* which man for a long time, for centuries, has built into it" (MuE 360 – italics added). And how has that come about? The source of technology's autonomy lies in the autonomy of thought. With Descartes, man sets himself up as the sole lord and master. He is autonomous. Subsequently this autonomous man becomes the victim of the technological development he has called forth. "Man is no longer in control of the mechanical uniformity which he himself has unleashed. This uniformity controls him" (PdT 197).

I agree with Jünger that both the absolutization of natural-scientific thought (in the form of technological thought, according to Jünger) and the all-controlling motive of the will to power in the background of technological development turn technology into a demonic power. "By means of technology, man is coercively injured in his freedom, for in technology the doctrine of mechanical functions pushes to the fore, and thereby also the conviction of a mechanical necessity to which man, too, is subject" (PdT 65). In technology *regarded* as autonomous, man becomes a machine; he loses his freedom. "The person becomes a component of the apparatus and organization" (MuE 307).[80]

Jünger makes a distinction between technological power and organization. In the latter, too, human freedom has disappeared; organization, he argues, is a direct result or a necessary side effect of technological development. Organization also has humanity completely in its clutches. "Organization here means all influences upon man that are caused by the proliferating effect of technology" (PdT 23).

The absolutizations that Jünger so clearly sketches have indeed had their effect in history. An absolutized technological type of philosophy issuing from spiritual decline and intellectual impoverishment is not concerned about the need for a clean environment and conservation. The resulting evil fruits are being plucked everywhere today. Only, the problems are not to be solved by throwing out technology along with the absolutization of technological thought, which is what Jünger does. It becomes apparent, then, that Jünger does not have a satisfactory view of technology. For him technology cannot be anything but autonomous technology. According to Jünger, this is a *necessity* given with technology itself. His perspective is inseparable from his answer to the question what technology is.

2.2.12.4 What is technology?

At the outset Jünger speaks of technology as *machine* technology. Later he aims at something more general, namely, the method and skill brought to expression in all manner of human activities, such as sports and music. Still, both machine technology and technology as method have a natural-scientific basis. Technology as method means for Jünger the technological-scientific method,[81] applied not only to inanimate nature but also to man and all his activities. Yet machine technology remains of supreme importance to Jünger; what he sees in it is the "mimicking" or "mirroring" of nature as nature is perceived by the natural sciences. The universal characteristics of natural science make technology a universal power.[82]

It may be concluded from all this that Jünger regards technology as applied natural science. Autonomous science becomes autonomous technology. Yet he winds up in conflict with his own ideas – without ever becoming quite aware of it himself – when he chooses two *inventions* as examples by which to explicate the character of modern technology as he understands it. Jünger is persuaded that death penetrates life through technology: he regards the clock and the wheel as precursors of technological development, which is necessarily accompanied by destruction and death. Dead mechanical time manifests itself in the clock, and the wheel guarantees that this time will be obeyed in technology.

The examples of the clock and the wheel should have made it clear to Jünger that there is no necessary relation between natural science and these inventions. An invention, after all, is an expression of something entirely new, something that does not follow from the natural-scientific basis of modern technology.[83] Jünger fails to perceive that the creative capacity of humankind can transcend a (given) technological development and thereby rise above the scientific basis of technology. He neglects invention as such. This means, of course, that he cannot help but fail to take its implications for technological development into account. The fact of the matter is that an invention is able to break through a stiffened technological development or a misdirected one to afford a new perspective.

Jünger's view of technology as autonomous technology precludes his setting limits to technology. His disregard of invention signifies that he does not perceive that human beings – who are free and who work with their fantasy – ought to lead technological development and that *they* are responsible for it.[84] The source of Jünger's difficulty at this point is his anthropology.

2.2.12.5 Jünger's anthropology

Jünger notices that technology threatens people in their freedom. He sees technological development as drawing toward a total technocracy or technological collective that affords no room for the free individual. "The collective that strives towards mechanical equality does not know what to do with freedom" (MuE 242). The ultimate result of technology is annihilation. "It is a dead world in which mechanism leads us: the quicker the automatons that help us make progress, the quicker the diffusion of death" (PdT 195).

Jünger arrives at this affirmation of necessity or inevitability through his acceptance of the *idea* of absolute technology, which is also applied in practice, and through his disregard of both invention and human responsibility for technology. He can therefore add that technological development is nihilistic in character. This does not mean that this development proceeds in the absence of a certain order. On the contrary: "Nihilism is not a process without order, not an orderless business. It is an orderly progress of affairs; it is planning and organization. The planning and organizing strives to tie everything to itself, and it becomes a goal in itself" (DVS 272).

I would, of course, affirm the ideal of the free person, which Jünger champions. People ought ever to be free, even in technological development. But Jünger denies the very possibility. He does so because he places the freedom of humankind over against technology. Jünger rises against the automatism of a technology sprung from the absolutization of technological thought to champion freedom as an absolutized freedom, as a freedom that has nothing to do with technology. This freedom that concentrates on itself apart from technology leads him astray with respect to technology. In my judgment, this is the basic reason for his failure to escape negativism in his reflection on technology.

In principle Jünger has broken the bond between humanity and technology. Over against an absolutized technology, he posits an absolutized freedom.

The great question is whether this opposition can be maintained consistently. Doesn't Jünger's whole case demonstrate that such a complete opposition of two absolutizations is impossible? Nevertheless, this opposition does entail the consequence that Jünger fails to find a therapy. There is a tension at work throughout his view, a tension between autonomous technology and free, autonomous humanity.

2.2.12.6 The tension between humanity and technology

According to Jünger, the rise of modern technology is due to technological thought. As he sees it, the thinking person stands at the beginning of technological development but is increasingly threatened by technology as it develops.

Although he realizes that people assumed the wrong attitude at the beginning of modern technological development, that is, that they absolutized technological thought so that technology attained a uniformity (*Gesetzlichkeit*) of its own to which humanity is subordinated, Jünger does not plead for a healthy relation between humanity and technology.

Because he neglects the creative capacity of people as it comes to the fore in inventing, because he does not see that humanity transcends technological development, and because he knows nothing of man's calling to disclose nature rather than misuse it, Jünger lacks perspective and cannot see beyond a technology that is predatory cultivation and diabolism.

In Jünger there is a tension between humanity and technological power. His initial attempt to flee before technology and to return to nature does not resolve this tension, any more than does his later relativizing of technology with the help of *Sprache*. In fact, his flight and his turning to *Sprache* are both outstanding examples of the tension. The difficulty is especially noticeable in his treatise *Sprache und Kalkül*. Technology arises from *Sprache*, only to turn upon *Sprache*, converting it into *Kalkül*. Yet it is this very *Sprache*-become-*Kalkül* that must somehow afford the possibility of deliverance. Just how contradictory this deliverance is becomes most apparent as one ponders the fact that Jünger contends that "automatism" – and thus also *Kalkül* – leaves no gaps. It is "*lückenlos*," which means that there is ultimately no room for freedom or *Sprache*.[85]

The same dialectical tension reveals itself in Jünger's view of the relations between man and machine, between man and demonic technology, and between property and technology.[86]

This dialectic is also clearly present when Jünger speaks of criticizing and resisting technological development. Criticism is necessary, he declares, although he knows all the while that it will be incorporated into the technological plan itself and thereby be robbed of its power.[87] Furthermore, Jünger wants people to resist technology even though he is conscious that their resistance only strengthens technology. Jünger has observed this in the resistance of young people to technology. "They [the youth] will be drawn to sabotage when they discern the exploitive character of the automaton, of which the exploitation of humankind is the ineluctable consequence. Yet the intellect of the plan is adequate to and superior to the impromptu assaults, and is even strengthened by them" (DVS 265).

The foundation of this dialectic is the opposition of two absolutizations. Technological power threatens humanity, but it needs humanity for the advancing development. Although absolutized human freedom excludes every technological tie, freedom is always prerequisite to technology.

From the dialectic as I have sketched it, it is apparent that an absolutization can never be carried through fully in practice. Even so, it cannot be denied that putting the *idea* of autonomous technological thought into practice entails enormous dangers and dislocations, since the direction in which technology develops must then be a wrong one.

An absolutization can never be actualized absolutely, because reality has the character of meaning. Although everything has a place of its own, everything points to everything else; nothing is self-sufficient. Our reality can be characterized as being-as-meaning. Everything is connected to the Origin. People can accept the truth, but they can also resist it by way of some absolutization or another. Yet their resistance always remains subject to being-as-meaning itself; indeed, it is parasitic upon it.[88]

In this light it is clear that so-called autonomous technological power cannot allow the free person to exit from the stage. We also see that the free person, in his pretension to autonomy, must chronically reevoke technological power. The egocentric subjectivism of Descartes brought humanity onto the road of collectivism through the power development of science and technology, contrary to what was intended. But the spirit of collectivism arouses opposition and generates the idea of the free person, who subsequently also proves to be characterized by egocentrism.[89]

Jünger has perceived that the dislocating effect of technology has a religious origin. He indicates often enough that humanity has fallen into the clutches of the modern ideologies – the religions of technology.[90] In general, modern man has placed his faith in technological development. Jünger, however, places his faith in the regular, everlasting movement in nature. Yet that faith of his is threatened, as everything is touched by the *Mobilmachung*.

The religion of those whom Jünger opposes gets its direction from technology, while his own religion is determined by the free, natural person. Both religions are secularized, and they are hostile to each other. Yet, they have much in common. In the people whom he opposes, Jünger discerns an absolutized thought that is at work in the background of technology and thus is directed outward. Eventually Jünger falls back on humanity – specifically, on thinking humanity. Yet, thought is directed inward, and its content is (still) unknown to him. Therefore his most recent publication on technology ends in great perplexity: "The thought that would be a match for rampaging technology has not yet been thought" (DVS 264, 265).

Jünger's religious certainty is in crisis because he believes that his certainty can be gained from something in our temporal reality, namely, the free person with his property. This certitude is shaken by dynamic technology.

Yet a religious reverence for technology has equally dangerous consequences: people place their confidence in technological development and yield to it. They reduce their responsibility for technology or lay their responsibility aside entirely, and that brings with it all manner of evil consequences.

For the moment we must leave the matter at this impasse. The same complex of problems occurs in other thinkers. I will attempt to penetrate the problems ever more deeply until, in the final chapter, using the christian faith as a point of departure, I will argue that a harmonious development between humanity and technology is possible after all.

2.3 Martin Heidegger:
In Quest of the Meaning of Technology

2.3.1 Introduction

Heidegger[91] has dealt with technology in many publications. His most extensive and comprehensive treatment of it is to be found in the lecture entitled "Die Frage nach der Technik" (The Question about Technicity), which was delivered in Munich in 1953 in the context of a series of discourses on "Die Künste im technischen Zeitalter." This lecture was republished in 1962, along with a previously unpublished essay about the "reversal," in a volume entitled *Technik und die Kehre* (Technicity and the Reversal). In this publication, what Heidegger has said elsewhere about technology is pulled together and worked out further.

Heidegger's view of technology cannot be understood in isolation from his entire philosophy of Being. The very title *Technicity and the*

Reversal indicates as much. It might even be argued that Heidegger's view of technology – especially his view of the nature of technology as *Ge-stell* ("that challenging claim which gathers man thither to command the self-revealing as fund"*) – constitutes the core of his "later" philosophy of Being. At any rate, the "reversal" occupies a central place in this "later" philosophy; as we shall see, the "reversal" is unbreakably connected with technology as the greatest danger of our time. Heidegger regards technology in its world-encompassing power position as the culmination of metaphysics, which seeks to account for all of reality by finding an ultimate basis and a first cause, in order to make possible the control of all being. It is in technology, in Heidegger's opinion, that metaphysics has achieved its most weighty form. Heidegger, being dissatisfied with metaphysics, devotes great attention to technology because he wishes to re-pose the question of the Being of being.

In this secton I will examine Heidegger's view of technology – or rather, his inquiry into technology (which I would wish to regard as a *search for the meaning of technology*) – and present it within the framework of his philosophy and its development.

I will begin by briefly summarizing the philosophy of the "earlier" Heidegger and the philosophy of the "later" Heidegger. Then I will take a close look at his definitive publications on technology. In a separate subsection I will investigate the unmistakable relation between Martin Heidegger and Friedrich Georg Jünger. (It should be noted that Heidegger does not mention such a relation in any of his publications.)

Then, before undertaking a critical analysis of Heidegger's search for the meaning of technology, I will pause to consider the confrontation between Friedrich Dessauer and Heidegger. Has Dessauer really understood Heidegger, and has he done him justice?

Heidegger, of course, is anything but easy to read. He often uses newly coined words, or words with a double meaning, or vague words that might better have been exchanged for clearer ones. Therefore the exposition will have to include more than the customary number of quotations.†

---

* Using Lovitt's translation. See the subsequent translator's note. – Trans.

† In this English translation, it seemed advisable to supply many of Heidegger's important German terms. I have placed them in parentheses after their English renditions. The English renditions vary, for two reasons. First, different renditions are called for in different contexts. Second, it is important to convey a larger area of meaning for certain German terms than any single English term would suggest. My renditions of Heidegger's German

2.3.2 The "earlier" Heidegger

Although Heidegger was a student of Edmund Husserl (1859-1938), he disagreed with him on some very essential points. Husserl was concerned to bring being in its being methodically to light in order to discover its "essential structures." Central for him was the transcendental thought-subject with its intentional "essential structures." His phenomenology is anchored in the transcendental ego, conceived of as an out-of-this-world and ultimately timeless subject.

In contrast, Heidegger inquired not after being as being but after Being as such, without either accounting for it from beings or founding it upon beings. He wished to reflect upon the "truth" of "Being itself." His starting point, therefore, is not the "I" with its consciousness as pole and as the ultimate point of reference of the phenomenological subject/object "correlation" ("cogito, *ergo* sum"). Rather, what is central is the existence that precedes thinking and is present in it. Heidegger begins by looking behind the positions of Descartes, Kant and Husserl. In them he discerns a dualism: the (thinking) subject stands over against the objects to be known. Husserl may have radicalized Descartes by moving from the "cogito" to the "transcendental cogito," but his philosophy continued to be governed by the "subject/object polarity." Heidegger wants to conquer this "polarity"; he seeks to emphasize the unity of all beings and tries especially to break through the rigid set(s) of this "polarity." Therefore he concentrates on "historical Being" (*geschichtliches Sein*) and poses the question of the meaning of Being.

In Heidegger's main work, *Being and Time* (*Sein und Zeit*, 1927), this question is dealt with extensively.

Since Heidegger does not give explicit attention to technology in *Being and Time*, I will limit myself for the present to the elements in that work which are of importance to my argument. My concern is not so much to give a synopsis of this first great work as to emphasize that Heidegger's central concern in writing it was There-being (*Da-sein*) as human existence. It is from this center that he both poses and attempts to answer the question of the meaning of Being. In the "later" Heidegger, Being itself, which reveals itself in human thought, assumes the foreground position.[92] Only then does Heidegger become intensively concerned with technology. Most of my treatment of Heidegger will therefore focus on the "later" Heidegger.

---

rely heavily – but not slavishly – on two authoritative English studies: W. J. Richardson's *Heidegger: Through Phenomenology to Thought* and William Lovitt's "A *Gespräch* with Heidegger," published in *Man and World*, February 1973. – Trans.

In *Being and Time*, Heidegger sees man as a being that occupies a unique place in the midst of all being: he is "There-being" as "to-be-in-the-world." This concerns the being itself of There-being in its being.[93] Prior to all reflection, There-being possesses an "existential comprehension of Being" (*existentielles Seinsverständnis*), by virtue of which it knows that it *is*. In its "comprehension of Being," There-being understands not only its own being but also being in general. The "to-be-him-self" of man is, after all, a "to-be-in-the-world." Man's understanding that he exists embraces himself, his being-in, and the world.[94] There-being as "to-be-in-the-world" is thus "comprehension of Being," which is to say that it is at once both ontic and ontological. The fundamental ontology resulting from this – which Heidegger calls *the* philosophy – ought to precede all other knowing and every science. As the hermeneutic of There-being, this fundamental ontology offers an "existential analysis of There-being."

This existential analysis, accordingly, would be an explanation of the elementary ways in which we have always understood ourselves in our commerce with beings. In other words, this analysis assumes the task of disclosing *that*, *how*, and *where-from* human existence in its being understands Being, and it issues in a clarification of the meaning of human-being, that is, of the *where-from* "to-be-in-the-world" exists.

The relation of There-being to being is differentiated as "preoccupation" (*Besorgen*) with "instruments" (*Zeug*), that is, as "referential dependence upon" or "immersion in" the world (*Sein bei*); and as "caring" (*Fürsorge*) in relation to other people, that is, as "being-with" (*Sein mit*). Both commerce with "instruments" and encounters with other people are implied in There-being as "to-be-in-the-world."

Being-structures are divided by Heidegger into "existentials" and "categories." The former are the structures of There-being, and the latter are the structures of "not-There-being-ness" (*nicht-daseinsmässige*). The "existentials" have primacy over the "categories"; the "categories" are subordinated, as it were, to the "existentials," or possibly even comprehended in them, since the understanding of the being of There-being as "to-be-in-the-world" implies the understanding of the "not-There-being-ness."[95]

There-being is clearly the center of Heidegger's conception in *Being and Time*. Therefore we need not pursue the structures of "not-There-being-ness," such as "(mere) entity" (*Vor-handenheit*) and "instrumentality" (*Zu-handenheit*). It is enough to simply take note of certain of the "existentials," observing their clear connection with the "categories."

Heidegger maintains that the being of man consists in his "power-to-be" (*Sein-können*). Man understands himself to be the free designer; in this designing as "power-to-be," the world has the primary meaning of "instrumentality." It is man who, in his designing, discloses the possibilities of "within-the-world-being." He takes that-which-is-to-hand in hand. In the serviceability of "instruments" (*Zeug*), it becomes apparent that man has a relation to the world which is antecedent to his assuming any theoretical distance from "objects," and that this relation is already a (primary) form of knowing, namely, "understanding" (*Verstehen*).

Yet, designing is not an unlimited possibility. The being of man is radically limited in its "power-to-be." Man is "thrown" into a "beginning situation" which profoundly affects every possible "project" or design. For this reason, the design in which man opens himself as his own power-to-be is a "thrown design." Hence "being-thrown" permeates design to the full. There-being is "to-be-in-the-world" as "designing-thrown being-in-the-world." Moreover, the last and highest possibility of designing coincides with "no-longer-having-power-to-be," with "no-longer-having-power-to-be as death" (*mit dem Nicht-mehr-sein-können als Tod*). Between the "to-be-thrown" and the "to-no-longer-have-power," There-being consummates itself in designing. Both the starting point ("being-thrown-ness") and the end ("being-unto-death") determine every project. The thrown design is finite; it faces death, "nothingness." The "thrown-ness of There-being" is a "being-unto-death."[96]

We can form an adequate picture of this being-unto-death as the outer limit of possibility of the thrown design if we imagine the thrown human-being to be like the trajectory of a bullet.

The initial conditions under which the shot is fired establish the conditions to be met by all the points along the trajectory. Even the final point, the point of impact (death), is comprehended in the conditions under which the shot is "thrown." The lot of man as a "being-unto-death" is likewise inevitable, since "human-being" is "thrown-There-being."

It is through "already-having-found-itself-there-ness" (*Befindlichkeit*) that man is aware of his own existence as "being-unto-death."

In his inauthentic being, in his everydayness, says Heidegger, man does attempt to escape his lot. He forgets who he is and therefore flees into the anonymity of the "they" (*das Man*). Yet this flight does not prevent the inescapable lot that we call death. It is therefore good to "accept" "human-being" as "being-unto-death." All human acting, planning and designing can only look forward to the "nothingness" of death, of no-more-There-being.[97]

What we have thus far is a summary, an extremely incomplete survey, of Heidegger's "analysis" of There-being. It is quite clear that everything revolves around There-being, that There-being is the hinge and joint of the whole conception.

It cannot be denied that Heidegger's terminology suggests the world of technology and man as the technician. Man is a being who designs, who in his designing wields "being-as-it-is-to-hand" (*Zuhanden-sein*) as "instrument" (*Zeug*), that is, as tools, implements, utensils. This designing is not, however, the free giving of form to a given material, but *self-designing*. Technology, accordingly, is not dealt with explicitly. To the extent that technology is implied, we must conceive, because of the centrality of There-being, of a subjectivistic or – to employ Heidegger's own later terminology – an anthropological ascertainment of the meaning of technology.[98] Since There-being is a "being-unto-death," every ascertainment of meaning arising from There-being ends in "nothingness," which is meaning-less-ness par excellence.

From Heidegger's later publications, it is clear that he rejects an instrumentalist or anthropological ascertainment of the meaning of technology. It is also clear that the central concern later is no longer There-being (which in its being addresses Being itself), but Being, which disposes over man and expresses itself in thought. Later, too, Heidegger gives explicit attention to technology as the most impressive phenomenon of our time.

2.3.3 The "later" Heidegger

In *Being and Time*, as we noted, the emphasis is placed on There-being, which discloses itself to itself in and through "comprehension of Being" (*Seinsverständnis*). Through the meditating of self, the structures of existence and of all being become visible. The point of departure is There-being, which is concerned in its being with this being itself. At the center of Heidegger's concern is human There-being as the path of admittance to Being. In contrast, the "later" Heidegger places the emphasis on Being, which expresses itself in the thinking of There-being, or comes to language in it. This Being is not static, as was so often the case in traditional ontologies, but dynamic. It is an original movement or event; it is "historical" and, as such, disposes over man. Man and Being belong together. The thinking person is the place where Being can come out into the open, and Language (*Sprache*), which is essentially bound to thought, is the "house of Being."

This turn to Being – this "reversal" – is accounted for by Heidegger by referring back to Parmenides. Following his line, Heidegger says: "Thought and Being belong together (*gehören zusammen*) in the same and out of the same."[99] Yet, little enough is clarified here about the mutual relation itself. How are we to understand the belonging to – or with – each other of Thought and Being? Heidegger replies: " . . . the hallmark of man consists in this, that he, as the thinking being, open for Being, is presented to Being, remains related to it, and answers it" (IuD 22). Being is "historical" (*geschichtlich*) and is expressed in "the-coming-to-pass-of-an-e-vent" (*Ereignis*). Human thought "responds" to this "appeal." It is in this sense that we are to understand Heidegger when he says at last: "Being and Thought belong in one identity, and essence is from that to-let-belong-together which we call e-vent (*Sein gehört mit dem Denken in eine Identität, deren Wesen aus jenem Zusammengehörenlassen stammt, das wir das Ereignis nennen*)" (IuD 31).

Although the essence of technology and science is to be found precisely in Being – for "the essence of technicity is Being itself"[100] – it must nevertheless be said that technological-scientific man has forgotten Being. Does man, then, no longer think? Heidegger maintains that "science does not think" and that "man has until now done too much and thought too little."[101]

The falling away from Being began as early as Plato; it was with Plato that the reduction of Being to imagined objectivity began. With Plato all understanding of Being is brought down to contemplating that realm of eternal ideas by virtue of which man, who is endowed with the capacity "to-think-in-ideas," possesses the possibility of comprehending his own knowing as an instrument of mastery. Since man possesses the capacity "to-think-in-ideas," he can design his world beforehand and make himself a design of himself. It is with the metaphysics of Descartes and its dualism of subject/object that this development really gets underway. All being is brought into relation to man as subject. From this subject all being is ruled. "Being as the over-against-ness (*Gegenständigkeit*) of objects (*Gegenstände*) is comprehended in the relation to the representing (*Vorstellendes*) subject. This relation between subject and object obtains from then on as a territory in which decision is had with respect just to being in its being, to being, that is, each time merely as the over-against-ness of objects, but never with respect to Being as such." "Thus there presents itself, just at this point, the possibility of that which we call modern science and modern technology."[102]

In our time technology has attained its fulness. Since Being has been reduced to being, Heidegger does not speak of "philosophy" after Descartes but of a philosophy corrupted into anthropology.[103] "Having become anthropology, philosophy destroys itself in metaphysics."[104] Human thought has become increasingly a "technological-calculating thought,"[105] and "calculating thought is no *meditating thought* that meditates *meaning*, which rules in all that is" (italics added).[106] Consequently, there is an ominous development taking place. "This thinking is itself already the explosion of a violence that could cast everything into *absolute negativity* (*ins Nichtige*). The rest that follows from such thought – the technological process of the functioning of machines of destruction – would be merely the final, dark sending forth of madness to *meaningless* being" (italics added).[107]

The technological revolution has so transfixed and bewitched man that he has forgotten to be a "meditating being" (*nachdenkendes Wesen*). Heidegger suggests a "remedy": "The step back into the abode of the essential man (*des Menschenwesens*) requires something other than progress toward the machine. Returning to where we already really are is the way to arrive on the way of thought that we now need" (UzS 190). Heidegger explains: "The most thought-worthy feature (*das Bedenklichste*) of our thought-worthy time is that we do not yet think" (WhD 63). Our time's dearth of thought gives us the most food for thought!

2.3.4 "The Question about Technicity"

Heidegger does not begin his inquiry into technology with an analysis of technology. He does not wish to spend much time looking at the manifold abundance that technology offers; he would rather push on and track down the essence of technology. "Technicity," he declares, "is not the same as the essence of technicity" (TuK 5). We are delivered bound to technology beforehand if we start by concerning ourselves exclusively with all that bears the stamp of technology. The path to a perspective on the essence of technology is then closed from the start. Similarly, says Heidegger, we shall be blind to the essence of technology if – as very often happens – we regard technology as a neutral means that man can either use or misuse.

In general, man sees technology as a means to some particular end or as a human doing. This anthropological definition, Heidegger believes, is *correct* (*richtig*) for the earlier technology as well as for modern technology. However, he thinks that it is more critical at this point to know precisely how the relation between man and the instrumental is to be further construed. As man tries to remain the master of technology and to "get (it) spiritually in hand," he finds that it slips through his fingers – the more so the harder he strives. Therefore the instrumental and anthropological definition of technology, although correct, is insufficient. Heidegger says: "The determination [of the instrumental definition of technology] of what is there (*das Vorliegende*), if it is to be correct, need by no means uncover it in its essence. Only where such uncovering occurs does the true take place" (TuK 7). Because modern man seeks certainty and security, he is already satisfied with the anthropological definition of technology.

However, Heidegger says elsewhere: "This correct masters the true and shunts the truth aside" (VuA 98). It is therefore necessary to inquire further, to inquire into the essence of technology. "We must ask: What is the instrumental in itself?" (TuK 7).

To show what it is, Heidegger returns to the traditional doctrine of the four causes: *causa formalis*, *causa finalis*, *causa materialis*, and *causa efficiens*. He says: "What technicity, presented as means, is, unveils itself when we carry the instrumental back to the fourfold causality" (TuK 8). The following questions then arise: Why are there precisely these four causes? What is the unity of these four, and just what is their origin? "So long as we fail to address these questions, causality, and with it instrumentality, and with this latter the current destination of technicity, remains dark and groundless" (TuK 8).

To find answers, Heidegger goes back behind Aristotle. "Causa" to the Greeks was "aition," which carried the connotation of indebtedness to something. This indebtedness or "guilt," however, embraced *more* than these four causes; as the unity of the four causes, it seems to have transcended them, as it were. Man is more than the *causa efficiens*, the fashioner; he reflects and is thereby in a position to bring together the three other ways of being indebted (*Weisen des Verschuldens*). This reflecting or pondering (*überlegen* in German, and *overleggen* in Dutch) is in Greek *legein*, *logos*. It is connected, in turn, with *apophainesthai*, and it signifies a leading forth or bringing to presence (*Her-vor-bringen*), according to Heidegger. He believes that this meaning has been lost in our discussions of causality and in our use of the word. The four sorts of causality in the sense of "indebtedness" (*Verschulden*) show their unity in this, that something comes forth into presence. "The four ways of indebtedness bring something forth into presence. They let it appear in coming-into-presence (*Anwesen*)" (TuK 10). "Being indebted" (*Verschulden*) is therefore to be understood both as "being occasioned" (*Ver-an-lassen*) and as "being led forth" (*Her-vor-bringen*). For "leading forth" (*Her-vor-bringen*) the Greeks also used the term *poiésis*, which signified leading forth as well as making, fashioning, doing, and creating, including artisanry and art.

"What is this 'leading forth,'" asks Heidegger, "wherein play the four ways of indebtedness?" (TuK 11). From his answer to this question it is apparent that he is not satisfied with what has already been discovered. Heidegger wants to delve still deeper. "The 'leading forth' is a leading forth from hiddenness into unhiddenness. Leading forth happens only insofar as the hidden comes forth into unhiddenness" (TuK 11). For this "revealing" (*Entbergen*) the Greeks used the word *alétheia*, which signifies truth.[108]

Following this way (of thought), Heidegger is able to arrive, via the instrumental and anthropological interpretation of technology, at a deepened understanding of technology: "This [*Entbergen*, revealment] obtains as the characteristic feature of technicity." In other words: "Technicity is not just a means. It is a way of revealment" (TuK 12). By taking technology as it is and delving into its origin, Heidegger gets on the track of the essence of technology: "Technicity comes to presence (*west*) in the region where 'revealment' and unhiddenness, where *alétheia*, where truth happens" (TuK 13).

Truth, for Heidegger, is an original *event*, a revelation-of-Being. Especially since Descartes, truth has been reduced to a *given*, to a correlate of representation or judgment: "Adaequatio rei et intellectus." This reduction of truth to correctness of representation fits in with the place of man in the thought of Descartes. Man as subject became the focal point of being. Man circled about himself in pursuit of certainty and security in "presenting-re-presenting" thought. The Being of being was lost in "objectiveness" (*Gegenständlichkeit*). The world thereby became an image of and for man.[109]

Since Descartes, technology has turned the world into a self-affirmation of man. Man has therefore lost the truth. He has gotten into the slipstream of himself and of being.[110] Instead of concerning himself with Being, from which all issues, man gives his attention in science and technology to being. Heidegger pleads for reflective consideration for Being or devotion to Being to show that the *origin* of technology is a *truth-event*.

This conclusion leads Heidegger to inquire still further into the original signification of the term *Technik*. It derives from the Greek *techné* and was used in two senses. In the first place it was used to denote artisanship, but also art. As such, *techné* belongs to leading forth, to *poiésis*. In the second place, before Plato *techné* coincided with *epistémé*, which connoted to ken, to have insight, to know. Accordingly, it is possible to maintain on the one hand that technology as "leading forth" is clearly distinguishable from a flower, for example, which appears as of itself, while on the other hand it is not dumbly exhausted in making and wielding (technology as means). Therefore it is etymologically correct to see the essence of technology in connection with "revealing in the sense of the four ways of occasioning (*Veranlassung*)" (TuK 13).

Heidegger goes on to inquire: Does what has now been said hold true as well for modern machine technology? That, above all, is what concerns us; it is that which gives us no rest. When we inquire into it, it is correct but nonetheless insufficient to answer that modern technology is based upon modern physics and stands in a reciprocal relation to it. Heidegger wants to penetrate still further.

Heidegger believes that modern technology, too, is a revealment – but not in the sense of the technology of artisanship, which is "a leading forth in the sense of *poiésis.*" Of modern technology he says: "The revealment holding sway in modern technology is a *challenging* that puts unreasonable demands to nature, in order that energy be supplied which can itself be challenged and stored" (TuK 14, using Lovitt's translation). He also says: "The revealment which reigns in modern technology has the character of a *setting-up* (*stellen*) in the sense of *challenging* (*Herausforderung*)" (TuK 16 – italics added).

This challenging, this provocation, as a *mandate* or *command of Being*, gives modern technology an entirely different character than the earlier technology. Revealment as challenge knows no surcease. It is dynamic and restless in character.[111] Revealment proceeds via the unlocking, the transforming, the storing of energy, through the re-dividing and switching over of the same, while the entire process regulates and secures itself at the same time.

After Heidegger has clarified what he means by providing a number of illustrations, he more or less summarizes the new situation of modern technology as follows: "Everywhere it is commanded (*bestellt*) to stand present at the place, and indeed so to stand as to be able itself to be commanded (*bestellbar*) for a further commanding (*Bestellen*). That which is so commanded has its own standing (*Stand*). We designate it the 'fund' (*Bestand*)"* (TuK 16). *When he uses the term "Bestand," Heidegger tries to bring to expression the way in which everything is affected by the challenging revealment.*

Heidegger goes on to inquire into the place of man in this challenging setup (*herausfordernde Stellen*). He says: "Man, it is true, can represent, fashion and execute this or that, thus or so. Only, man does not dispose over the unhiddenness in which the real ever and again shows forth and withdraws itself" (TuK 17). Rather: "Only to the extent that man on his side is already challenged (*herausgefordert*) to challenge the powers of nature can this commanding revealment (*bestellende Entbergen*) happen" (TuK 17). An original challenge precedes the commanding (*Bestellen*) effected by man. Because he answers the various ways of revealment, man knows himself in the unhidden (*Unverborgene*). This means that man is neither fund (*Bestand*) nor origin of the challenging (*Herausforderung*). He is rather, it might be said, a mediator between them. In contrast to anthropological views of technology, Heidegger can therefore say: "Modern technology as 'commanding revealment' is accordingly no purely human affair" (TuK 18).

---

*See Lovitt, "A *Gespräch* with Heidegger on Technology," where *Bestand* is translated as "fund" and is said to denote a store or supply as "standing by," as a "ready-reserve-that-endures." – Trans.

That which transcends human doing, the original provocation, provides man his place in the commanding (*Bestellen*).

The *"that which gathers together (das 'Versammelnde'), from whence there come forth the ways by which we feel thus or so"* (TuK 19 – italics added) is *"the heart of man (das Gemüt),"* according to Heidegger. And "the challenging claim which gathers man thither to command the self-revealing as fund," that is, "the essence (*Wesen*) of modern technicity (*Technik*)," he calls the *Ge-stell*.* Of the *Ge-stell* Heidegger says, in summary: *"The Ge-stell designates the gathering of that setting (stellen) which people set, that is, challenge (herausfordern), to reveal (entbergen) the real in the way of the commanding (Bestellen) as fund (Bestand). Ge-stell designates the way of revealment which rules in the coming-to-presence (Wesen)† of modern technicity, and is itself not technical (technisch)"* (TuK 20 – italics added).[112]

Heidegger states, furthermore, that in the term *Ge-stell* there resides not only the sense of the "provocative commanding" (*herausfordernde Bestellen*) appertaining to modern technology but also – still – the sense of that "leading forth into presence" (for *hervorbringende Her-stellen*) which was a function of the older technology. Modern and classical technology "do indeed differ fundamentally in character and yet continue to be essentially related. Both are ways of revealment (*Entbergen*) of the truth (*A-létheia*)" (TuK 20). Only, in modern technology man is challenged to revealment (*Entbergung*). This had already become the case with the advent of modern, exact physics. "Its manner of representing sets nature up as a power-coherence calculable in advance" (TuK 21), making possible the total control of nature.[113]

Mistaken conclusions must be avoided at this point. Historically (*historisch-gerechnet*), modern technology first arises when it can utilize the mathematical natural sciences, Heidegger says, but *"historically thought (geschichtlich gedacht), that fails to reach the truth (das Wahre)"* (TuK 21 – italics added). Heidegger continues as follows: "The modern physical theory of nature is not the preparer of the way for *Technik* but for the essencing (*Wesen*) of modern technicity (*Technik*)" (TuK 21 – italics added). As Heidegger says elsewhere: "Modern science is grounded in the coming-to-presence (*Wesen*) of technicity"

---

*This important Heideggerian term is untranslatable, according to Lovitt. He describes its meaning as follows: "a peculiarly self-perpetuating and self-intensifying revealment," a "way of revealment" which "gathers together all the modes of the challenging revealment which are built on the verb *stellen* ...." – Trans.

†"Essencing," see Lovitt. – Trans.

(WhD 155). Even if it was not perceived to do so, the "presencing" of modern technology governed the rise of modern physics. Heidegger is therefore able to say: "...that which, according to historical establishment, is the later, that is, modern *Technik*, is, in view of the presencing governing within it, the historically (*geschichtlich*) earlier" (TuK 22). It is for this reason that Heidegger considers it incorrect to speak of technology as applied natural science; technology only makes use of this science. "Since the essence (*Wesen*) of modern *Technik* lies in the *Ge-stell*, it must therefore make use of the exact natural sciences" (TuK 23).

From what has now been said, it is clear that Heidegger rejects any instrumental or anthropological definition of technology. Heidegger rejects the idea that man can enter into an independent relation with technology (and science). Technology is not merely a human affair; after all, man is originally (*ursprunglich*) *placed* on the way of revealment (*Entbergen*).[114] "As the so-challenged one, man stands essentially on the terrain (*im Wesensbereich*) of the *Ge-stell*" (TuK 23). Revealment does not happen only and nakedly through man, for man should then forget that Being claims him for that purpose. It is the *Ge-stell* that "sends" man "into a way of revealment." "*That gathering 'mittance' (Schicken) that brings man upon a way of revealment we designate as 'sending' (Geschick)*"* (TuK 24 – italics added). From all this we may not draw the conclusion, says Heidegger, that "the historic destiny of revealment (*das Geschick der Entbergung*)" is a compulsory fate, or that technology as a whole is fraught with fatality. No, man is in fact *originally* free "insofar as he belongs on the terrain of the sending (*in dem Bereich des Geschickes*); he is thus not a pedial slave (*ein Höriger*) but one who really hears (*ein Hörender*)" (TuK 24). Man must keep unremittingly in view the fact that he is in the truth whenever, having been commanded by the *Ge-stell*, he enters upon the way of revealment (*Entbergen*). However, he may not subsequently surrender himself to the revealed (*Entborgene*) and thereby let it determine his acts; in other words, man may not derive his norms from technology as if he has been delivered helpless into its hands. That danger does threaten: "Brought between these possibilities, man is threatened from out the sending (*Geschick*). The sending of revealment is there as such in all ways (*Weisen*) and is therefore inevitably a danger" (TuK 26).
  Why does the revealment assume a negative value and become a

---

* "Historic destiny," see Lovitt. – Trans.

danger in modern technology? On the one hand man turns aside from Being to regard everything as if it were in the dimension of that which can be fashioned or worked, so that everything becomes fund (*Bestand*). On the other hand there is inherent in the destiny of Being (*Seinsgeschick*) the danger that man will turn aside from the revealment (*Entbergen*).

Heidegger wishes to say that, time and again, whenever the sending (*das Geschick*) brings us upon the way of revealment, this way veils or shuts off the origin of revealment: every revealment implies a concealment of the origin. "All revealment belongs to hiding and hiddenness" (TuK 25).

In this light it becomes clear what Heidegger means when he says that the moment of greatest danger is precisely when "destiny (*Geschick*) rules in *Ge-stell* fashion (*in der Weise des Ge-stells*)" (TuK 26).[115] Not only does man move at such times at the brink of an abyss of pure fund, which threatens to swallow him up, but he is also no longer aware of himself as the commander of the fund (*Besteller des Bestandes*) called to disclosure of the truth – and this is so just at the moment of his pretension to be lord and master of the earth. "It must seem to him as if man encounters everywhere only himself" (TuK 27). Everything bears the mark of having been made by man. "*Meanwhile, man meets himself nowadays nowhere in the truth*, that is, in his essence" (TuK 27). Man has forgotten that he is in the service of the happening of truth (*das Ereignis der Wahrheit*); instead, he rules. But again, if it is the danger of technology that concerns us, we must take into consideration more than just technology and the man-of-technology: "The *Ge-stell* distorts the appearing and ruling of truth" (TuK 27).

Heidegger's view that man does not (any longer) encounter himself in modern technology and science seems to contradict Heisenberg's notion that man can encounter no one but himself when he is involved with an apparatus or a machine or when he wanders in a landscape thoroughly marked by technology.[116] Heisenberg claims that it is not only in technology but also, for example, in microphysics that man comes upon structures which he has himself called forth. Knowledge of atoms and of their movements as such, that is, independent of experimental investigation, is impossible. Heidegger does not want to deny this; what he wishes to emphasize is the degree to which being is reduced in such a view to object, to *Gegenstand*, to being to be ruled.

The same is true even of man, who no longer listens to Being but has become a slave to the destiny or mittance of Being (*Seinsgeschick*), who departs from Being to go forth upon the way of technological control. "Progress!" Yet man is reduced to *gestellter Mensch*, to fund (*Bestand*), and he loses himself in it.[117]

The dichotomy in Heidegger's view of technology will engage our attention repeatedly. On the one side danger threatens as destiny of Being (*Seinsgeschick*), and on the other side man no longer listens to Being. Heidegger does not try to escape the dichotomy by proposing that technology is ruled by darker, demonic powers. The danger does not arise, in his view, "from afar." Heidegger does not seek a cheap way out. He even acknowledges that we are *allotted* to technological objects, that we cannot (any longer) do without them, even to the extent that we must constantly be improving them. In contradiction to Heidegger's Yes to technology stands his No, since technology bends and distorts the essence of man, confounding and finally destroying him. "I should like to use an old word to name this simultaneous maintenance of Yes and No towards the technological world: resignation (*Gelassenheit*) with respect to things" (Gel 25).

Heidegger's acquiescence is not so complete as to reduce him to silence. The essence of technology as *Ge-stell* may, as a destiny or sending of revealment (*Geschick des Entbergens*), indeed be the greatest danger. All the same, he asks how the danger may be overcome and where deliverance may be found. He joins the poet Hölderlin in saying: "Where the danger is, there grows the saving power also" (TuK 28), in order the better to inquire to what extent deliverance from technology may be concealed within its very essence. And then he discovers the possibility of deliverance by reflecting on the meaning of the word *essence* (*Wesen*). This word, Heidegger says, should not be taken in the usual sense of *essentia*, that is, as a word for a genus, such as that which makes all trees trees. Rather, it should be taken in the sense that one finds expressed in words like *Hauswesen* and *Staatswesen*, where there is an implication of the manner in which "house" and "state" are governed, of how they direct and develop themselves. Heidegger believes that this signification of the term *essence* is best expressed by the verb *keep* (*währen*), which means *last*, *endure* and *persist*. "Sojourn, continue, endure, is of course the old meaning of the word *being* (*Sein*)," says Heidegger elsewhere (SvG 207).

From this analysis it is quite clear that Heidegger does not wish to construe the essence of modern technology as a genus or noun, as something which is. He sees it more as a verb, as being that is coming-to-being, or becoming. There is more, however, for essence (*Wesen*) is more than that which endures. "How technology 'is' (*west*) allows itself to be understood only from out that enduring in which the *Ge-stell* happens as a sending of revealment (*Geschick des Entbergens*)" (TuK 31). Therefore Heidegger, following Goethe's example, would rather speak not of *währen*, which means *enduring*, but of *gewähren*, which has a still more original signification. *Gewähren* means to stand as a surety for something, to tender something. "*Only the granted endures (Nur das Gewährte währt)*" (TuK 31). In short, Heidegger would say that if the danger is the greatest just when man is fully consumed by the provocation or challenge of revealment (*Herausforderung des Entbergens*), then it is precisely the essence of technology that summons the same man and at the same time stands as surety for him, in order to "return in the worthiness and space of his essence" (TuK 32). "Every sending of revealment happens from out the granted (*Gewähren*), and as such. For this carries for man only that part of the revealment (*Entbergen*) which the happening of the revealed (*Entbergung*) requires. As the so used, man himself comes to belong to the happening of truth. The granting power (*das Gewährende*), which disposes (*schickt*) in the revealed in one way or another, is as such the delivering power (*das Rettende*)" (TuK 32). But the precondition for this is "that we begin doing our part, which is to heed the essence of technology." Therefore everything depends on our "reflecting upon the arising (*Aufgang*) (of the saving power) and devotedly tending it" (TuK 32).

Of this forked nature of technology as danger or deliverance there is little more to say; it is at bottom a *mystery*. It must be clear in any case that technology itself cannot bring deliverance, that only *reflection* on the essence of technology affords perspective. We can nourish the growth of deliverance only by paying constant attention to the most extreme danger. "The closer we approach the danger, the more brightly lit become the paths to the delivering power, and the more inquisitive we ourselves become. For inquiry is the devotion of thought" (TuK 36).

2.3.5 "The Reversal"

In his essay "The Reversal" ("Die Kehre"), Heidegger is principally concerned with the *forked meaning* of the *essence* of technology, and he adds to what has been brought up in "Die Frage nach der Technik." On the one hand the essence of technology is danger (hiddenness in the revealment), while on the other hand it is deliverance (revealment in the hiddenness). According to Heidegger, as we saw, the equivocal character of this definition of the meaning of technology is at bottom a mystery, and the essence of technology accordingly cannot be grasped scientifically or anthropologically – as if technology as peril might be escaped by grasping it in such a way, with salvation then brought within the gates.

It seems that Heidegger sometimes strongly emphasizes Being as the "origin" of the danger of modern technology, and that at other times he emphasizes man who forgets Being. This forgetfulness-of-Being, furthermore, is itself a destiny-of-Being (*Seinsgeschick*).

The dichotomy between Being and man permeates not only Heidegger's view of technology as danger but also his view of the possibility of deliverance from it. Originally man "set" technological reality, after having been called to that task be Being in e-vent (*Ereignis*). The essence of modern technology, the *Ge-stell*, gradually preempts man for itself. Man is altered from a mediator between *Ge-stell* and fund (*Bestand*) and, becoming fund himself, a mere *gestellter Mensch*, he is estranged from Being. However, man is again essential to deliverance from forgottenness-of-Being, for he serves, as it were, to create the preconditions for the reversal.[118] The reversal itself, furthermore, is a Being e-vent (*Seinsereignis*) that disposes over man.

On the one hand man who has fallen into forgottenness-of-Being is essential to initiating a turn for the better, but on the other hand Being uses man to arrive at a new Being-e-vent. Man must begin to give reflective consideration, or devotion, to Being. Thereafter man will be liberated from himself as ruling subject, by the grace and favor of Being; he will be liberated to the height, the breadth, and the depth of Being, to the innermost world (*Weltinnenraum*).[119] In thoughtfulness for Being, man will find himself on the way back from modern science and technology.[120]

Before we pursue Heidegger's view of the reversal any further, it is important to note that he nowhere speaks of man as responsible for nihilistic technological development or culpable because of it. This is of great importance in connection with the way of escape he finally proposes. What Heidegger seeks is the way back; he is ultimately at a loss as to what to do with the meaning of technology and of technological development. He flees from technological development; only in flight, in turning to Being, does he find any hope of deliverance from nihilistic technological development. It is for this reason that the reversal attracts considerable attention.

Now, although "the *Ge-stell* comes to presence as danger" (TuK 37), says Heidegger, it does so without the danger being noticed: the danger hides behind all the activities, "through the commanding of the *Ge-stell*," by which all being is made fund (*Bestand*). Nevertheless, the fact that the danger remains hidden is its "most dangerous feature" (TuK 37). The source of the danger is man's constantly continuing to construe technology as a means at his disposal. "In truth, however, the essence (coming-to-presence) of man is so commanded that it abets the essence of technology" (TuK 37). To understand this we must recall that the *Ge-stell* is a sending-of-essence of Being (*Wesensgeschick des Seins*). All too frequently, however, we are disposed to pay attention only to what lies at hand, thereafter drawing our conclusions on the basis of historical givens, for example. Thereby we commit a grave error: "We set history (*Geschichte*) on the terrain of that which has

96

happened (*des Geschehens*) instead of considering history in its provenance as a coming-to-presence (*Wesensherkunft*) out of sending (*Geschick*)" (TuK 38). Beyond this, the additional error we can commit is to construe the mittance (*Geschick*) statically. What we must see is that the destiny of Being (*Geschick des Seins*) is being itself as that e-vent or as that movement in which all that is, appears.

Being disposes in the e-vent (*Ereignis*) and therein disposes over man. Therefore Heidegger can say: "Technology, of which the essence is Being itself, never allows itself to be conquered by man, for that would imply man as the master of Being" (TuK 38).

Is man not, then, the victim of Being, asks Heidegger, if the essence of technology is Being? No, for although the essence of man belongs to the essence of Being, it is at the same time true that "the essence of Being needs the essence of man." Therefore we may conclude that "the essence of technology cannot be guided in the course of its sending without the help of the essence of man" (TuK 38). Clearly Heidegger means that there is an essential relation between man and technology. This must still be brought into sharper focus. "We must think the particular coming-to-presence [essence, *Wesen*] of man so that it belongs to the coming-to-presence of Being and is thereby used to keep (*währen*) the coming-to-presence of Being in its truth" (TuK 39). It is therefore necessary before all else that we "think the coming-to-presence of Being as being worthy of thought." But Heidegger inquires: "How must we think?" (TuK 40). What is necessary is that we prepare (*bereiten, bauen*) a place for Being in the midst of all beings, instead of exorcising it, as happens in technological thought. On the other hand, the Language (*Sprache*) of Being still has to precede thought. "Language is the initial dimension in which man is really able to answer Being and the summons of Being and, in that answering, to belong to Being. *This first answering*, deliberately completed, *is thought*" (TuK 40).

The essence of the *Ge-stell* is danger, as we saw, but within the danger *hides* the possibility of a reversal, in which the forgottenness-of-Being so turns itself "that with this reversal the truth of the essence of Being returns to Being" (TuK 40). This reversal e-ventuates (*ereignet sich*) only when the danger comes to light *as* danger. Accordingly, this reversal is a reversal within the Being-event itself. Perhaps we are already standing in shadows cast forward by this reversal. "When and how this mittance (*Geschick*) will happen, no one knows" (TuK 41). The last word – notwithstanding "that man must first open himself to the essence of technology" (TuK 39) – belongs to Being! "Only when man as the shepherd of Being keeps watch over the truth of Being can he be on the watch for a coming of the destiny of Being

(*Seinsgeschick*) without falling into pure willing-knowing (*Wissenwollen*)" (TuK 41). Therefore the words of Hölderlin – "But where the danger is, there grows the saving power also" – Heidegger claims, can better be made to say: "Where the danger is as danger, there grows too, already, the saving power" (TuK 41). Danger and deliverance are identical, on the condition that the danger come to light *as* danger. "When danger is as danger, there e-ventuates (*Ereignis*) with the reversal from forgottenness the coming-to-light [*Wahrnis*, keeping] of Being" (TuK 42). Therefore Heidegger can also say: "In the essencing (*Wesen*) of the danger there comes-to-presence (*west*) and dwells (*wohnt*) a favor, namely, the favor of the reversal from forgottenness of Being into the truth of Being" (TuK 42). As Being is everywhere and nowhere, in other words, lacking all definiteness, it can never be said when the reversal will come. "The reversal from the danger occurs suddenly" (TuK 43). That is all that can be said of it.

Do we not thereby come close to the notion of blind fate? No, for then we would be forgetting once again that the essence of man belongs to the essential space of technology, which is Being. "The *Ge-stell* is, although veiled, still outlook – no blind sending in the sense of a totally opaque lot" (TuK 45). "If insight e-ventuates, then man is struck in his essence by the lightning of Being. Man is the one discovered in the moment of insight" (TuK 45). Only then does man experience what freedom is! "Only when the human-essence in the e-vent of his insight as the thereby discovered one is denied human self-will and projects for himself the insight to himself and away from himself does the human in his essence satisfy the summons of insight" (TuK 45).

Elsewhere Heidegger has cast the turning toward Being in a more or less different light, namely, that of Language (*Sprache*). "Language conceals within itself the treasure of all the real" (HdH 33), says Heidegger. Language and e-vent are closely related. "The e-vent speaks. Likewise speaks Language, after the manner of the e-vent's revealing and concealing itself" (UzS 262).

Language in its fulness is relegated increasingly to the background in modern technology, because of the everyday hustle and bustle and also because of technological-scientific thought. Language is reduced more and more to a technological instrument. This can be seen in machines for calculating, thinking, and translating. "The translation machine (*Sprachmaschine*) is – first and foremost – a way in which technology disposes over the nature and world of Language as such" (HdH 36).[121] Technology has preempted Language for its own use and has thereby become a sinister force that overpowers the essence of man. Through habituation man fails to perceive the alien-subversive-*ergo*-ominous (*das Unheimliche*) character of this technological power. "The *Ge-stell*, the all-controlling coming-to-presence (*Wesen*) of modern technology, commands the formalized language, the way of giving word, by virtue of which man is fitted into the technological-calculating essence (*Wesen*), that is to say, is controlled; step by step, man surrenders 'natural language'" (UzS 263). In other words, technological-calculating thought introduces a language which drives out "natürliche Sprache" step by step. Yet there is a residue left over "for want of formalizing." This residue offers the possibility of deliverance: "What, when utterance [*Sage*, created by natural language], instead of merely disturbing the destroying power of information, shall have already recovered it from out the 'uncommandable' of e-vent? What, when e-vent – nobody knows when or how – shall become in-sight (*Einblick*), from which the clearing bolt (*lichtender Blitz*) strikes that which is, and is kept for, Being? What, when e-vent, through its introspection (*Einkehr*), shall withdraw every that-which-is-present (*Anwesende*) from pure commandability (*Bestellbarkeit*) and bring it back to itself?" (UzS 264).[122]

Thus the basis here for the hope of a turning and wending toward Being is Language (*Sprache*), which cannot be fully formalized. That is because nature-as-objectiveness (*Gegenständlichkeit*) is only one of the ways in which nature exposes (*herausstellt*) itself.[123] Nature is more than objectiveness. Nature should sooner be understood in the sense of the Greek thinkers: "*Physis*: the coming and going to its presence and absence of all that-which-comes-to-presence (*alles-Anwesenden*)" (HdH 28).

Being, Language, Nature – these seem to be synonymous for Heidegger. To investigate this matter further, however, would carry us too far afield. It is now clear that deliverance from the danger of technology is not to be found, according to Heidegger, in the technology that lies at hand.

To consider and inquire into the essence of technology is not to analyze the technology at hand and is not to become preoccupied with technology; rather, it is to answer the clearing summons of the truth of Being in the commanding of the *Ge-stell* (*Bestellen des Ge-stells*). Only, the sheer willing and doing of man and his quest therein for the future represents the set of technological-calculating re-presenting, in which the summons of Being gets lost. It is for this reason that Heidegger prefers not to give us a description of our own age, for such a description would necessarily remain superficial and afford no real perspective. "As long as we do not thoughtfully experience what is, we cannot belong to that which shall be" (TuK 46).

Along the way from the technological future back into the past, according to Heidegger, the way to the future lies open. In the silent, thoughtful experience of the danger *as* danger – in resignation, then – there is the beginning of possible deliverance from technology and its power. Man ought to repent from taking himself as subject, as the point of reference of all being. Then Being might possibly be inclined to disclose itself anew and bestow upon us a new dwelling place – a home (*Heim*), in the Language of Being. There is nothing that can be said of this with any certainty. And in this respect Heidegger is consistent in his break with the notion of truth as certainty. We remain standing before a mystery ("*das Ge*-heim-*nis*"), but then there is hope that Being will reveal the hidden meaning of technology.

Heidegger points to the essence of technology, the *Ge-stell*, as the danger, but he also speaks about the hidden meaning of technology. Apparently he means by this the *deliverance* from technology as danger. Of the meaning of technology he says, thinking not only of the danger but also of the possible deliverance: "There reigns, then, in all technological processes, a meaning which claims all human doings, a meaning which man has not invented and made." Also: "We do not know what the alien-subversive (*unheimlich*) aggrandizing power of atomic technology has in mind. *The meaning of the technological world hides itself.*" Thus the meaning present in technology-as-danger is hidden. By setting ourselves open – "openness for the mystery" – there is perspective; there is a "vista of a new, solid ground" (Gel 25).

2.3.6 Martin Heidegger and Friedrich Georg Jünger

From what we have seen above, it is clear that various of F. G. Jünger's ideas turn up again in Heidegger. While this matter is not directly related to the subject of this study, it does seem desirable to say something about the possible connections between Heidegger and Jünger.

As far as I have been able to determine, there is no indication that Heidegger ever mentioned any of Jünger's publications, let alone that he engaged him in discussion. Heidegger did address the thought of *Ernst* Jünger, in *Zur Seinsfrage*. F. G. Jünger, on the other hand, was clearly influenced later by the "later" Heidegger, as we have seen. Moreover, he published a poem in the anniversary volume for Heidegger's seventieth birthday.[124]

In general, Heidegger is correctly regarded as an original thinker. In his ideas about technology, however, the influence of Jünger is present, in my opinion. Heidegger often uses terminology reminiscent of Jünger, who wrote about technology earlier. Nor is Heidegger's contact with Jünger surprising, considering his clear sympathy for poets. Heidegger frequently cites Goethe, Rilke, Hölderlin, and Hebel, a line to which Jünger also belongs.

The conjecture that Heidegger learned from Jünger is well founded. To support it, I will point to a number of transparent similarities. In the first place, the expressions *Perfektion der Technik* and *Automatismus* were already being used by Jünger in 1939, with the former even serving as the title of his first publication about technology. Heidegger, and many other authors after him, also used this terminology, and invested it with the same signification, namely, that modern technology leaves nothing undisturbed as it comes to completion. In and through technology, man as subject and the world as object reach completion.[125] When Jünger addresses the historical dynamic brought about by technology, he speaks of *totale Mobilmachung* (total mobilization), a term which Heidegger appropriates and places between quotation marks![126]

Jünger asserted that modern technology turns the university into a *Technikum* in which all the sciences and ancillary disciplines are converted to the service of technological development. The practice of modern science has thereby become an industry. The same ideas appear in Heidegger.[127]

Like Jünger, Heidegger holds the view that modern technology brings forth the modern, total state, and that it leads to the organization of world public opinion. In short, man is delivered into the hands of powers which he can neither govern nor control.[128]

Jünger claims that the destroying power of technology remains hidden because technological thought remains intact in the face of the devastation. There is agreement between Heidegger and Jünger on this point, too, since Heidegger states that the greatest danger of technology is that the danger remains hidden.[129] Jünger believes that with technology, death penetrates life. Technological thought attempts to conceal this; death and suffering are bureaucratized. According to Jünger, death is thus made manifest in the most striking manner, inasmuch as people are bereft of all feeling. For Heidegger, technology as the negation of death can do nothing to alter the fact that there is nothing more certain than death. That is why the epoch of technology as human self-protection is the epoch of unprotected, threatened humanity.[130] Beyond this, Heidegger developed these ideas along the lines of Rilke's view of death.

When the destroying power of technology reaches full completion and therein attains the victory, it is at the same time defeated. The victory of technology means death to itself. "That is the point at which man is thrown back upon new thought" (MuE 352). The same idea appears in Heidegger when he says in connection with the reversal: "Where the danger is as danger, there grows too, already, the saving power" (TuK 41). The technological-calculating thought of technology yields to reflective thought.

Jünger, as we saw, appeals for a return to nature. "The earth needs man as a caretaker and shepherd" (MuE 363). In Heidegger, the nourishing, guarding and tending of nature become the nourishing, guarding and tending of Being. Although differences are clearly present here, there is also agreement in that Heidegger has used the term *Nature* to refer to Being.[131]

In conclusion, I believe it can be argued that Heidegger's view of technology is somewhat less original than has often been thought. Jünger's ideas about technology appear to have influenced him. Could it be that it is partly because of this influence that Heidegger abandoned the notion of human There-being (*Dasein*) as self-designing There-being, a notion important in *Being and Time*? That the reversal in Heidegger's philosophy has something to do with his view of technology is a conjecture I have raised before.[132] In the light of Jünger's view of technology and the influence which it may have had upon Heidegger, the question whether there is a connection between technology and the reversal can be answered affirmatively.

On the other hand, it is evident from Jünger's publication *Sprache und Kalkül* that the "later" Jünger was in turn influenced by Heidegger. He mentions Heidegger by name.[133] Now Jünger is no longer satisfied to simply maintain technology as predatory cultivation and diabolism; he is out looking for a basis for the hope of deliverance. Heidegger is of the opinion that technology cannot be understood as a means in the hands of people, since technology attempts to overpower humanity. He rejects the notion of a diabolism of technology since such a view never comes more than halfway to the "way of thought." Jünger took the lesson to heart.

2.3.7 Dessauer's "confrontation" with Heidegger

Heidegger is not an easy thinker to understand. Unless one makes an exceedingly careful study of his philosophy, erroneous interpretations lie at the door. I will now raise the question whether this might be the case in connection with Dessauer's treatment of Heidegger. The concomitant advantage of this line of investigation is that it brings up the major elements of Heidegger's thinking on technology again, placing them in a new context.

Friedrich Dessauer is perhaps the greatest of the philosophers of technology. He vigorously propagated his philosophy of technology for many years (1924-1960). In one of his last publications, *Streit um die Technik* (1958),[134] there is a chapter entitled "Die Technik in existenzialphilosophischer Schau" (pp. 311-68), in which he considers Heidegger's view of technology ("Martin Heidegger über die Technik," pp. 348-68). In order to understand and evaluate this critical analysis, I will first offer a brief statement of Dessauer's philosophy.

In order to get on the track of the meaning of technology, Dessauer, as a good philosopher, tried to avoid getting lost in the great welter of technological things and facts. He sought the source and origin of all that is technological, and he believed he had found it in invention. He believed that with this insight he had provided a foundation for the independence of technology over against economics and the natural sciences, and that he had once and for all done away with the so-called instrumentality (*Mittelhaftigkeit*) of technology.

Yet Dessauer does not escape difficulties in connection with his idea that the source of technology is to be found in invention. Construction and fabrication, for example, also belong to technology, but they are clearly distinguishable from invention. Dessauer later believes – correctly, in my opinion – that he can avoid this difficulty by analyzing invention in further detail. He then distinguishes between pioneering invention (*Pionierserfindung*) and developmental invention (*Entwicklungserfindung*).[135] With this distinction Dessauer seeks to indicate that there is indeed a difference between the more or less unexpected invention on the one hand, in which the creative character is clearly primary, even though experimentation, research, and thought may have been involved, and the developmental invention on the other hand, which rests upon the methodical labor of a number of engineers and inventors. The former depends primarily upon technological fantasy, and the latter upon technological science. In the latter case, the invention is developed not with great rapidity but step by step, according to a work schedule or design and with the help of amassed technological know-how. The surprising element in pioneering invention is missing in developmental invention.

These kinds of invention are definitive for the further development of Dessauer's philosophy of technology. Yet he is aware that he still needs to discover the origin of invention. Only if he is given the knowledge of this origin will he be able to speak of the meaning of technology – and he does speak of a religious-metaphysical delimitation of meaning. From his definition of technology, this much is apparent: "Technology is real being from ideas (*reales Sein aus Ideen*), through final forming and fashioning of naturally given material" (SuT 234). Dessauer ascribes a place of importance to technological ideas: together they constitute an independent realm that possesses the highest grade of reality.

With the introduction of this realm of technological ideas, Dessauer, inspired by Plato, has taken the meaning of technology to be something referring beyond technology.[136] By means of inventions, the ideas are realized. Dessauer already speaks of this in one of his earliest publications: "This that is new [invention], transcendental, qualitatively other, loosing itself from the origin, autonomously working further, coercing man, altering the form of the earth, lies in the deepest essence of technology."[137] Man is called to actualize these ideas. Dessauer cites Genesis 1:28: "Subdue the earth." Akin to this view is the notion that through the struggle to actualize technological ideas, technology will produce progress – first for technology itself, but thereafter for human society too.

On the basis of what we have seen above, we are in a position to understand Dessauer's religious-metaphysical interpretation of technology as it is summarized by Klaus Tuchel: "All technological realizations are preceded by, and the realm of pre-established solutions is founded upon, the plan of God, which is the real ground and presupposition of all technology. Its essence can be transparent only to one who reconciles both the naturally given and the final fashioning of it to the Biblical act of creation, and who understands technological inventions in the sense of a *creatio continua*."[138]

While a great deal more might be said about Dessauer's philosophy of technology, the preceding is sufficient to develop an understanding and evaluation of his confrontation with Heidegger. The question, really, is whether Dessauer does Heidegger justice, and whether it is accurate to speak of an authentic confrontation. This is especially so in view of Dessauer's not seeking the meaning of technology in dynamic Being, let alone entertaining the notion that the meaning of technology might be hidden in Being. Dessauer finds it rather in a fixed realm of ideas. His optimistic expectations contrast strongly with Heidegger's view of technology as *the* danger. The important point to watch for is whether Dessauer perhaps proceeds too hastily to interpret Heidegger in the light of his own view of technology. I will make a point by point comparison as follows: What is technology? Is there a difference between old-fashioned technology and modern technology? Is Dessauer's interpretation of revealment (*Entbergung*) correct? Is the confrontation superficial, as is perhaps illustrated by Dessauer's perception of the danger of technology?

1. Dessauer gives clear evidence of an interest in existentialist philosophy. "We are interested to know if existentialist philosophy has attained an enriched, deepened, broadened, more insight-full, clearer view of technology" (SuT 357). He appears to take cognizance of what to his mind are the positive elements in Heidegger's view of technology, while not hesitating to point out the errors and reject them. Thus Dessauer is of the view that Heidegger understands modern technology too much in the sense of "machine technology," and that Heidegger is accordingly oblivious to the rich diversity afforded by modern technology. Do not synthetic medicaments, electric light bulbs, automobiles, telephones, and so forth belong to its terrain? Now, all these

things mentioned by Dessauer are not really technological things at all; even if they are technologically "founded," they are not destined for use in technological forming. Apart from this, however, it is clear from this argument that Dessauer has not understood Heidegger very well. Heidegger, as we noted, saw the essence of modern technology in the *Ge-stell*; the dominion (*Herrschaft*) of the *Ge-stell* is the origin of everything's being made into fund (*Bestand*), of everything's coming to bear the stamp of technology. Machines and gadgetry may be the most striking examples, but for Heidegger, fund (*Bestand*) embraces even more – for instance, the forest ranger[139] and the travel bureau.[140] "Fund (*Bestand*) characterizes nothing less than the way in which everything comes to presence (*anwest*) that is struck by the challenging revealment (*herausfordernden Entbergen*)" (TuK 16). It is therefore legitimate to question whether Dessauer realized just how comprehensive technology is for Heidegger, or whether he fathomed Heidegger's deepest intentions.

2. The conjecture that this is not the case is confirmed by Dessauer's disinclination to accept Heidegger's distinction between classical and modern technology. Heidegger's error, in Dessauer's opinion, is that he speaks of challenge (*Herausforderung*) with respect only to modern technology and regards it as being absent from the older technology. Dessauer asks: "Was it not a 'challenging' intervention when man rooted out the forests, even to excess, to make pastures?" (SuT 358). And is there now really an essential difference between the old windmill and the modern hydroelectric works? Do not both cases involve producing energy through a "setting up (*Stellen*) in the sense of challenge"? Dessauer will not hear of a "preserving bringing forth" (*bewahrendes Her-vor-bringen*) in contrast with a "challenging commanding" (*herausforderndes Bestellen*). He objects to any real distinction. "The difference concerns the degree, the mastery, the yield of this intervening, not its essential character" (SuT 358). This difference with Heidegger is to be explained on the basis of the fact that Dessauer interprets his *Herausforderung* too much along the lines of Toynbee's "challenge."[141] He thereby fails to understand Heidegger correctly. Heidegger's challenge (*Herausforderung*) is intended to show the difference between classical and modern technology as being a difference of destiny of Being (*Seinsgeschick*). For classical technology, the e-vent of Being (*Ereignis des Seins*) had a very conservative, limited, we might even say, a more or less static character. For modern technology, there is talk of an all-affecting, all-embracing dynamic character, to which Heidegger gives expression when he speaks of challenge (*Herausforderung*).

Raby is correct when he observes that Dessauer, in contrast to Heidegger, makes only a quantitative – and not a qualitative – distinction between classical and modern technology. Heidegger has given more consideration than Dessauer to the fact that modern technology utilizes scientific method and – I would hasten to add – to the fact that it is based upon science. The old technology made use of natural possibilities (the windmill); in modern technology, nature is dissected according to a scientific plan and then transformed. The desired results are foreseen and provided for in this plan. "Man approaches nature with this pre-established plan and forces her to carry it out."[142] Heidegger has seen the qualitative difference with the technology of craftsmanship. "In setting up the conditions in modern technology, there is added, in contrast to the older technology, the mathematical plan."[143]

3. Dessauer would interpret Heidegger's frequently used concept of "revealment" (*Entbergen*) as follows: "In Heidegger's language, the 'leading forth' (*Hervorbringen*), the 'revealing' (*Entbergen*), the 'happening of the true' signify, after all, that peculiar capacity of technological invention to take the forms of solutions for human problems (in the framework, that is, of problems solvable in the natural order) and – while not just contriving them arbitrarily, but while having respect instead to their pregiven form, that is, to their so-being, their *Quiddität* – to proceed to reveal (*entbergen*) them from out hidden (since only potential) being (they have not been actualized by nature itself; they are not "at hand" or "vorhanden," not "to hand" or "zuhanden," but just possible); to reveal (*entbergen*), that is to say, to find out (*auffinden*), to actualize, and to make visible" (SuT 355-6). How transparent it is at this point that Dessauer has read Heidegger through his own glasses! He interprets revealment (*Entbergen*) as a connection between the realm of ideas and technological reality, while Heidegger means to express the im-parting of the sending (*Schickung des Geschicks*). Moreover, it seems to elude Dessauer that Heidegger rejects Plato's doctrine of ideas, believing it to be the source of "technological-calculating thought," the kind of thought that makes us forget Being. How much more, then, should Heidegger have rejected the Plato-inspired thought of Dessauer, which is colored by technology in every respect! It is impossible for Heidegger to speak of "potential" being, since he rejects all determination of Being on the ground that every determination of Being is forgottenness of Being!

While Dessauer awaits the progress of mankind through technology, Heidegger, by contrast, regards destruction as its most extreme consequence and therefore calls for *thoughtful reflection*. Now, this does not mean that Heidegger never thought of invention in connection with revealment. More than one passage indicates the contrary. Heidegger values technology positively where invention is concerned, since in invention the truth manifests itself and freedom becomes visible. Heidegger is negative, however, toward the technological development into which invention is taken up and converted to fund (*Bestand*). The freedom of the invention is destroyed again in technological development. This dichotomy has far-reaching consequences for Heidegger's total view of technology. I will return to this matter in my critical analysis.[144]

4. It is obvious from the preceding that we cannot speak of a genuine *confrontation* between the two thinkers. Dessauer draws hasty and ill-considered conclusions from Heidegger's writings. Given his own ontology, he is unable to understand Heidegger's framing of the problem. He does not trouble himself to undertake an investigation of Heidegger's deepest motives, which go beyond Heidegger's thought and yet constantly come "to language" in it. Had Dessauer undertaken such an investigation, Heidegger's point of departure and the ineradicable stamp set upon his philosophy by his nonphilosophical religious choice would not have continued to seem arcane to him. Nor would Dessauer have accepted one portion of Heidegger's philosophy of technology after interpreting it to suit his own views while leaving another portion unaltered, including, for example, the "reversal." Thus there is reason enough to accuse Dessauer of a "superficial" interpretation of Heidegger. This is again apparent when Dessauer discusses Heidegger's "obscurantist way of saying things" (*verküllender Sprachweise*). Dessauer thinks Heidegger uses difficult terminology to discuss familiar matters! The advantage to this, in Dessauer's opinion, is that old problems acquire new life. In the meantime, however, he brushes Heidegger's problematics off the table.

Dessauer writes: "That technology is more than human handling and means is not true of modern technology alone; that has frequently been said, for decades. As for the metaphysical or religious declarations lying behind it, these are exchanged by Heidegger for the concepts *Gemüt* and *Gestell*, in which the problem of the transcendental, or more-than-human, is concealed (*eingehüllt*) and hidden (*eingerätselt*). If that is legitimate, then it is also legitimate, and perhaps more fruitful, to consciously attempt the step toward the metaphysical declaration and the religious analogy" (SuT 361). Of course Heidegger would have opposed this interpretation, saying, as he does: " . . . the anthropological determination of technology does not allow itself to be completed by a metaphysical or religious declaration merely placed behind it" (TuK 21).

5. Dessauer acknowledges that great perils are concealed in modern technology especially, but in contrast to Heidegger, he regards them as likewise present in classical technology. In modern technology, specialization of craft threatens man with the loss of his contact with the backgrounds and the eternal. To prevent such a loss, it is necessary to be reminded where and how technology arose and remains possible. It

is in this sense that Dessauer understands that the *Ge-stell* can be both danger and deliverance. Instead of Heidegger's obscure way of speaking, he would then rather say: "How simple and clear it would be if man ... would only consider that about twelve hundred years before Christ the books of Genesis were written, in which the command was given to subdue nature and in which the possibility of actually carrying out the command was suggested by the likeness between created man and the Creator" (SuT 366). Now, this interpretation could never have gained Heidegger's approbation. Heidegger does not wish to appreciate technological development anew; rather, he wants to draw back from it, in order to command all devotion for the great mystery of Being. Dessauer speaks of the plan of God for this world and of the clear mandate to man to actualize that plan. That such language is alien to Heidegger is abundantly clear from the following statement, in which Dessauer's philosophy is rejected and his incomplete and incorrect interpretation of Heidegger is illuminated: "That God lives or remains dead is not decided by the religiosity of man and still less by the theological aspirations of philosophy and natural science. Whether God is God *takes place (ereignet) from out the constellation of Being and within it*" (TuK 46 – italics added).

At bottom it eludes Dessauer that Heidegger is not a realist. Heidegger rejects the fixity and eternality of technological ideas in Dessauer, not as an imperfection in Dessauer's thought but rather as a matter of principle. Heidegger's dynamic Being stands *over against* Dessauer's realm of ideas. Heidegger's revealment (*Entbergen*) implies the realization of possibilities, but not in the sense of aristotelian potential or in the sense of pregiven platonic ideas. Moreover, the irrationalism of Heidegger cannot be reconciled with the rationalism of Dessauer.

2.3.8   A critical analysis of Heidegger's search for the meaning of technology

2.3.8.1   Introduction

Heidegger realized that there is much about technology that demands philosophical reflection. One of the greatest services he rendered was to approach technology philosophically as a nontechnological person.

What is especially positive in Heidegger's philosophizing about technology is that he has given considerable attention to the relation between science and technology; that he has noticed (regardless of how he eventually interpreted it) the autonomy of technology; that he has inquired into the ground and possibility of technology; and that he has demanded attention for the crisis of meaning everywhere apparent in our technological epoch.

Heidegger, as philosopher, rightly and necessarily addressed himself to technology, since "our" world is no longer thinkable without it. It might well be argued that the beginning and elaboration of Heidegger's philosophy of Being received a stimulus from the development of modern technology. A first requirement for understanding his philosophy would then be to investigate what he has said about technology. Is it not modern technology as danger that leads Heidegger to speak of a reversal essential to bringing deliverance? And it is this reversal that gives substance to Heidegger's philosophy of Being.[145]

Although Heidegger speaks of his own reversal as a reversal in the history of philosophy itself, it does not represent a complete break with that history. Heidegger often gives attention to the history of philosophy. He emphasizes that western metaphysics reaches its completion in technology, and that it displays its most extreme possibilities in technology. If we wish to escape this technology as *the* danger, then we must discard our metaphysical attitude. Heidegger intended his *thoughtful reflection* on technology to be the beginning of this escape.

I have endeavored above to sum up some of the things Heidegger observed in his search for the meaning of technology, in order to show how his remarks on the meaning of technology are to be understood in the context of his general philosophy and its development. I will now proceed to a critical analysis of his thinking about technology. The central question is whether Heidegger's philosophy, in which technology is seen as the great symptom of the crisis of meaning, affords a new perspective for the future. The question of special importance is whether Heidegger's philosophy can offer deliverance from the all-embracing power of technology.

I will proceed as follows in my critique. First I will deal with Heidegger's view of technology and its development and his view of the relation between science and technology. Next I will explain explicitly how Heidegger understands the philosophical backgrounds of technological development. After having shown that Christianity, in Heidegger's opinion, has abetted the advance of the destructive power of technology, I will summarize the problems he faces. After that I will try to shed some light on Heidegger's rejection of several current "definitions" and also on the dichotomies that run through his thought. Finally, once the dialectic between Being and Thought and the parallel dialectic between technology and man have been sketched, I will raise the question of the consequences of Heidegger's standpoint for the future of technology. Is Heidegger successful in pointing out a way in

which deliverance from the nihilistic development of modern technology – as he sees it – is possible?[146]

2.3.8.2 Technology and its development

For Heidegger, modern technology has an absolute and imperialistic
character. It has set an ineradicable mark upon everything. "What
now is, is stamped by the power of the essence of modern technicity, a
power which is already visible in every area of life by virtue of such
notable marks as functionalizing, perfection, automating, bureau-
cratizing, and information" (IuD 48). Although man *ex origine* can
never be transformed into fund (*Bestand*) and can never be transformed into a component of technology, this technological dominion as
destiny of Being (*Seinsgeschick*) is, as we saw, a threat to man.
Technology as revealment (*Entbergen*), in which man experiences his
freedom, stands over against technology as fund (*Bestand*), which
contains a threat to – and finally the destruction of – the freedom of
man and of man himself, and which every "revealment" "necessarily"
brings along with it. "The powers which universally and persistently
claim, chain, drive, and cramp man in one form or another of technological installation or arrangement – these powers have long since
grown beyond man's will and decision-making capability, because they
are not man-made" (Gel 21). Man is steadily transformed into "gestellter Mensch."[147] This ominous development is not yet at an end. In fact,
the opposite is sooner the case: "We presently know, without properly
understanding it, that modern technology irresistibly presses forward
with its arrangements and all that it has brought forth toward the
most all-embracing and greatest possible perfection. This perfection
consists in the completeness of the calculable securing of objects
(*berechenbaren Sicherstellung der Gegenstände*), of the calculating
with these objects, and of the securing of the calculability of the
possibilities of calculating" (SvG 197-8).

According to Heidegger, technological development is ominous. It
grows beyond man's capacity to contain it. Even man himself is
embraced by technology as it comes to perfection and is made a
calculable quantity subject to manipulation.

It will become apparent that the origin of this development is
twofold. On the one hand man yields to technology, allowing it to
establish his comings and goings for him, while on the other hand
technology comes upon man as a power in itself. The essence of
technology, the *Ge-stell*, works itself out as a destiny of Being
(*Seinsgeschick*) fully within history, leaving nothing untouched. History is for Heidegger the history of technology.

From this survey it is clear that Heidegger brings everything under the rubric of technology, while missing the specific meaning of technology as freely giving form to material. He does not distinguish technology in this specific sense from techné as a general method. Moreover, for Heidegger not only the "made and used" but also the "needs and purposes served" belong to technology.[148]

In short, Heidegger does not establish any demarcation between technology and other areas of culture, to say nothing of speaking in a normative sense about the relation between these areas and technology. Heidegger wields only that one norm that obliterates all differences – the dominion of the *Ge-stell* as destiny of Being (*Seinsgeschick*).

Next we shall examine the consequences of all this for the relation between science and technology.

2.3.8.3 The relation between science and technology

Heidegger acknowledges that modern technology is unthinkable apart from the antecedent development of natural science. In opposition to many others, however, he has concluded that even the natural sciences were already governed by the essence of technology. Viewed "historically," it is true that natural science did precede technology, but if this matter is thought through from the perspective of the history of Being (*Seinsgeschichtlich*), the order is reversed. Modern technology, of necessity, had to employ natural science. According to Heidegger, there is no distinction of principle between them.

Heidegger has seen – correctly – that modern technology can only come to development on the basis of natural science. However, this insight has at the same time put him on the wrong track. When the difference of principle between science and technology is obliterated and natural science is absolutized, a proper evaluation of technological development becomes impossible.

Related to this is the fact that Heidegger views technology too much as machine technology.[149] (Admittedly there are passages in his work where one finds a different emphasis; see his handling of Dessauer,[150] for example.) The machine, he says, is the most visible expression of scientific thought. Scientific thought regards nature as a calculable power coherence. Writes Heidegger: "Modern physics is not experimental because it uses instruments to investigate nature. Rather the reverse: because physics, and then already as pure theory, sets up (*hinstellt*) nature to present itself as *a pre-calculable coherence of powers*, therefore the experiment is commanded (*bestellt*), namely, to inquire whether nature as it is thus set (*gestellt*) announces itself, and how" (TuK 21 – italics added). The strong emphasis on the calculability of the power coherence facilitates seeing the essence of the natural sciences as being taken up in the essence of technology. "Re-presenting" (*vor-stellen*) turns into "setting" (*zu-stellen*) and "placing" (*her-stellen*), says Heidegger. He has thereby obliterated the boundaries between natural science and technology.

In the same connection, Heidegger pays little attention to inventions in modern technology. The revealment of Being is indeed the source of this technology, but its course thereafter is by way of natural science. Heidegger believes this is precisely what is new about modern technology. As I see it, he thereby fails to do justice to the character of inventions, which also occur in modern technology. It is true that they have physics as their basis, but they cannot be explained from physics alone. It is human creativity or productive fantasy that transcends the scientific basis and that at a given moment "sees" a new invention.

Because the character of invention in modern technology eludes Heidegger, he also fails to grasp the particular signification of technology as a *free* forming of material on the basis of physics. Sometimes that precise basis of physics is not even present yet. This is true, for example, of certain sectors of chemical technology. People working in that area wrestle with the tremendous problem of introducing exact calculation, which is extremely difficult. The main emphasis in this sector of technology is on conducting experiments the results of which cannot be established in advance. However, great advances have been made by conducting numerous experiments, as the manufacture of synthetic materials demonstrates.

Even where Heidegger does construe revealment (*Entbergen*) as the expression of human creative potential, he refers only to the origin of modern technology and not to its development. This development is controlled instead by the *Ge-stell* as destiny of Being (*Seinsgeschick*). Man is subordinated to it; he cannot break through the *Ge-stell*. Heidegger has thereby taken the continuity and the discontinuity which are intrinsically interrelated in modern technology, as we saw in Chapter 1, and separated them, setting them after and over against each other.

With this view Heidegger has impoverished both natural science and technology. Their independence of one another, as it arose through the historical differentiation process, is sacrificed to unity of essence. I am willing to admit that when natural science and technology fructify each other, the boundary between them can become obscured. Yet we should continue to distinguish between (1) technology, which aims at the free giving of form to given materials, (2) technological science, which formulates the laws for what is thus formed, and (3) physics, which charts the laws for the nature-side of reality. In research, experimentation, and designing, the culture terrains of technology, technological science, and physics encounter and fructify each other.

Because of Heidegger's identification in principle of science with technology, he is apparently oblivious to the fact that there have been some nonscientific influences on technological development. He thereby forces technological development into a straitjacket, overlooking the influence of economic and political powers and such stimulants for technological development as have previously been produced by wars.

In the following section we will see that these misconceptions and oversights are consistent with Heidegger's way of looking at the history of philosophy, which, according to him, forms the background of modern technology.

2.3.8.4 The philosophical background

We have seen that Heidegger regards technology as breaking its way through natural-scientific thought.

In order to place the character of that thought in the proper light, Heidegger reveals its philosophical origin. That origin is given with the subject/object dualism that has ruled western thought since the philosophy of Descartes but cannot be charged to his account, since his philosophy, too, was subject to the destiny of Being (*Seinsgeschick*). The *Ge-stell* ruled his thinking. It is in this light that we must seek to understand what Heidegger says in his essay "Wissenschaft und Besinnung" (Science and Thought), where we read: "The objectiveness (*Gegenständigkeit*) becomes the mere constancy of fund (*Beständigkeit des Bestandes*) determined from out the *Ge-stell*. Only then does the subject/object relation attain its pure relating, that is, commanding (*Bestellung*) character. This does not mean that the subject/object relation disappears. Quite the opposite: it arrives at its supreme power as pre-established by the *Ge-stell*. It becomes a fund (*Bestand*) to be commanded (*bestellt*)" (VuA 53). The theoretical subject/object relation is predestined from out the *Ge-stell* to be *used* in technological dominating. Scientific thought attains completeness in technology. It is for this reason, as has already been said, that western metaphysics is not dismissed. Instead it plays its last and highest trump, according to Heidegger, in the order of technology.[151]

Thus the rise of modern technology is intrinsically connected with the history of metaphysics as the thought "on this side of human-being-ness" (*diesseits des Menschseins*).[152] Then Heidegger also calls modern technology "completed metaphysics" (*vollendete Metaphysik*).[153] He finds the first traces of this kind of metaphysical thinking as

113

early as Plato. Via Descartes, Leibniz, and Kant, it has developed fully into an *eingleisiges Denken*,[154] something we might speak of as "one-dimensional thought."[155] This thought sees everything in the light of control and is therefore followed, of necessity, by technology.

Man has gradually come to stand at the center as the "representing-placing" (*vorstellend-herstellend*) subject. Heidegger regards this as an expression of the modern freedom idea as *Selbstgewissheit*.[156] The advancing objectification of all being implies the increasing *self-assurance* and *self-affirmation* of man.[157] Heidegger counts it one of Nietzsche's contributions that he perceived this development as fatal and drew the consequences from this insight. Heidegger goes right along with him in regarding the dominion of modern technology and of the *nihilism* ("God is dead") that flows from it as the consequence of "the completion of western metaphysics" and not, if one may put it that way, as a coincidental traffic accident along the way.[158] In the will to power, man increases and establishes his power over being. He has become the pivot upon which everything turns. Given this central position of man in technological development, truth has increasingly become a matter of the value of being to man – thus, an element of his certainty.[159] With this reduction of truth, *the* truth as revealment of Being is ignored, and this in turn provides the basis for nihilistic technological development.

Modern man has grown mighty in the exercise of power. He rises up violently to subject all things to his own dominion. In so doing he has forgotten Being, from which, by which and unto which all that is, *is*; he has become a nihilist. Heidegger says: "The essence of nihilism, in the sense of the history of Being (*das seinsgeschichtliche Wesen*), is aban-donment of Being (*Seinsverlassenheit*), to the extent that in nihilism there e-ventuates (*sich ereignet*) Being's abandonment of itself to making" (VuA 83). The world stamped and controlled by technology has thereby become an "unworld of errance": "In the sense of the history of Being, it is a wandering star (*seinsgeschichtlich der Irr-stern*)" (VuA 89). Human activity in technology has thereby become devoid of meaning.[160]

A dichotomy runs through Heidegger's presentation here. He regards the subjectivism of Descartes as an abandonment of Being, but at the same time he claims that this subjectivism and the technological development ensuing from it are a destiny of Being (*Seinsgeschick*). We shall have to look more closely at this tension between guilt and dispensation of Being.

I agree with Heidegger's claim that when subjectivism is the driving motive, tremendous dislocations can arise in and through technology. The examples are legion. But surely there is more to be said! Does not the faith of Christians, for example, signify the rejection of subjectiv-ism and the rejection of the reduction of *the* truth to self-affirmation and self-assurance? And has that christian faith not had an inspira-tional effect on technological development?

2.3.8.5 Technology and Christianity

Heidegger opposes subjectivistic humanism because in that humanism everything revolves around man. For the sake of his own will, man subjects the world of beings to himself through technology. Does Heidegger's rejection of humanism indicate a choice in favor of Chris-tianity? In Christianity, is it not God as the "Transcendent One" who should be at the center instead of man?

Heidegger sees that Christianity has been a clear stimulus for technology. However, in the light of his view of technology as a power that is at once broken loose from Being and a destiny of Being (*Seinsgeschick*), Heidegger arrives at a rejection of christian belief. According to him, that belief is equally an expression of subjectivism. The Christian, too, especially the protestant Christian, is interested in himself; everything revolves around his personal eternal salvation as he strives for the personal assurance of salvation.[161] In Protestantism, which is directed toward the transcendent, truth is still reduced to certainty. Whatever differences there may be, that is the congruence with humanism.[162] Both revolve around man.

According to Heidegger, Nietzsche saw through the egocentricity of Christianity too. For Nietzsche, man had become the autocrat, the one ruling alone, determining his own values. To be consistent, if God is dead there would also have to be a transformation of all values (*Umwertung aller Werte*). Heidegger claims that Christians have de-stroyed and are still destroying the transcendental foundation just by *speaking* of the "foundational" and the "transcendental." Subjectivistic Christianity advances the exorcising of God (*Entgötterung*).[163] Man is thrown back upon himself and out upon the way to technological world conquest.[164]

It cannot be denied that Christianity bears some marks of a serious decline. "Christian" subjectivism does indeed revolve around man – around *his* earth, *his* heaven, and even *his* god. But that ought not to be, and often it isn't. Selfish self-interest is really Enemy Number One for the Christianity in which God is central. Heidegger's view of technology as the great danger is such that he deprives himself of the possibility of doing Christianity integral justice. He believes that technological power has arisen because man has always assumed the central place in humanism and Christianity. The idea that Christian-ity ought to revolve around *serving* God accordingly eludes him also.

115

Even if Heidegger should invoke certain utterances of Christians to support his interpretation of Christianity, he would still only be noting *facts* and failing to speak of the biblical *norms* for christian believing.[165] Do those norms not signify that the life of a Christian is not to be consumed in technology, but that technology ought to have its own subordinate, richly meaningful place?[166]

2.3.8.6 Heidegger's problems

Heidegger's rejection of modern technology and, with it, of scientific thought, is placed in an extremely clear light when he says: "This thought [calculating thought] itself is already the explosion of a violence that could drive us all into nothingness (*ins Nichtige*). All the rest that follows from such thought, the technological process of the functioning of the machine of destruction, would only be the last dark sending of madness into the meaningless" (UzS 190).

Heidegger declines to concern himself with problems of a secondary order, though they demand attention everywhere. For example, the question whether atomic energy is to be used for peaceful purposes or waging war does not, at bottom, reach *the* problem. Heidegger is of the opinion that the danger of the misuse of atomic energy is subordinate to the danger of modern technology as such. In modern technology, nature is reduced to calculable nature and truth is reduced to the certainty of the human subject. And this reduction of truth is a threat to man even if he should escape the incidental danger of an atomic war.[167] Heidegger's view, accordingly, is that thought runs aground short of a safe haven; it warns of the dangers of atomic energy, but it neglects to inquire after the ground upon which and from which this new energy is discovered and released in nature by scientific technology. More important is the question: "What does it mean to say that a period in world history is marked by atomic energy and its release?" (SvG 199).[168]

For Heidegger it is not the incidental problems that matter, but the *structural* problem of modern technology. And to get this properly in view, a different attitude is necessary than the attitude taken by science and technology. "Science does not think." The technological man calculates, makes plans, and equips an enterprise, but he does not think things through either. According to Heidegger, technological development is controlled entirely by scientific thought, which is not thought at all. Heidegger would like to get at the essence of modern technology by following the way of *reflective thought*; he wishes to follow this route in reverse, as it were, to see where technology goes astray and what its meaning is – the meaning "that rules in all that is" (Gel 15).

It may seem surprising that Heidegger should inquire concerning the meaning of technology since he gives such a central place to its meaninglessness. However, Heidegger's real quest is not for the meaning of technology after all – although he would like to catch sight of its origin – but for deliverance from the destroying power of technology. The question that faces us is whether Heidegger succeeds in freeing himself from subjectivism. We saw that the "earlier" Heidegger of *Being and Time* lacks the possibility of doing so. The end of each (thrown) human existence as (thrown) design is *death*, meaninglessness *par excellence*. It is true that the "earlier" Heidegger is concerned with There-being (*Dasein*) as admittance to Being (*Sein*), but human being itself always precedes any mention of Being.

In *Being and Time* Heidegger still finds himself in the subjectivistic and anthropological tradition of metaphysics.[169] If the meaninglessness of technology is to be broken through, if a meaningful inquiry into the meaning of technology is to be undertaken, and if deliverance from its destroying power is to be made possible, then a *reversal* in thought is necessary. "This reversal is not an alteration of the standpoint of *Being and Time*; rather, in this reversal, that 'tried' thought arrives at the place of the dimension from out of which *Being and Time* was experienced, and then experienced out of the ground experience of the forgottenness of Being (*Seinsvergessenheit*)" (PH 71).[170] In this reversal, then, continuity with the "earlier" Heidegger has been preserved even while it affords new possibilities that are not yet present in *Being and Time*. In *Being and Time* Heidegger is still underway toward his later philosophy of Being. Thought ought to be ruled by Being and ought to be directed toward Being. This offers the perspective of escape from the turmoil, frenzy and destructive urge of modern technology. "Must we not find ways in which thought can respond to the thought-worthy, instead of always being bewitched by calculating thought into passing the thought-worthy by? That is the question. It is the universal question of thought. What will become of the earth and of the There-being (*Dasein*) of man upon this earth depends on the way this question is answered" (conclusion of SvG 211).

Victory over nihilism and deliverance from it cannot be attained through the same attitude that convinced man he could rule being through knowledge in the first place. It can only be attained through the withdrawal of this claim. In taking this step, Heidegger rejects humanism construed as the attitude with which man-as-the-control-center of reality thinks and from the potential of which he makes himself his own master. The "remembrance of Being" turns man around and is sufficient to man's becoming open to Being.[171] This remembrance is "...the changeover from departure to introspection in the widest compass of what is open" (Hol 285).

Partly because of the etymology of the word *technology*, Heidegger believes that he need only journey along the way of thought in order to find what he is seeking. *Techné*, *poiésis*, and *epistémé* mean the same thing. Our differentiation of technology, art, and science respectively was unknown to the Greeks. Heidegger wishes to return to this unity, and he sees it in thought. He joins Parmenides in claiming that Thought and Being are the same. In Thought, Being reveals itself as truth. In the meantime, the problem of their identification and differentiation remains unresolved. And that, as we shall see below, calls up dichotomies in Heidegger's view. Through the thinking of Being, Heidegger believes he can overcome the separation of Being, thought, and technology while at the same time incorporating and building upon it. It is as if he hopes, through his thinking, to get upon the track of the unity, coherence and diversity of all being and its origin. It is thought that reveals the essence of the meaning of all that *is*.

Once again I would point out that Heidegger exposes subjectivism as the fundamental ailment of modern technology. In the thinking of Being he believes he has found the remedy. In the following subsections, I will take a critical look at the question whether Heidegger's attempt to conquer subjectivism has been successful. In other words, is deliverance from the danger of technology possible through the thinking of Being?

2.3.8.7 Rejection of current definitions of technology

Heidegger's desire is not to orient thinking to what is at hand, not to run aground in beings, and above all not to derive his norms from them. He also desires, in contrast to what happened in traditional ontologies, not to reduce all that has being to some highest being. Rather, he would see in Being the unity of all that "is" (*west*).

Because he requests reflective consideration for Being as it is *present* in everything, Heidegger is able at the same time to reject a number of real misconceptions about the nature and meaning of technology. Not only does he reject the notion of technology as applied science, he also opposes an instrumental and anthropological definition of technology. Such a definition makes technology an instrument of human power. Man then stands at the center. Yet he is the same man who, through technology, is estranged from himself in forgottenness-of-Being (*Seinsvergessenheit*): "In the meantime man currently can encounter himself precisely nowhere in truth, that is to say, in his essence!" (TuK 27). "Man does not hear the claim of Being which speaks in the essence of technology" (IuD 26). In short, the views mentioned fail to do justice to Being; they set man at the center and are devoid of any awareness of the forgottenness-of-Being.

Heidegger similarly rejects a number of other conceptions of technology: technology as neutral, technology as demonic, and technology as a fate. In these misconceptions, says Heidegger, justice is not done to man and his freedom.

I can willingly join Heidegger in rejecting these incorrect definitions or "determinations of the meaning" of technology. Certainly, technology is something more and something other than applied science. Moreover, the meaning of technology does transcend human-being. To "flee" into the notion that technology is demonic or to accept the notion of technology as a fate is indeed to obscure the fact that man is both free and responsible for technological development. Furthermore, when man regards technology as neutral, he does turn a blind eye to the dangers that may be concealed within it.

All the same, I cannot agree fully with Heidegger, for we must consider the way in which he arrives at his rejection of these misconceptions. Heidegger wants to let Being speak in technological development, but at the same time, on the ground of the unity of Thought and Being, he wishes to do justice to the position of man. In the "conquest" – this word should be set off in quotation marks because we must still investigate whether the "conquest" is real – of subjectivism, Heidegger is determined not to forget the human subject.

Meanwhile, there is undeniably a dichotomy in Heidegger's rejections. On the one hand Being rules technology, but on the other hand this ruling of Being is revealed by the *thoughtful reflection* or *thinking of Being* of Heidegger himself. Who has the last word – Being or Heidegger? If indeed it is Being, does human freedom not vanish – swallowed up in the destiny of Being? And if it is Heidegger, is subjectivism not triumphant?

2.3.8.8   Dichotomies in Heidegger's thought

As we have seen, Heidegger's search for the meaning of technology ends in questions. Along the way he seems to have said quite a few things that might afford some "anchorage." Let us now inquire into their real significance.

On the one hand Heidegger declares: " ... *the essence of technicity comes to presence (west) in the happening of the truth*" (TuK 35 – italics added). On the other hand, it is thought that leads to this pronouncement. After all, Heidegger also says: "Therefore there is a difficulty upon the terrain of thought, to think through the first thought still more originally – not the foolish will to renew the past, but sober preparedness to be surprised at the coming of earliness" (TuK 22). While the conclusion might be drawn from all this that Heidegger's own argument is open to relativizing, I should like to call particular attention to the fact that what Heidegger states here is that it is thought that lays bare the essence of technology – while earlier it was Being that opened up this essence in e-vent (*Ereignis*). Now it would seem that Being as historicity (*Geschichtlichkeit*) and mittance (*Geschicklichkeit*) is sometimes all-controlling, and that thought must join Being, while at other times thoughtful reflection as such is required for the understanding of forgottenness of Being as a destiny of Being (*Seinsgeschick*). Thus Being bestows thought upon man and presides over it, but at the same time it is thought that must bring all that into the open! There is a dichotomy here between Being and thought.

There is a parallel dichotomy between Being and man. A destiny of Being (*Seinsgeschick*) brings man upon the way of the revealment (*Entbergen*). On the other hand: "Whenever we *open ourselves* to the essence of technology, then we are unexpectedly confronted with a liberating summons" (TuK 25 – italics added).[172] If Heidegger often creates the impression that man needs Being and receives from Being his freedom – "All revealment comes out of the free, goes into the free, and carries into the free" (TuK 25) – he just as frequently leaves the impression that Being needs man. Being reveals itself by making use of man. In this revealment Being conceals itself. If man, who due to the concealment of Being has come onto the way of the forgetfulness of Being, wishes to grasp this revealment, then he must himself be active in thought – that is, *open himself (eigens öffnen)*. Man, thinking, approaches Being, which concealed itself in the revealment. In thought, Being finally lets itself be found again. Heidegger says: "Every sending (*Geschick*) of a revealment (*Entbergen*) happens from out the granted (*Gewähren*), and as such. For this carries to man only that portion of the revealment (*Entbergen*) that the happening of the revealment *(Entbergung) requires (braucht)*. As the so *used (Gebrauchte)*, man belongs to the happening of the truth" (TuK 32 – italics added).[173] Later, Heidegger even speaks of the value of man consisting of his guarding (*hüten*) both the hiddenness and unhiddenness of all beings on earth. He speaks also of "the guarding that is by way of remembrance" (*andenkend Hüten*).[174] Man is " ... the one who is required (*der Gebrauchte*) to be the keeping (*Währnis*) of the essence of truth" (TuK 33). Being is the guarantor of revealment, true enough. Yet, man is still clearly indispensable, for *brauchen* means not only to use but also to require in the sense of need.[175]

The dichotomy thus sketched is also present in the essay "The Reversal." Being embraces everything that there *is*. Only, without mankind it cannot come to mention. Man must prepare a place for Being in the midst of all beings. Only thereafter does Being speak and does man respond.[176] The dichotomy between Being and man is evident in connection with the reversal when Heidegger states that Being reverses itself and therein disposes over man. But that is possible only when man, through thought, reverses himself and wends his way toward Being.

The dichotomies mentioned between Being and Thought and between Being and man are not without significance for the relation between man and technology. Technological development is a destiny of Being (*Seinsgeschick*) in which Being has abandoned technology, but then it is also true that man in technology has abandoned Being. Technology has become a power over against man because in technological development, Being has been lost. In the revealing (*Entbergung*), man and technology were still united; now man is subordinated to technological development. Although technology ignores death, there is nothing quite so certain as death.[177]

Heidegger wants to resolve all these dichotomies by speaking of the mystery. Man must bring about deliverance from technology as danger through his reflecting, but at the same time the possibility of his getting a grasp on Being must be denied him. Otherwise all of Heidegger's effort would be in vain.

I must acknowledge that something of the real state of affairs comes to expression in Heidegger's thought, namely, that Being and man are indeed not self-exhaustive. They point beyond themselves to the Creator and thus are certainly not self-sufficient. Yet, it seems to me that the dichotomies mentioned, which express the predicament of Heidegger's thought, are called forth by Heidegger himself – partly to camouflage his powerlessness with respect to the development of technology, and partly because "Being" (in my judgment) is a speculative product of Heidegger's own thought. Yet, that would mean that Heidegger has failed to overcome subjectivism, even though it was his express intention to do so.

2.3.8.9 Heidegger and dialectic

Gradually the standpoint from which Heidegger philosophizes is becoming clear to us.

Heidegger's intention was to overcome subjectivism, which had forgotten Being. The question that forces itself upon us is how this should ever be possible if subjectivism is at the same time a destiny of Being (*Seinsgeschick*). How does Heidegger himself manage to escape that sending (*Geschick*)? That is indeed his claim, is it not? Does he not pronounce a *judgment* concerning the ruling technological development? And does that judgment not presuppose that Heidegger occupies a place outside technology as a destiny of Being (*Seinsgeschick*), or at least that he has been privileged to receive something that transcends technological development? Upon what ground does he assume this exceptional position? Furthermore, is it Heidegger *himself* who calls for a return to Being, or is it Being that speaks by means of Heidegger? If the former is the case, then Heidegger has taken an extremely exclusivistic and subjectivistic standpoint. And if the latter is the case, Heidegger is subordinate to a destiny of Being (*Seinsgeschick*) in which Being is necessarily hidden, since every revealment of Being implies a simultaneous concealment of Being. That, however, cannot be the case, for Being is the crux of the matter for Heidegger. Therefore only the first possibility remains, namely, that Heidegger is in an exceptional position. Yet, this means a *subjectivistic Heidegger* who is himself not subordinate to Being, let alone to a destiny of Being (*Seinsgeschick*), but is lifted up above it.

Despite Heidegger's appeal to Parmenides, I believe it is proper to conclude that Being is itself a product of Heidegger's thought, a speculative product whereby room is created for the mystery of the reversal. Thought becomes stranded in the posing of questions of its own making: "For questioning is the piety of thought" (TuK 36).

Because of this subjectivism, Heidegger becomes enmeshed in a dialectic. On the one hand there issues from Being the necessary destiny of Being (*Seinsgeschick*) to which man is subordinate and because of which he is thus not free. On the other hand, Being must give man freedom. Heidegger's idea of Being is intrinsically contradictory.

Through this fundamental dialectic, the dichotomies mentioned previously become dialectical dichotomies. Being as such excludes "thought" (man) at the other pole; yet, for Being to be brought to language, "thought" (man) is again required. Viewed in reverse, "thought" as such has to understand forgottenness of Being, but at the same time this forgottenness of Being is a destiny of Being (*Seinsgeschick*).[178]

The last question that remains to be answered is: What do Heidegger's subjectivism and his dialectic mean for the future of technology?

2.3.8.10   The future of technology

Technological development is dangerous because it estranges man from Being. Deliverance is possible, according to Heidegger, when man turns to Being. "Yet human *reflection* can *think* that all that delivers (*alles Rettende*) must be of a higher but at the same time of a kindred being as that which is threatened (*das Gefährdete*)" (TuK 34 – italics added). We would do well to take a look at the real meaning of the reversal to Being.

The origin of technology as a Being-revealment is a truth-e-vent, says Heidegger. But truth implies untruth at the same time. Has not this same technology sent man, via concealment of Being, along the wrong track, to forgottenness of Being? The inherent danger can be escaped if a "revealment as self-keeping of Being" brings deliverance. Nevertheless, given Heidegger's philosophy, this deliverance, too, is necessarily doomed to become a danger. Does not every revealment of Being imply a concealment of Being?

Moreover, a reversal is required for deliverance – a reversal from forgottenness of Being as a destiny or sending of Being (*Seinsge-schick*). The question arises whether this reversal can bring about a new *Seinsgeschick* that will be stronger than the sending (*Geschick*) from which it extricates itself *and* that, as sending (*Geschick*), will be the precondition of reversal, of deliverance. And the question remains whether forgottenness of Being can thereby be overcome. If it should be overcome in this way, however, what is then meant by the notion of the forgottenness of Being as a destiny of Being? In other words, if technology as danger is a disposition of Being, then no deliverance is possible; but if deliverance is possible, it is not a destiny of Being. This dialectic leads Heidegger's thought into a vicious circle: danger and deliverance exclude each other and evoke each other.

Heidegger, following Nietzsche, has seen that technological culture is a crisis culture, since man as the ruler of all that has being is the sole center. Man finds his fulfilment in the will to power. Although techno-logical man seeks security for himself in the power of technology, he is in principle without protection (*schutzlos*). In and through technology, man has become unprotected and threatened. He smashes himself to smithereens in nihilism, as it were, because technology has gotten him in its grasp.

Heidegger argues for a fundamental change in the *basic attitude* of humankind. He rejects modern subjectivism, but we have seen that he could not overcome subjectivism. The turning to Being that Heidegger proposes is not only a return to Parmenides but equally a turn within thought itself. As a turning around, it is at the same time a turning inward. Thought takes a turn, all right, but in this reversal it remains *thought* – and subjectivism continues. Only its direction is altered. Over against subjectivistic thought directed outward toward techno-logical development stands Heidegger's subjectivistic thought, which is directed not outward but inward.

Heidegger has made it clear that the autonomous thought in the background of technology leads to great perils, and even to the destruction of culture. Although he rejects the autonomy of thought, Heidegger cannot extricate himself from it. Autonomy cannot be overcome by giving thought a different direction.

Meanwhile, this turning in autonomy contains dangers as great as those about which Heidegger is concerned. While he desires a return to Being, which he sometimes colors as a return to Nature,[179] technological development continues its dynamic advance. Heidegger states that technological-calculating thought, as a destiny of Being (*Seinsgeschick*), is legitimate, just as the attendant forgottenness of Being (*Seinsvergessenheit*) in which he detects nihilism as the great peril is legitimate. Because it is a destiny of Being, however, Heidegger can offer no resistance in principle from his espoused standpoint. Even with his consideration for Being, technological development continues to advance. Indeed, as a sending (*Geschick*), it must develop to its full unfolding, with all the attendant consequences – increasing derangement of technological power and an all-permeating nihilism. The question arises whether Heidegger's expectation of a reversal is not a *flight* from human responsibility for technological development. Is it not the case that his dialectic is an outstanding medium for concealing not only the evil of technology and the sin in it but also the responsibility for it? Heidegger's "way out" is no way out. That same autonomous technology-broken-loose-from-man which Heidegger fears so much will only be strengthened if all mention of human responsibility (and irresponsibility) is omitted. The only result of such a silence will be an aggravation of the dialectic between Being and man, technology and man.

If this philosophy of Being, which will have nothing to do with the calculating thought of science and technology, should become accepted, would that secure the future of technology? Does the turning toward Being not rather imply the *end* of technology, of cultural consciousness and cultural calling and task? This philosophy of Being wishes to return to nature;[180] it discharges mankind from work and denies responsibility and calling. If it were possible, this philosophy of Being would even bring forth chaos and destruction. Zuidema has correctly stated[181] that Heidegger's philosophy of Being legitimizes the thistle and death. Are the consequences of Heidegger's thought for the future of technology not identical with those of nihilistic development, of which the only ultimate certainty is death?

Even when we reject the consequences of Heidegger's thought, we cannot say that we are finished with him. His problem is all too much *the* problem of our culture. Heidegger has contributed a great deal to the exposure of this problem. The great questions of our current technology-marked culture should concern *us* just as they did Heidegger. The unfolding of technology and the growth of material welfare are associated with a growing contradiction in human existence, human freedom, and human well-being. Western culture finds itself in a crisis that affects everyone.

A culture that is seen only as an affirmation of human dominion robs man of his freedom and drags him down toward his final demise. Man as ruler becomes the functionary, the slave, the victim of technology.[182] Heidegger is correct in regarding the imperialistic urge for expansion that has been exhibited by moderns since Descartes as one of the roots of our culture. The subjectivistic anthropology has steadily increased its influence. The belief in man as a man-of-culture and as a *technicus* has even given our age the character of an "age without God." In positivism, materialism, and marxism, man is the ruler of being; being is nothing but material for human fashioning and control. Again, these influences are real. Heidegger has absolutized them, and in so doing he has kept himself from being able to conceive of a meaningful development of technology.

What is essential is not a conversion to Being but a conversion to God, the Creator, the Redeemer in Jesus Christ. In this conversion there is no hint or shadow of flight, for man comes to stand once again in the place to which God originally appointed him – not at the center, not as the ruler of reality, not as man sufficient to himself (whatever the direction in which autonomy may work itself out), but as a bondservant called by God to engage in the forming of culture, including the development of technology, and accountable to Him for every act of commission and omission. This confession is liberating in the face of the worshippers of science and technology on the one hand, and in the face of those who would flee science and technology on the other.

2.4 Jacques Ellul: The Technological Society

2.4.1 Introduction

The central theme in Jacques Ellul's[183] publications on technology is his contention that our present society has become a technological society. In the technological culture, he says, everything has originated through technology, everything exists for technology, and everything *is* technology.[184] He is especially concerned about the position of mankind in this technological society.

There are two reasons why a consideration of Ellul is indispensable in such a study as this.

In the first place, Ellul is a Christian who, *as* a Christian, presents a view of technology which is considerably different from the views of other Christians. He is compelled to take a stand against technology, he believes, precisely because he is a Christian. Technology tends always to direct people's attention to the things of this world, he asserts, instead of pointing to the world that rises above material reality. He sees the christian religion as standing over against the religion in which technology is honored as a god.

In the second place, Ellul has made an important preliminary contribution to the discussion of what is now generally called technocracy. He shows that technology as a scientific method of control assumes an all-embracing character. According to Ellul, economics, psychology, sociology, ethics, law, and politics are all component fields of one and the same phenomenon – modern technology. On the one hand he sees the state as a concentration of technology, but on the other hand the state has itself become a technological state. The principle of efficiency has become so commanding that human freedom is disappearing and the world is being transformed into one great concentration camp.[185]

In Ellul's judgment, modern technology has inevitably become an autonomous power through its scientific basis. Modern technology influences human existence profoundly; it robs man of his freedom. This effect issues first of all from material technology, he claims, but many psychological and sociological "techniques" play a contributing role as well. These "techniques" may confer upon people a certain feeling of happiness, but they do, finally, deprive them completely of their freedom: " ... the absolute disparity between happiness and freedom remains an ever real theme for our reflections" (TO 399).

Because he has accorded the connection between science and technology a central place, Ellul has been able to offer a description of a number of characteristics of technology which are of the greatest importance for understanding the place and significance of technology in contemporary culture.

Ellul is anything but optimistic about the future. Every well-intentioned solution to the growing problem turns out to be a scheme in which technology again occupies the central place. Accordingly, the power of technology in culture intensifies.

Ellul does not wish to be labeled a pessimist or a fatalist with respect to the growing tension in culture occasioned by the increasing influence of technology. Although he often cites Jünger sympathetically, he does not propose to flee to the past, for such a response is as impractical as it is impossible.[186] What he desires is to see man take his bearings correctly in culture. And the first prerequisite for this is that man must understand the forces that govern contemporary culture. Only then will he be able to understand the extent of the illness that has come upon our culture.

In his first publication about technology, Ellul manages only to suggest a diagnosis. Later he undertakes the therapy and formulates a number of conditions that might offer some prospect of healing. It remains a matter of great uncertainty for him, however, whether man will ever be freed from technology and its influences.

I will present Ellul's treatment of technology by following his own themes: what technology is; the origin and development of technology; characteristics of modern technology; technology, economy and the state; mankind and technology; technology and the future (the diagnosis); technology and the future (the therapy).

2.4.2 What is technology?

To Ellul, technology is a cultural phenomenon – a phenomenon that embraces a great deal indeed. This determines his whole view.

Technology does have something to do with machines, Ellul says, but we would greatly underestimate the power and possibilities of technology if we restricted our consideration to them alone. Machines represent only a small portion of technology. For Ellul they are really not technology at all; technology, as he sees it, is method, a method of controlling things and people: " ... technique is not only the ways in which one influences things, but also the ways one influences persons" (TO 394). Moreover, this technology concerns all human activity – not just human activities in production, as Fourastié, for example, incorrectly thinks. Furthermore, since technology as method is often followed by technology as machine, technology is often wrongly identified with machines: "For, wherever a technical factor exists, it results, almost inevitably, in mechanization: technique transforms everything it touches into a machine" (TS 4).

The machine is just one more result of technology as method, according to Ellul. It escapes him at this point that what the machine does is to take over human work in forming. The machine is never separable from technology as method. There may be something to the notions that the machine is no friend of the worker (Jünger) and that the machine is antisocial (Lewis Mumford); yet Ellul thinks that these

127

notions arise from an incorrect view of technology. Technology-as-method will do away with the unhealthy consequences of technology-as-machines. "The metal monster could not go on forever torturing mankind. It found in technique a rule as hard and inflexible as itself" (TS 5).

Technology preceded science; therefore it cannot be applied science. Technological development did get up a good head of steam through the influence of science. Under that influence, technology as a method of control became a scientific method of control. Modern technology has made more and more use of science: " ... science has become an instrument of technique" (TS 10). Through the influence of technology, science has become pragmatic and has come to stand in the service of technology-as-control.

It is important to note that for Ellul, technology is not the giving of form to inanimate nature with the help of tools. Technology is always a method of control. To be precise, technology, for Ellul, is the technological-scientific method as central in designing.[187] Nor is technology limited just to designing or to the control of technological forming processes. Ellul sees technology as the scientific method of control at work in organization, in business, in politics, in the economy, in education, in medicine, in eugenics – in short, wherever people interfere with nature in order to alter it.

Technology as a method of control is determined by *efficiency*, says Ellul. The means employed by technology are therefore more important than the ends toward which they are directed: " ... the means ... are more important than the ends" (TS 11). Mathematics must be used to find the best available means: "It is really a question of finding the best means in the absolute sense, on the basis of numerical calculations" (TS 21).

All the elements in Ellul's conception of technology mentioned above are contained in the following definition: "The term technique, as I use it, does not mean machines, technology, or this or that procedure for attaining an end. In our technological society, *technique is the totality of methods rationally arrived at and having absolute efficiency* (for a given stage of development) *in every field of human activity*" (TS 2).[188]

This perspective on technology has far-reaching significance for the rest of Ellul's view. That he should finally encounter difficulties ought to come as no surprise, for whenever the so-called technological-scientific method is utilized not only for material technology but also to influence and control people, then people will be perceived as unfree, as things. An absolutized technology banishes human freedom. In modern technology, says Ellul, things, animals, and people are standardized.

128

2.4.3 The origin and development of technology

Ellul's ideas concerning modern technology and its influence on culture – an influence so astounding that culture, in his view, has become a technological culture – can be seen in a correct light when we consider what he has to say about the origin and development of technology.

Within primitive technology Ellul draws a distinction between magic and material technology. Both are controlled by religious motives, by reverence, which also characterizes modern technology.[189] Both magic and material technology submit to precise rules. Magic as an intellective technology mediates between humanity and the higher powers, while material technology assumes a place between humanity and nature. Both sorts of technology are concerned with subjection. In the case of magic, the higher powers as such are subjected, while in the case of material technology it is nature as controlled by the higher powers that is subjected.

Magic knows no development according to time and space; as a folk group perishes, it disappears. "In manual technique we observe an increase and later a multiplication of discoveries, each based on the other. In magic we see only endless new beginnings, as the fortunes of history and its own inefficiency call its procedures in question" (TS 27).

Both sorts of technology are eastern in origin. Insofar as the Greeks (who concentrated all attention on the contemplative, on pure thought) were obliged to use technology, they introduced eastern material technology. As much as possible, they then relegated it to slaves. "The great preoccupation of the Greeks was balance, harmony and moderation: hence, they fiercely resisted the unrestrained force inherent in technique, and rejected it because of its potentialities. For these same reasons, magic has relatively little importance in Greece" (TS 29).

Social technology was first introduced by the Romans. In the great coherence of Roman culture, individual freedom and responsibility were guaranteed by juridical techniques. The objective was a harmonious society.

After the fall of the Roman empire, Christians opposed material technology. From Augustine they had learned that their attention should be directed not to this world but to the other world. Given this attitude, it was a long time before technological development could get underway.

In the twelfth century, under the influence of trade with the East, a feeble technological development commenced. This development increased in tempo as the influence of Christianity began to wane.

The fact that nature was de-apotheosized by Christianity and slavery was abolished is often unjustifiably construed, according to

Ellul, as a fostering of technological development. Technology developed strongly in Egypt despite slavery. Therefore the abolition of slavery cannot be a sufficient cause of the rise of technology. The de-apotheosizing of nature *hindered* technology rather than abetted it, for people wished to leave the divine order undisturbed. And what truly mattered to Christians? "It was wiser to be concerned with eschatology than with worldly affairs" (TS 37). According to Ellul, the quest for righteousness before God and the preoccupation with the vision of another world were "the great obstacles that Christianity opposed to technical progress" (TS 38). The influence of the Reformation on technological development is usually greatly exaggerated as well. "The age of the Reformation, in its effort to return to the most primitive conception of Christianity, broke down many barriers. But, even then, it was not so much from the influence of the new theology as from the shock of the Renaissance, from humanism and the authoritarian state, that technique received a decisive impetus" (TS 38).

After an initial blossoming of technology in the fifteenth century under the influence of navigation, the succeeding centuries brought very little progress. The reason for this is that people at that time did not address their scientific attention to the concrete and factual but to the universal. "The intellectual ideal was universality..." (TS 40).

The technological revolution of our times is usually limited in people's minds to the industrial revolution. It is then forgotten, according to Ellul, that technology has gradually affected every sphere of life. Ellul regards the French Revolution, too, as a technological revolution: " ... it might be said that technique is the translation into action of man's concern to master things by means of reason, to account for what is subconscious, make quantitative what is qualitative, make clear and precise the outlines of nature, take hold of chaos and put order into it" (TS 43). The union of scientific research and technological ingenuity has brought technology into such full bloom that science itself is now governed by "a technical state of mind."

It is transparently impossible to summarize all the factors that made the rapid acceleration of technological development possible in the last century. Certainly, materialistic philosophy alone was not responsible. "The optimistic atmosphere of the eighteenth century, more than this philosophy, created a climate favorable to the rise of technical applications" (TS 47). While the myth of progress held people in its tightening embrace, a number of conditions necessary for giving concrete shape to technological development were also present. Lengthy technological experience had reached ripeness; the population had grown so much that needs and wants were complemented with a suitable work force; a stable economic situation had arisen, which limited the risk of utilizing technological means for production; a

130

formidable technological attitude had displaced Christianity, eliminating various taboos; and people generally did not object any longer to meddling with the established natural and social order. In the nineteenth century, all these factors came together for the first time. They had all been present at one time or another in the past, in isolation, but they had never proved capable of stimulating technological development.[190]

That technological development could continue when technology was fully in the hands of the state and of the bourgeoisie and when the workers had set themselves strongly against it is attributable in no small measure to Karl Marx. He preached that technology would liberate the workers from the oppression of the capitalists. This motivated the workers to yield to it completely. The later resistance of the intellectuals, of whom Kierkegaard is a forerunner, had scarcely any restraining influence, even when it increased after the two world wars. Despite those two wars, or perhaps precisely because of them, technological development continued. Technology brings ease and pleasure for everyone, and poverty and human suffering are diminished through the shorter work periods and increased social security which technology delivers.[191] And the end of this prosperity is not in sight. All obstacles to a heaven on earth must be overcome. "When man finds the foe who stands in his way and who alone has barred Paradise to him (be it Jew, Fascist, Capitalist or Communist), he must strike him down, that from the cadaver may grow the exquisite flower the machine had promised" (TS 191).

Both the myth of destruction (the old must be done away) and the myth of paradise (the new must come) motivate people to turn to technology for the fulfilment of the promises that are not yet redeemed. "This promise restores to man the supernatural world from which he had been severed, an incomprehensible world, but one which he himself has made, a world full of promises that he knows can be realized and of which he is potentially the master" (TS 192).

2.4.4 The characteristics of modern technology

Before Ellul proceeds to an analysis of the characteristics of modern technology, he summarizes the traits of classical technology in the context of its relation to the society of its time.

The earlier society was scarcely influenced by technology: "Society was free of technique" (TS 64). Only the slaves concerned themselves with technology: "Technique was not part of man's occupation nor a subject for preoccupation" (TS 66). Technological means were extremely limited. New tools were hardly sought after; existing ones were improved. There was scarcely any technological development, for technology was still bound by time and place. "Just as one society is not interchangeable with another, so technique remained enclosed in its proper framework . . ." (TS 69). In short, people were free of technology and had control of it; technology did not upset the social balance but allowed room for fantasy, beauty, and irrationality, which were in fact dominant.[192] "There was reserved for the individual an area of free choice at the cost of minimal effort. The choice involved a conscious decision and was possible only because the material burden of the technique had not yet become more than a man could shoulder" (TS 77).

In modern technology, the means have assumed such significance that one can speak of a qualitatively altered state of affairs. Moreover, the new technology is founded upon science—especially upon quantifying mathematics. The scientific influence has even led to calculating machines and fully automated factories. The place of people in relation to technology has been fundamentally changed: they have become slaves of technology. Technology has transformed culture and made it one great technological culture in which everything is connected to technology.

The characteristics of modern technology are summarized by Ellul, as it were, when he says: "In fact, technique has taken substance, has become a reality in itself. It is no longer merely a means and an intermediary. It is an object in itself, an independent reality with which we must reckon" (TS 63).

Ellul is correct in seeing the foundation of all of technology's distinctive characteristics in the scientific basis and the scientific method of modern technology. He believes, moreover, that these characteristics, when taken together, lead necessarily to the autonomy of technology.

We shall now take a look at some of these characteristics, namely, rationality, artificiality, automatism, self-reinforcement, monism, universalism, and autonomy.

The *rationality* of technology has altered both its intrinsic and extrinsic traits. And the latter, especially, have set their stamp upon culture. "It is not, then, the intrinsic characteristics of techniques which reveal whether there have been real changes, but the characteristics of the relation between the technical phenomenon and society" (TS 63).

That the relation of people to technology has been strongly altered by science becomes obvious when Ellul says summarily: "This rationality, best exemplified in systematization, division of labor, creation of standards, production norms, and the like, involves two distinct phases: first, the use of 'discourse' in every operation; this excludes spontaneity and personal creativity. Second, there is the reduction of

132

method to its logical dimension alone. Every intervention of technique is, in effect, a reduction of facts, forces, phenomena, means, and instruments to the schema of logic" (TS 79).

While technology was once imbedded in nature, technology has now brought forth a completely *artificial* world. "The world that is being created by the accumulation of technical means is an artificial world and hence radically different from the natural world. It destroys, eliminates or subordinates the natural world..." (TS 79).[193]

Technological development runs *automatically*; it is "self-directing." People no longer possess freedom of judgment and choice in modern technology. Rather, "It [technique] is self-determining in a closed circle" (TO 394). Capitalism may in some sense impede this development, but capitalism will ultimately be overthrown. Then Marx will be vindicated on this score: "...it is the automatism of technique, with its demand that everything be brought into line with it, that endangers capitalism and heralds its final disappearance" (TS 82). This means that technology will bring everything under its control as efficiently as possible.[194]

From this it follows that technology *reinforces itself*. Lewis Mumford believes that material technological development has already arrived at the end of its possibilities. Ellul agrees with him that material technology is limited by the physical world, but he disagrees with Mumford's contention that physics, as the basis of material technology, is no longer in development. From recent developments in the field of nuclear research and nuclear energy, it would appear that Ellul is quite clearly correct. Moreover, Ellul believes that in such areas of nonmaterial technology as politics, administration, economics, and social engineering, the possibilities of technology will accelerate rapidly, especially as the computer begins to play a large role.

According to Ellul, this self-reinforcing technology conforms to two laws. The first is that technological development is irreversible. The second is that its progression is not of an "arithmetic" nature but of a "geometric" nature. It is all-embracing: "A technical discovery has repercussions and entails progress in several branches of technique and not merely in one" (TS 91).[195]

Ever more people are involved in this process, but their influence on it is constantly diminished. Eventually humanity will be completely subordinated to technology. This means, too, that technology can no longer be predicted and that people will no longer be able to set goals for technology. "The evolution of techniques then becomes exclusively causal; it loses all finality" (TS 93).[196] Even though the technological phenomenon is blind to the future, it increases in intensity.

It is clear that in following this line, Ellul qualifies technology as a *monism*. "The human hand no longer spans the complex of means, nor does the human brain synthesize man's acts. Only the intrinsic monism of technique assures cohesion between human means and acts. Technique reigns alone, a blind force and more clear-sighted than the best human intelligence" (TS 93-4). On the basis of this monism, Ellul rejects any weighing of the advantages and disadvantages of technology in Mumford's vein. Ellul does not even share the view that technology in itself is good, though often misused, for this view assumes that technology is dependent upon moral considerations. Nothing could be further from the truth. "But a principal characteristic of technique ... is its refusal to tolerate moral judgments. It is absolutely independent of them and eliminates them from its domain. Technique never observes the distinction between moral and immoral use. It tends, on the contrary, to create a completely independent technical morality" (TS 97).[197]

The consequences of technological development can no longer be foreseen. This means that we can no longer determine whether the newest technological possibilities will be utilized for goals of war or peace. In general it can be said that every new possibility will also become reality. Therefore the atom bomb *had* to be used, says Ellul. But after the first application, oddly enough, technology has the intrinsic tendency to eliminate the harmful effects. Ellul accordingly denies the possibility of extremely disastrous consequences that allow of no correction.

This "optimistic" view of Ellul is probably connected somehow with the fact that in contrast with Jünger, he does not believe that technology will end in disaster. "On the contrary, technique has only one principle: efficient ordering. Everything, for technique, is centered on the concept of order" (TS 110). The order of technology is really not the kind of order that allows human freedom maximal chances; rather, it is the order of a concentration camp. Ellul finds the automatism and monism of technology more dangerous than the atomic bomb because they are not particularly striking and also because they give people a feeling of safety, certainty, and happiness.[198] Technology, it would appear, will still resolve all harmful side effects. In short, "It is an illusion, a perfectly understandable one, to hope to be able to suppress the 'bad' side of technique and to preserve the 'good.' This belief means that the essence of the technical phenomenon has not been grasped" (TS 111).

Ellul locates the connection between the self-reinforcement and the monism of technology in the *plan*. With the plan, another characteristic becomes clear, namely, the *universality* of technology. The plan can be executed repeatedly and is interchangeable. Through the agency of trade, wars, technological assistance, and the influence of modern means of communication, technology has attained worldwide scope

134

and universal significance. Technology has erased established barriers between nations: " ... technique leads to technical identity in all countries" (TS 120). Even the various world religions are nearing extinction because of the universal power of technology. "To their transcendental religion a 'social' religion is opposed, a religion which is but an expression of technical progress" (TS 121).

By the universality of technology, Ellul means not only geographically universal technology but also qualitative universality. Because science aims at the universal itself and because it is the basis for technology and its method, it has assured that technology, too, has become universal. "Technique depends upon a science itself devoted to the universal, and it is becoming the universal language understood by all men" (TS 131).

In summary: "Geographically and quantitatively, technique is universal in its manifestations. It is devoted, by nature and necessity, to the universal" (TS 131).

All the above-mentioned characteristics are to be accounted for from the influence of science on technology, and they find their ultimate unity in the *autonomy* of technology. Technology is an organism, says Ellul, an organism that is entirely dependent upon itself. It allows ethical, economic, political, and social considerations no influence. "It is autonomous with respect to values, ideas, and the state" (TO 394). Everything has been transformed to satisfy the wishes of autonomous technology. "Technique has become a reality in itself, self-sufficient, with its special laws and its own determinations" (TS 134).[199]

Autonomous technology must respect the limits of physics and biology, of course, but within these limits it is the sole controller. Human beings are totally removed from their places by autonomous technology and are reduced to technological animals or things.[200]

If someone were to object by saying that the conditions for automatic progress were established by people, Ellul would apparently have to disagree. As Ellul sees it, technology is so autonomous that people can no longer set "goals" for it, for "technique is formed by an accumulation of means which have established primacy over ends" (TO 394).[201] Ellul anticipates that in the future technology will develop itself entirely without human participation. At present human beings are still the slaves of technology. They still resist it. In the future, however, they will simply be reckoned as among the technological objects. The remarkable thing about these technological persons will be their possession of an imagined freedom. "True technique will know how to maintain the illusion of liberty, choice, and individuality; but these will have been carefully calculated so that they will be integrated into the mathematical reality merely as appearances!" (TS 139).

The autonomy of technology can also be discerned in the religious veneration of technology. Modern machines, especially, are trusted, admired, and feared like gods. "Nothing belongs any longer to the realm of the gods or the supernatural. The individual who lives in the technical milieu knows very well that there is nothing spiritual anywhere. But man cannot live without the sacred. He therefore transfers his sense of the sacred to the very thing which has destroyed its former object: to technique itself" (TS 143).[202] Marxism is the technological religion *par excellence*, especially for workers: "Technique is the hope of the proletarians; they can have faith in it because its miracles are visible and progressive" (TS 144).

In short: "Technique has become autonomous; it has fashioned an omnivorous world which obeys its own laws and which has renounced all tradition" (TS 14).[203] This autonomous technology behaves like a tyrant, transforming culture into one great, totalitarian whole. "*Technical civilization* means that our civilization is constructed *by* technique, *for* technique, and *is* exclusively technique" (TS 128).

It cannot be denied that Ellul often sketches the characteristics of modern technology correctly. The great question is whether he perhaps presents them too much in the light of his own opposition to technology, setting them in an incorrect framework, striking the wrong accent – in short, offering a distorted image. In my critical analysis of Ellul, I will give these characteristics fresh consideration[204] and try to account for them on the basis of my own analysis of modern technology.[205]

2.4.5 Technology, economics, and the state

As we have seen, technology, for Ellul, is in the first place the scientific method of control, and in the second place technological tools, machines, and so forth. As to coherence, the technological method of control strives to employ machines to the maximum extent possible.

Ellul shows that technology also controls economics and is concentrated and led by the state.

The dependence of economics on modern technology is apparent from the increased use in economics of statistics, mathematics, opinion polls, models, and computers. The influence – in fact, the dominance – of technology may also be observed in the increasing concentration of capital, through which technological development is fostered and the newest inventions are more rapidly and efficiently realized.

Through planning, modern economics has become a technique. "The principal criterion of the planned economy is rationality (or efficiency): in a word, technique" (TS 201).[206] The technological plan is an ever-expanding one; this expansion is coupled with force, whereby its success is assured. "Planning is inseparably bound with coercion" (TS

183). The various plans are steadily integrated. All the talk about decentralization might lead us to believe otherwise, but that would be to succumb to an illusion, for decentralization can only be brought about when we assume a central unity: " . . . decentralization is not possible unless there is a powerful *planned organization for decentralization*" (TS 199).

Technological economics as technique has achieved anything but the enlargement of human freedom. "Man indeed participates in the economy, but technique causes him to participate not as a man but as a thing" (TS 216). People are transformed by the technological economy into producing-consuming beings. "All human functions are mobilized in the 'production-consumption' complex" (TS 224). Marx erred rather badly, then, in believing that an economy based on technology would bring freedom. Nothing could be further from the truth. Technology has conquered economy and has gradually made humanity subordinate to itself. Furthermore, this is a logical development – not a dialectical one.

Technology as the scientific method of control levels itself not only at the economy but also at the military, political, administrative, and social sectors of culture. The *state* assumes responsibility for the integration of these techniques. "The basic effect of state action on technique is to co-ordinate the whole complex. The state possesses the power of unification, since it is the planning power par excellence in society" (TS 307). This development is not a fortuitous or detached development subject to possible exchange for some alternative one. "Technique, once developed to a certain point, poses problems that only the state can resolve, both from the point of view of finance and from that of power" (TS 237).[207]

From the foregoing one might conclude that technology is not autonomous and that it does not run automatically. But that would be to fall into self-deception, for the modern state has become a technocracy. The state itself is fully controlled by technology. "The state becomes a machine designed to exploit the means of the nation" (TS 265). Present-day democracy is bogus democracy. Yet the myth of democracy conceals this truth. "What technique wins, democracy loses" (TS 209). The technocracy is led by specialists, who are a new aristocracy. And they, just like the masses, are fully subordinated to technology. Autonomous technology rules both the new elite and the masses.

Differences between national states, including differences of ideology, are abolished by the power of technology. In practice, the "state-machine" functions in the very same way in so-called democratic states as it does in dictatorially governed states. Ellul accordingly concurs with Wiener when Wiener argues that modern technology makes the world state unavoidable, and that this state will have a totalitarian character.[208] Parliamentary discussions, elections, and protest movements cannot stem this development; rather, they take their place within the development. Human beings cannot resist technology; they stand naked and unarmed before it. The only posture left to humanity is submission to technology and acceptance of its benefits, such as they are.

Of the relation between technology and the state, Ellul says in summary: "In view of what has been said, it may be affirmed with confidence that, in the decades to come, technique will become stronger and its pace will be accelerated through the agency of the state. The state and technique – increasingly interrelated – are becoming the most important forces in the modern world; they buttress and reinforce each other in their aim to produce an apparently indestructible, total civilization" (TS 318).[209]

The fact that Ellul views technology as autonomous prevents him from indicating a meaningful perspective for technology and forces him to depict technology as a frightening monster and tyrant.[210] Under the tyrannical power of the state as a great machine, human beings are bereft of their freedom; they are reduced to things.

The unavoidable question is this: What power renders the technological-scientific method of control independent of people, so that they are made slaves?[211] I will return to this question later.

2.4.6 Humanity and technology

While the theme of humanity and technology has often been alluded to thus far in our discussion of Ellul, its centrality suggests that we should examine it in a separate subsection.

People may have had technology in hand once, but now technology controls people, in Ellul's opinion. "The reality is that man no longer has any means with which to subjugate technique...." "Man is unable to limit it or even to orient it" (TS 306). Technology no longer has room for people as individuals. "The individual in contact with technique loses his social and community sense as the frameworks in which he operated disintegrate under the influence of technique" (TS 126). The natural relation of people to one another is replaced by a technological relation. "Technique has become the bond between men" (TS 132).

Not only the workers but also the civil servants and the specialists have become slaves of technology. None of them are in a position to exercise any influence on technological development. The technological monster puts pressure on all people and at the same time completely absorbs them: "...it engages the whole man and supposes that he is subordinated to its necessity and created for its ends" (TS 320).

Ellul is not looking at the derailments; he believes that the position of a person in modern technology is necessarily that of a machine. In and through technology, people are estranged from their original environment. They live in a technological world for which they are not suited. "The human being was made to breathe the good air of nature, but what he breathes is an obscure compound of acids and coal tars. He was created for a living environment, but he dwells in a lunar world of stone, cement, asphalt, glass, cast iron, and steel" (TS 321).[212] Since Ellul believes that technology has devastated nature, he concurs with Jünger when the latter says: "Only rats and men remain to populate a dead world" (TS 321).

Although technology has liberated people from physical coercion, their situation has not been ameliorated. Not just in their work but in their whole lives they have become subject to the rhythm of dead clock time. Ellul agrees with Jünger and Mumford that the clock is the most important and most dangerous machine of our culture. "Today the human being is dissociated from the essence of life; instead of *living* time, he is split up and parceled out by it" (TS 329).[213]

In this unavoidable development, people begin to conduct themselves more and more like each other, to the point that they can even replace each other where necessary, yet without needing each other. The person has become a mass person. Because *community* is broken, this mass person is very lonely: "The process of massification corresponds . . . to the disappearance of anything resembling a community" (TS 333).

It is clear that when Ellul seeks the source of this development which is so harmful to humanity, he focuses his attention on *technology*, and not on humanity itself. Therefore he rejects technology not in its derailments but *as* technology.

Ellul believes his view is confirmed when he discovers that the consequences of technology harmful to humanity can only be corrected by more technology – a process in which the terrifying aspects of technology remain hidden. Whenever people attempt to escape technology and flee before it, technology refuses to accept their resistance and flight. It suppresses that resistance, and it pursues them. Thereby technology's grip on humanity is tightened. "The purpose of the techniques which have man as their object, the so-called human techniques, is to assist him in this mutation, to help him to find the quickest way, to calm his fears, and reshape his heart and his brain" (TS 334).[214] The "human techniques" strive for the assimilation of people by technology; moreover, they seek to protect people from technology and to give them the feeling that they are above technology. Humanity is thereby subjugated in double measure. "We must find solutions to the problems raised by techniques, and only through technical means can we find them" (TS 340).

The "human techniques" have their origin in the "human sciences"; these, especially through the use of mathematics, have become techniques. "Biometry, psychometry, sociometry, and cybernetics have become the chief intermediaries for creating these techniques" (TS 342).

The ultimate goal of the "human techniques" is that human beings should once again be able to face the problems evoked by technology and become anew the lords of technology. "The grand design of human techniques is to make man the center of all techniques" (TS 337). In addition, the "human techniques" are to render humans superhuman. "It is instructive to see how many intellectuals hope to find in the creation of superman the solution of all the otherwise insoluble problems posed to the common man by the technical world in which he lives" (TS 338).

Ellul demurs, believing that these "techniques" only aggravate the situation; the result is simply that people become increasingly encapsulated by technology. "Man feels himself to be responsible, but he is not. He does not feel himself an object, but he is" (TS 224). In short, the "happiness" of modern man is a mere appearance, for his freedom has vanished.

The basis for all this is the fact that the "human techniques" are not focused on the individual, as techniques in art are. These "human techniques" have a universal character. Furthermore, they are objective, impersonal, lasting, and repeatable. All these characteristics, taken together, generate the fundamental problems.

Ellul explains this by referring to various "human techniques." Thus *child-rearing technique* must assure that the growing child will later conduct himself as a technician. "It makes men happy in a milieu which normally would have made them unhappy, if they had not been worked on, molded, and formed just for that milieu" (TS 348). *Career counselling* serves to give people pleasure in their future work – but still, and above all, to make them serviceable in that work to technology and technological progress.[215]

Every conceivable form of *advertising* and *propaganda*, usually coordinated by the state by way of the mass media, is similarly directed. Any vestigial critical attitudes of individuals are stifled through the creation of collective needs and desires. Propaganda fosters a mass public which is easy to manipulate. The technology of the *organization*, in combination with the psychological techniques, must provide people with the feeling that they are one with their work and their working conditions.

All these methods, complemented by *amusement* and *sports,* assure that "the human being becomes a kind of machine, and his machine-

140

controlled activity becomes a technique" (TS 383). They have the character of drugs. The myths of "democracy," of "happiness," and of "progress" overcome the last vestiges of resistance, so that people cease to oppose technology and revere and worship it instead. "In short, man creates for himself a new religion of a rational technical order to justify his work and to be justified in it" (TS 329).[216]

Ellul's view of the relation of humanity and technology is consistent. His argument follows from his view of technology as an autonomous power that subjects even man in religious worship to itself. Wherever there are signs that might seem to belie the terrifying faces of this technology, such as material prosperity and the liberation of people from their natural limitations, Ellul (justifiably) does not neglect to point out their contradictions.

That man can become more human in and through technology is only an illusion to Ellul. The opposite is in fact the case. Humanity's freedom is a chimera. That, basically, is Ellul's reason for denying, whenever the question arises, that technology can have a meaningful future, and that is why he is unable to indicate a way of escape from the real, dramatic development that transforms humanity into a technological thing.

2.4.7 Technology and the future: the diagnosis

Considering Ellul's view of technology, it should come as no surprise that he envisions a somber future. It is also apparent that he has difficulty offering a meaningful alternative and a perspective for technological development.

As the power of technology increases, so does humanity's weakness. "Enclosed within his artificial creation, man finds there is 'no exit'; that he cannot pierce the shell of technology to find again the ancient milieu to which he was adapted for hundreds of thousands of years" (TS 428). Here Ellul misses two questions: Did humanity's natural milieu not impose limitations? And is the juxtaposition of nature and technology in such a fashion justifiable?

For Ellul it is idle to contend that humanity can lead and control technology. Humanity is the prisoner of technology. As we have seen, every attempt at liberation results in the aggrandisement of technology. "The first solution hinges on the creation of new technical instruments able to mediate between man and his new technical milieu" (TS 429). The computer fulfills this mediating function; when people can no longer cope with the problems of technology, they utilize the calculating machine. Yet this technology of the second degree reinforces technological power. "The second solution revolves about the effort to discover (or rediscover) a new end for human society in the technical age" (TS 430). People find this effort necessary because in the great march of technology, the means have come to occupy the place of the "ends." It would be a mistake, Ellul says, to believe that the new ends are of a nontechnical nature. "If ends and goals are required, he [the technologist] will find them in a finality which can be imposed on technical evolution precisely because this finality can be technically established and calculated" (TS 431). This means that people in technology are ever less human. Every intervention in technology is of necessity a technological-scientific intervention. "Everything in human life that does not lend itself to mathematical treatment must be excluded – because it is not a possible end for technique – and left to the sphere of dreams" (TS 431). The consequence of this is that the human being as a technological object comes to lead a still more abstract existence: "Excluding all but the mathematical element, he is indeed a fit end for the means he has constructed" (TS 432).

Accordingly, the technologists who concern themselves with the future do so in a manner that is in many respects superficial. People give their attention one-sidedly to that which is technologically possible, but they forget to give due consideration to the path along which the promises will have to be realized. For example: " ... automation does not result in labor saving favorable to the workers, but is expressed through unemployment and employment disequilibriation" (TO 416).²¹⁷ Moreover, to the extent that people do pay heed to the interim period, they attempt anew to get its problems under control with technology.

In short, technological development is of such a nature that technology constantly extends itself, becomes stronger, and becomes an all-embracing monster, " ... a world-wide totalitarian dictatorship which will allow technique its full scope ..." (TS 434).²¹⁸ Not only is man threatened; he has in fact become fully the prey of technology. Meaning is lost; meaninglessness rules everywhere. "The 'wherefore' is resolutely passed by" (TS 436).

In *The Technological Society,* Ellul does not go beyond establishing a diagnosis; he does not suggest a therapy that could offer the prospect of healing. To his mind, the sickness of the technological society cannot be healed. In his subsequent study, *The Technological Order,* he does try to approach a therapy, after summarizing his diagnosis. But even then he is conscious that his attempt has little chance of success, for "where freedom is excluded ... an authentic civilization has little chance" (TO 403).

Suggesting a possible solution to the problem of the technological society is extremely difficult, Ellul states, because one's point of departure must be the present situation. A flight to the past is a flight to dreamland, and it offers no chance of improvement. "It is our duty to find our place in our present situation and in no other" (TO 403). And the present situation is so heavily colored by technology that we scarcely know how to step outside of it in order to judge it, in order to be able, eventually, to limit technology.

It is self-evident that Ellul's difficulty must follow from his judgement that technology is an autonomous power. If this power is truly autonomous, how, indeed, can there be any escape from it? If technology ushers in the opposite of freedom, if it is "an operation of determination and necessity" (TO 402), then the obstacles in the way of escape simply cannot be overcome. It is futile even to mention the possibility. That Ellul still hopes for escape suggests he does not take his own pessimistic conclusions quite so seriously. Perhaps he has accentuated them to shake up his readers. On the other hand, it must be said that he has not escaped the influence of those conclusions.

Of course Ellul rejects the cheap solutions of the technologists, scientists, and politicians who say that the problems will automatically resolve themselves. The myths that give expression to such expectations, such as the myths of progress and of happiness, of democracy and of a new golden age, deceive people. That deception is powerfully present in the religion of communism. "The Russians have gone farthest in creating a 'religion' comparable with Technique by means of their transformation of Communism into a religion" (TO 398).

Ellul also disagrees with Fourastié, who states: "Technique produces the foundation, infrastructure and superstructure which will enable man really to become man" (TO 405). According to Fourastié, the human intellect, the spiritual and the moral life, is full-grown only when it has as its basis the complete satisfaction of material needs, such as security, safety and assurance against hunger and sickness. Fourastié believes that it is correct to say that "progress occurs automatically, and the inevitable role of Technique will be that of guaranteeing such material development as allows the intellectual and spiritual maturation of what has been up to now only potentially present in human nature" (TO 405).

Likewise, Ellul rejects the notions of philosophers and theologians who say that technological development poses a great challenge for people and possibly requires a great alteration in them. According to Ellul, such thinkers have not grasped the scope of the real problem, and they are still too optimistic about technology. Bergson contends, for example, that technology bestows freedom on humanity on the condition that the technologically enlarged body of humanity receive the necessary complementary filling of the soul or the mystical. Bergson says, among other things: " ... technique will never render service proportionate to its powers unless humanity, which has bent it earthwards, succeeds by its means in reforming itself and looking heavenwards" (TO 406). Ellul claims that many theologians, especially Roman Catholics, believe that things go wrong with technology when mankind turns away from God, for "it is God Himself who through man is the Creator of Technique, which is something not to be taken in itself, but in relation to its Creator" (TO 407).

Teilhard de Chardin takes a position in the middle; he adapts Christianity to technological development. This development is for him a christian development. "Chardin holds that in technical progress man is 'Christified' and that technical evolution tends inevitably to the 'edification' of the cosmic Christ" (TO 408). [219]

In all these views, says Ellul, technology is accepted with an innocent unawareness of its all-destroying character – a quality that demands to be taken into account. People fail to penetrate beneath the surface appearances, and every attempt they make at finding a solution, however well intended, has a contrary effect.

2.4.8 Technology and the future: the therapy

As we have noted, Ellul himself does not point to any way of escape from the difficulties engendered by technology. Yet he finds it more necessary than ever to be looking for one. "The further technical progress advances, the more the social problem of mastering this progress becomes one of an ethical and spiritual kind" (TO 408). Ellul, however, never gets beyond the suggestion of a number of conditions essential to the direction in which the solution must be sought.

In the first place, people must come to see that they are being robbed of their freedom by both the material techniques and the "human" techniques. People must reject these so-called "human techniques."[220] Once they have done that, they may become conscious of the dangers of material technology as well.

Secondly, people cannot avoid the fact that humanity is subject to a "historical law or order," and that the modern myths allege the contrary. It is a lie that technology is "an instrument of freedom, or the means of ascent to historical destiny, or the execution of a divine vocation, and the like" (TO 410). In order to (re)gain the mastery, "desacralization" and "de-ideologization," at the very least, are necessary. In addition to certifying that technology should never be more than material technology, Ellul maintains that even the positive sides of material technology, such as leisure, hygiene and prosperity, are not worth the difficulty of having people spending their whole lives pursuing them. All adoration of technology must be rejected, since it implies expectations that cannot be realized. "As long as man worships Technique, there is as good as no chance at all that he will ever succeed in mastering it" (TO 411).

Thirdly, only when people have reflected adequately on the first two conditions will it become possible for them to achieve the distance from technology that is essential to regaining power over it. There will be obstacles to this approach, for "to affirm that these things have no importance at all in respect to truth and freedom, that it is a matter of no *real* importance whether man succeeds in reaching the moon, or curing disease with antibiotics, or upping steel production, is really a scandal" (TO 411).

Fourthly, a different philosophy will be required. Modern philosophy has directed itself too much toward separate givens. "How, in the nature of things, can a philosophy which is nothing more than a research into the meaning of words, get any grip on the technical phenomenon?" (TO 411). Only "a *truly* philosophic reflection will be necessary" (TO 411). What "truly" means, Ellul does not specify. Given some of his other remarks, it may be assumed that he means a philosophy in which the unity of thinking and acting is central. He regards Greek philosophy as an example of it.[221]

Finally, a solution to the problem of technology will require some sort of dialogue "between technique's pretensions to resolve all human problems and the human will to escape technical determination" (TO 412).

All these conditions, if taken together, afford some hope.

Strangely enough, Ellul never explicitly mentions his christian convictions about life as he discusses these conditions – to say nothing of pleading for christian faith as a condition for mastering technology. Given his position, this is understandable. Technology, as he sees it, is not advanced by Christianity. Christianity has no interest in ruling this world. This is probably also the reason why he sees no possibility of saying something more – from the vantage point of his christian faith – about the *meaning* of technology. The conditions he lays down serve only to keep autonomous technological development within certain paths so that humanity remains free outside those paths. Ellul, it would seem, believes that freedom can only be present outside technology and never manifests itself within technology. Technological development as necessary, autonomous development cannot be resisted; at best it can be led from without. What is required, therefore, is room for freedom outside technology. Technology destroys freedom. Yet it is the freedom of humanity that must establish the limits of this freedom-destroying technology. With Ellul, the basis for all this remains uncertain. At best his statements allow us to conclude that a limited technology makes room for freedom and the religion of humanity – the *christian* religion. Technology and Freedom compete; technology and the christian religion compete.

In short, Ellul believes that today's vanquished must become tomorrow's victor.

2.4.9 Critical analysis

2.4.9.1 Introduction

Ellul has shown what becomes of a culture whenever technology as the scientific method of control becomes all-determining and people surrender to it in religious trust. A technocracy arises wherever technology is relied upon to solve all problems. People become technological objects.

When people absolutize technology as the scientific method for controlling practice, they soon learn from experience that absolutized technology is an autonomous power. Ellul has given eloquent expression to this experience. At the same time, since he has correctly understood the connection between science and technology, he has been able to offer an excellent description of the characteristics of modern technology.

Although I have a high regard for Ellul's view, I feel I must offer some criticism of it. Ellul regards the tendency of technological development toward an absolute technocracy as unavoidable. Its autonomy, he believes, is given with modern technology itself. This conviction is the basis of his notion that technology and secularization necessarily go hand in hand. Technology comes to occupy the place of the christian religion. Now this does often occur, of course, and we must admit that it can indeed lead to a technocracy that stifles life. I have no quarrel with Ellul on this score. What I do object to, however, is the idea that technology and Christianity cannot go together, and that christian faith is reserved for people who feel they must turn away from this world, in world avoidance.

We should note that Ellul is not consistent in his viewpoint, since he still seeks a solution to the problem of the increasing power of technology. Moreover, even as he searches for a solution, he knuckles under to the autonomy of technology. In the final analysis, as we saw, he is unable to indicate a way of escape.

First I will attempt to shed some light on Ellul's philosophical background. We shall see that on the one hand he allows himself to be led by positivism in examining the facts at hand, while on the other hand he clearly reflects existentialism – specifically christian existentialism. He disqualifies observable and describable facts because he wants to preserve human freedom.

Once the philosophical background is clear, we will critically examine aspects of Ellul's view of technology, of its characteristics, of the problems centering on autonomous technology and the individual, and of the relations respectively between technology and religion, and technology and freedom.

2.4.9.2 The philosophical background

Approached philosophically, Ellul is difficult to place. Nevertheless, an explication of his philosophical background is essential to an understanding of his analysis of contemporary culture and of the conditions he sets for achieving a limitation of the power of technology.

On the one hand, Ellul gives the impression of joining the positivists, for he says that he wishes to describe the facts as they are.[222] Experience is the point of departure for his philosophy.[223] In his foreword to *The Technological Society,* Ellul states: "I am concerned only with knowing whether things are so or not" (TS xxiv), and "I ask only that the reader place himself on the factual level and address himself to these questions: Are the facts analyzed here false?" (TS xxvi). On the basis of this positivism, Ellul proposes to refrain from making value judgments: "I make no value judgments" (TS 217). "The real problem is not to judge, but to understand" (TS 189). He fears metaphysical judgments.[224]

On the other hand, when Ellul describes facts he is of the opinion that technology deprives man of his freedom and threatens him with destruction and death. Here he is clearly in the existentialist camp. The world in which people live has become so colored by technology and is so much in the grip of technology that Ellul repudiates the structures (which for him are technological structures) and "saves" freedom as a transcending freedom. Wilkinson says of this: "His concept of the duty of a christian, who stands uniquely (is 'present') at the point of intersection of the material world and the eternal world to come, is not to concoct ambiguous ethical schemes or programs of social action, but to testify to the truth of both worlds and thereby to affirm his freedom through the revolutionary nature of his religion" (TS xxiv).[225]

Ellul's philosophical position is perhaps most easily clarified with a metaphor. On the "ground floor" he follows positivism in his description of facts. He consciously declines to utilize the distinction between fact and norm. "Upstairs" he is an existentialist, advocating human freedom. These two "floors" are not separated; they are connected. Indeed, one can speak of a contradiction between his positivism and his existentialism. On the "ground floor" he refuses to speak normatively about technology, but "upstairs" he repudiates technology because it threatens human freedom. First he sees facts as being unnormed; later, "upstairs," he interprets them on the basis of norms and concludes that they are a threat to existential freedom.

In the following section it will become apparent that this philosophical dualism carries great weight in Ellul's judgment of our culture, in which technology plays so great a role, and also in his search for a possible way of escape from the problems generated by the march of technology. This fundamental conflict in Ellul's view of technology prevents him from indicating concrete possibilities for avoiding the dangers of technology. As he himself admits, his "solution" to the difficulties must inevitably remain theoretical and abstract, because it never has any basis in existing facts.[226]

2.4.9.3 Technology

For Ellul technology is the scientific method in contact with things and even with people. The criterion this method meets is that of efficiency. The method involves the optimum introduction of machines. Ellul refers to machines as material technology, in distinction from technology as a scientific method of control. In the meantime, it escapes him that "material" technology, technology in the narrow sense, embraces more than machines. In the framework of his thought, that "more" belongs to the category of technology as method.

According to Ellul, technology as method has become all-embracing because of its scientific basis and its scientific character. "In fact, nothing at all escapes technique today" (TS 22).

Ellul brands as technology every intervention in the realm of praxis in which science plays some role, regardless of whether machines are used. The reasons for this broad characterization are his failure to analyze technology, his orientation toward factual language, and his identification of "techné" and technology. It is true that in the control of human activities and of people themselves, so-called technological-scientific methods are used. However, people are then construed not as free persons but as the mere equals of things that can be controlled technologically.

In summary, technology for Ellul is not only the panoply of tools and instruments and the technological-scientific method of designing; even science and organization fall under the heading of technology, according to him.

In our philosophical analysis of modern technology, we have seen that the character of modern technology differs strongly from that of classical technology, since designing is carried out on a scientific basis and with scientific method.[227] This technological-scientific method as the method of analysis, abstraction, and synthesis reflects clearly the relation between science and technology. Yet these reciprocal influences may by no means be allowed to lead to a denial of their different destinations. Involved in science is the *knowing* of reality, in a general sense, in terms of its normed character. Involved in technology is the possibility of *giving form* to physical things and facts with the help of tools and instruments.

Thus the differences between science, technology, the technological-scientific method, and the various scientific methods that are utilized to alter or influence reality, as is the case in organizing and in the "human techniques," are neglected by Ellul. He fails to take cognizance of the diversity in what he calls technology. He does not properly appreciate the boundaries and limitations, for he has established no norms.

Whenever the norms are not taken into account, great difficulties are sure to arise. Nature-things and nature-facts can be made objects in technology. In organization and in the "human techniques," however, people are cosubjects. This means that the scientific method may only be employed there if human freedom and responsibility are honored as limitations of principle. Scientific methods ought then to guarantee and give maximum room to the freedom of the person as cosubject. This is also true for "material" technology to the extent that people cannot (yet) be dispensed with in the execution. Even then, the technological-scientific method has limitations of a normative character.[228] If norms are ignored, people are reduced to technological things. Ellul justifiably opposes such dehumanization; yet, in doing so he also rejects technology as such. Not until the real character of technology is taken into account does it become meaningful to talk about the problems and dangers of modern technology which are certainly pervasive enough.

In short, technological development and the various "techniques" must be distinguished and, furthermore, assessed in the light of norms. This opens the possibility of achieving a correct view of technological development and of the various "techniques." The question then arises whether a broad, general utilization of the word *technology* is satisfactory, and whether the differences in the objects (things or persons in their various relations) that are to be controlled or influenced, together with the pertinent differences in norms, do not require a more highly differentiated terminology. A terminology containing more nuances might obviate unnecessary misunderstandings and erroneous conclusions. Furthermore, technology in the proper sense might no longer simply be rejected out of hand; rather, its independence and its "internal" normative structure within established boundaries might be recognized, and room might be made for an inquiry into its coherence with norms of a (primarily) nontechnological nature. Overestimation and underestimation of technology can be avoided only if technology's boundaries and limitations are properly acknowledged.

It should be noted that whenever Ellul seeks a way of escape from threatening technological development, he is of necessity driven back to material technology – to machines – and is forced to reject technology as method. But in taking this recourse, he ignores the fact that machines owe their existence to the technological-scientific method in designing. Similarly, he rejects the "human techniques" on the grounds that these "techniques" destroy the human being as an intellectual and ethical subject: "... in every domain, Technique has established stricter and stricter domination over the human being" (TO 409). It would therefore seem appropriate to stop calling these human methods "techniques." This should not be construed as implying that science is not important for other, nontechnological cultural activities. These other methods, however, ought to be normed differently than the technological-scientific method. Technological *designing,* which is always directed to the future, arises by *analogy* in other sectors of culture, such as the social, the political, the economic, and so forth. The same can be said of the technological-scientific *method* of designing. Different norms arise, depending on the qualification of the cultural sector for which the appropriate scientific method is expected to suffice. In contemporary futurology, this matter should be a burning issue.[229]

It will become evident in the next subsection that Ellul's indiscriminate use of the word *technology* has something to do with his view of science as autonomous as well as his identification of the autonomy of technology with the autonomy of science. Here, too, he appeals to facts. Indeed, science's *pretension* to autonomy is an actuality that does have dangerous consequences for technology, which really is founded upon science. Whenever science and technology are addressed in a way that is not normative – and Ellul does not attain normative insight from his positivistic viewpoint – then there will eventually issue from "upstairs" an all-embracing yet powerless condemnation of technology, for technology, so conceived, destroys human freedom.

2.4.9.4 The characteristics of modern technology

In his description of the characteristics of modern technology, Ellul makes a number of excellent observations. That he means material technology in one place and method in another is extremely confusing. Yet, when we examine his characteristics of technology in the narrow sense, it becomes apparent that he has perceived how modern technology displays the stamp of science.

Ellul mentions the following characteristics of technology: rationality, artificiality, the automatism of technological development, self-reinforcement, monism, and universalism.[230] Taken together, these characteristics indicate that technology has become an independent organism subordinating everything to itself. People no longer have any influence over it but are subjected to it as by a blind force.

Ellul has understood the influence of science on technology. Undoubtedly, the hallmarks of science are mirrored in technology.[231] Yet, Ellul absolutizes these hallmarks. Thereby he loses sight of the fact that people (whether singly or communally) ought to lead technological development and correct any wrongly aimed development – in short, that people should act in freedom and responsibility to set the right course.

Ellul says that technology is rational. What this means is that technological development is governed, according to him, by the iron laws of logic. Abstract scientific knowledge renders modern technology entirely artificial. The advancing development of scientific knowledge surges on in the automatism of technology. As scientific knowledge perpetually expands, so does technology. Autonomous scientific development assures that technological development likewise cannot be hindered by influences external to technology, and because science is universal in both the geographical and the qualitative sense, so is technology.

From all this it is apparent in the first place that Ellul conceives of science as autonomous science and that he accordingly recognizes no norms or limits for science. Consequently he states that scientific method is technology par excellence. Thus technology, too, is autonomous.

Here Ellul gives every appearance of not knowing technology from the inside. Otherwise he could never have failed to reflect upon the central significance of invention in technology. Furthermore, if he really knew technology from within, he would have perceived that the tendency of technological development towards continuity is not only shot through with discontinuity, but also that continuity, by means of invention, can be broken through in an astounding manner resulting in new perspectives for technological development.

That the development of modern technology is of such a nature that individuals have increasingly less influence upon it may indeed be the case, but this does not exclude them. Rather, it leads to their participating in cooperation with others. This situation can prove to be highly inauspicious. The human community may lack the power and capacity to shoulder its increased responsibility in a truly neighborly and

151

communal way. The individual may then be left with the possibility of transferring his responsibility to an impersonal group, a collective or an autonomous organization. The anonymity that accompanies such a transfer may be determined by the structure of modern technology. Yet all this is never actually necessary. It depends entirely upon the individual and the community. Where a consciousness of norms is wanting, what Ellul has perceived will most assuredly occur. It cannot be denied that secularization can lead to just such a nihilistic development of technology. Where people do think normatively about technology, however, it can be argued that the influence the individual exercises on technology through cooperation with others attains a higher plane. But since Ellul does not think and speak normatively about technology, he can hardly agree. Moreover, his opinion is based upon facts that often seem to vindicate him.

For the rest, dynamic, complicated and comprehensive technology does provide ample cause for concern. Technological development constantly places people before new situations that require normative judgments. Traditional norms no longer suffice. On the basis of immutable normative principles that are beyond arbitrariness and indicate the right way to go, norms must be positivized so that they may be in effect for the new situation. The positivization of norms requires a tremendous effort on the part of people, and it highlights their responsibility. Where that responsibility is not positively exercised, or where people do not have (or no longer have) the capacity to assess all the results of technological development, serious consequences are inevitable. In this present age of environmental pollution, which is overwhelming mankind, it would appear that people are much too late in beginning to reflect on the possibilities and thus on the dangers of technology. People have not adopted appropriate measures soon enough; in other words, people have not set norms that would both conserve nature and preserve the life of man as well as beast.

Ellul perceives the threatening dangers of technology on the one hand, but on the other hand he believes that the very logic of technology will somehow resolve these dangers. He is convinced that there is some kind of an order intrinsic to technology, an order that strives towards the abolition of all the dangers of technology. The price to be paid for this, he states, is the high price of the abolition of human freedom.

For Ellul, the monism of technology is self-evident. Humanity can no longer set goals for technology; technology goes its own way. The question that arises at this point is whether Ellul does sufficient justice to the actual state of affairs. In point of fact, is it not the case that technological development is stimulated for the most part by military and industrial powers, and that many experience this situation as an inescapable necessity? Should Ellul not take these powers into consideration? It would then become apparent quickly that the development of technology is not nearly as automatic as Ellul would have us believe.

The same oversimplification of the complex question of technological development arises again when Ellul addresses the tendency of technology toward uniformity and universality. It becomes apparent once more that the idea of autonomous science is at work in the background of his idea of autonomous technology. And once again it is undeniable that Ellul is correct in noting that integration into a world culture is progressing strongly under the influence of technology. Even the realization of a totalitarian world state through the possibilities of technology has gained ground. Only, this development is not inevitable, as Ellul believes it to be. Human responsibility ought to function in such a way that people resist this development. Advancing integration should be accompanied by advancing differentiation. The norm for technological development is not only that of integration but also that of differentiation.[232] If the latter norm is forgotten, people are satisfied with advancing integration. To say this, however, is not yet to say that technology must inevitably usher in a totalitarian world state.

Ellul fails, furthermore, to do justice to certain facts that might have served to make it clear to him that the trend toward such a world state is something less than inevitable. Advancing integration entails stiffness and diminished flexibility. Large concerns are much less able to adapt to the latest developments, for example, than are modest enterprises. As a result, the latter will often be faced with unexpected opportunities. Nor can it be denied that integration often entails a high degree of dependence of small concerns upon large ones. This points to the possibility that integration and differentiation do indeed go together.

By the same token, we must reject Ellul's notion that technology controls the economy and the state. The very difference in terminology ought to have rendered him more cautious at this point, for in the usage itself we can discern the differences in the destinations of technology, the economy, and the state. Once again, it is not to be denied that there is a growing interplay between these areas and that one gets the impression that technology is going on to control everything. Viewed normatively, however, we should be cautious here; we should guard against oversimplified conclusions, no matter what kernel of truth is concealed within.

2.4.9.5 Autonomous technology and the individual

All the characteristics mentioned by Ellul find their unity in what he calls technology. The dynamic development of technology (to which the individual can now contribute only marginally on his own), the influence of science upon technology in elevating it to a universal power both geographically and qualitatively, the independent action of relatively autonomous technological operators – all this can indeed engender in people the sensation that they have become *completely* dependent on technology as an autonomous power. For Ellul, however, this is not just a feeling but reality itself. "The more technical actions increase in society, the more human autonomy and initiative diminish" (TS 402). Technology overwhelms mankind. People can no longer discern whether technology is accompanied by a good or an ill, a just or unjust development; they no longer possess the capacity to set "goals" for technology. "It is formed by an accumulation of means which have established primacy over ends" (TO 394).[233]

Ellul claims that problems posed by material technology are resolved by "human techniques," giving people the feeling that they are again the masters of technology. In the meantime, all these techniques converge and make people totally dependent on technology *itself* as the unity of its various forms of manifestation: " ... the human being becomes completely incapable of escaping from the technical order of things" (TS 397). This involves his work, his leisure, and his relations with other people. Real fellowship and community disappear, to be replaced by some artificial substitute. The person as mass person is totally integrated and in the grip of technology. Every truly human initiative proves abortive. "Man is caught like a fly in a bottle. His attempts at culture, freedom and creative endeavor have become mere entries in technique's filing cabinet" (TS 418). In short, Ellul presumes that this autonomous technology will usher in an absolutely impersonal dictatorship. Technology attempts to give people a feeling of freedom through the "human techniques," but people are not free. While they think they face technology in a manner worthy of themselves as human beings, they are reduced to technological objects. Freedom and responsibility are merely an illusion. The true nature of the illusion remains hidden by the influence issuing from such modern myths as democracy and progress. Anyone who can see behind the myths will perceive, however, that autonomous technology brings a universal concentration camp in its train, and that it is a blind force for the future.

While I must concede that a narcotic influence issues from the nonmaterial "techniques" when the norms that obtain for them are not accorded due consideration – which means that Ellul is correct in a certain sense – I am still convinced that he has gone wrong. He proceeds from the notion that technology is autonomous. Ellul ought to be concerned not with technology and its power but with people who build technology and who ought to guide technology on the basis of nontechnological norms. The illusion is not man's freedom and responsibility; the illusion is that technology is autonomous. But when it is not recognized that the alleged autonomy of technology is an illusion, there is no room left for human freedom and responsibility in the practice of technology. The source of the whole question is not technological power; it is people and their vision of science and technology. Whenever science and technology are taken to be autonomous, people surrender to them. And that is only possible if technological power is regarded as a neutral power rather than as a normed power. Ellul bows to this autonomy on the basis of his positivism. Then he condemns this state of affairs from his existentialist "upper story."

Meanwhile, it should not be forgotten that Ellul will be vindicated if progressive contempt for norms advances. Our world will then assume the form he has described. Yet it is not at all inevitable that it will – at least, not if people assume their central role in the philosophy of technology and in technological practice. But if this is not done, humanity itself will have to shoulder the blame for the devastating power of technology.

When free, responsible people are conceded their central place, another misunderstanding can be removed, namely, that the starting point for technology must be the individual. As we have previously noted, modern technology has elevated human responsibility to a higher plane – that of responsibility in cooperation with others. When this, too, is acknowledged, the danger that the individual in technological development will vanish is lessened. The individual receives his proper place in modern technology in dependence upon others. And that place need not be an unfavorable one. Certainly, it seems unfavorable to those who regard technological development from the viewpoint of an absolutization of the individual; however, it is impossible to achieve a harmonious view of technology and its development if all one can offer in opposition to the autonomy of technology is an absolutized place for the individual.

2.4.9.6 Technology and religion

According to Ellul, the power of technology is so great that people fall down before it in religious reverence. A technological religion arises. "Technique is the god which brings salvation" (TS 144).[234] The reason there is such a clear power of attraction in technology, Ellul claims, is that in contrast to theology and philosophy, it is unambiguous and demonstrable.[235] In other words, technological development is inevitably accompanied by reverence for technology.[236]

Just as in the previous section, it would appear that here, too, Ellul has a distorted view of technology, of man, and of the relation between the two. In fact, the autonomy of technology originates with people, who impute autonomy to technology after they have declared themselves autonomous. People suppose that in technology they have gained the opportunity to establish their own power independently. Precisely because they are unable to realize their pretension to autonomy, however, they project their (supposed) autonomy into the impersonal power of technology, of which they subsequently become the victims.[237]

In Ellul's opinion, secularization inevitably accompanies technological development. Yet he has the roles reversed here, for it is because of secularization that people construe technology as an autonomous power. People themselves have made an idol of technology; they have placed their confidence in it, and they expect salvation from it in the end. While Ellul does not share that confidence and that hope, he, too, has run aground on the presupposition of the autonomy of technology and has failed to examine this presupposition critically.

Because he sees technological development as accompanied necessarily by secularization, Ellul goes on to proclaim that christian belief is separate from technology. This disclaimer comes close to countermanding God's mandate to man to unfold the creation and to make and keep it livable. Is Ellul perhaps giving lopsided consideration to the dislocations wrought by technology while passing too lightly over its fruits?

Ellul's interest in the Greeks is understandable. The Greeks – Ellul views them in the same light as do Jünger and Heidegger – still recognized the unity of thinking and acting.[238] However, could it be that Ellul associates Greek life and thought too closely with the christian perspective in this regard?

There is another question that must be raised, namely, whether the impossibility of integrating christian belief and technology will not lead to secularization's being aggravated by an advancing technological development. Surely that is the last thing Ellul desires!

Ellul's point of departure is mistaken, as we saw earlier, because he orients himself by way of facts that he fails to perceive as normed facts. When he speaks of the future he does introduce norms, but then not as norms that are to be employed by people in technology. Instead he appeals to norms that should prevent people from touching technology. That Ellul should even think this to be possible is an indication he does not really regard the autonomy of technology as absolute, for then every appeal to people to resist technology would be idle, and only acquiescence would remain possible. Or is he ultimately appealing for acquiescence after all when he says: "The weight of technique is such that no obstacle can stop it" (TS 106) and "The human being is delivered helpless, in respect to life's most important and most trivial affairs, to a power which is in no sense under his control" (TS 107)? He emphasizes this acquiescence once again when he says that his ideas concerning a possible way of escape must remain theoretical and abstract: " . . . they are nowhere apparent in existing facts" (TS xxix). Later he says: " . . . except in print, I see no sign of any modification of the technical edifice . . ." (TS 116).

Because Ellul places the christian religion over against technology, it is in principle impossible for him to arrive at a responsible, christian view of technology. The christian religion must be satisfied with just a portion of human life, a portion that is permanently contested by the religion of technology. I reject this view because the christian faith is a radical and integral, saving and liberating faith. On the basis of that faith, perspectives open for the whole of culture – including technology and scientific method.

2.4.9.7 Technology and freedom

Technology and freedom cannot coexist, according to Ellul. He states repeatedly that modern technology destroys individual freedom. It becomes a matter of some importance, then, to know what Ellul means by freedom. Freedom is not " . . . an immutable fact graven in nature and on the heart of man." "As a matter of fact, reality is itself a combination of determinisms and freedom, and freedom consists in overcoming and transcending these determinisms" (TS xxxii). Now, technology is absolutely determined, Ellul says. Accordingly, it stands in an adversary posture toward human freedom. Ellul, in order to preserve freedom, therefore repudiates the technological structures which imprison humanity. He denies that material technology should ever be able to function as a basis for an increased and enlarged freedom.[239] "We must look at it [at freedom] *dialectically,* and say that man is indeed determined, but that it is open to him to overcome necessity, and that this *act* is freedom. Freedom is not static but dynamic, not a vested interest, but a prize continually to be won. The moment man stops and resigns himself, he becomes subject to determinism. He is most enslaved when he thinks he is comfortably settled in freedom" (TS xxxi). Freedom is not a gift but an assignment. It commences when people become conscious of determination. Modern technology robs people of their freedom because it lulls them like an opiate: " . . . Technique can never engender freedom" (TO 402).

Technology and freedom, according to Ellul, exist in a permanent condition of dialectical tension. Transcending – or existential – freedom cannot do without the determined; the possibility that lies open to this freedom is to transcend technology, but at the same time this freedom is encapsulated by determined technology.

It is striking that Ellul should thus reduce freedom to a "transcending" freedom. I would not wish to deny that transcendence is an aspect of freedom – perhaps even its most central aspect. But I cannot agree that freedom is entirely "transcending" freedom, for freedom is also a freedom bound to "determined" structures. Those structures can form the space in which freedom is possible, even while freedom "transcends" them. In any case, the space of freedom is not exhausted by the space of those structures. Stated in reverse, freedom is more than a freedom outside structures and more – contra Ellul – than a freedom outside technology. Freedom can surely be threatened by those structures when they fail to conform to the norms that obtain for them; freedom can even be relinquished by people if they take distance from its source, that is, if they turn away from God. In that case, freedom in the sense of choosing and deciding – the so-called modal freedom – can still function, but freedom as central, "transcending" freedom is perverted. This must have consequences in turn for modal freedom, namely, for its direction.

What is at issue is not the retarding of technological development as such in order to make room for a "transcending" freedom and to offer it additional possibilities. What is really at issue is whether people will accept their freedom in responsibility. What this means is that people must positivize norms for material technology and other "techniques," and that they must be mindful of derailments. Then freedom need no longer stand in dialectical opposition to technology, for the dialectic is resolved.

The task is a difficult one, certainly, considering the problems before which dynamic technology and its comprehensive character place us. Yet, the task is a christian one.

2.5 Hermann J. Meyer: The "Technologized" World

2.5.1 Introduction

The remaining transcendentalist who requires our attention is Hermann Meyer.[240] In a number of ways, he is linked with the others we have looked at. To begin with, he was inspired by Heidegger's view of technology. Because of this connection, I will deal with him only in a summary manner. Yet there are two reasons why we cannot pass him by. In the first place, Meyer's contribution is his attempt to understand the crisis of modern culture from its source. He shows that the subjectivism of Descartes and the natural-scientific method of Galileo led to a "mechanization of the world picture," and that this finally led to the autonomy of technological power. In the second place, Meyer is the first of the transcendentalists to discuss norms and yardsticks with an eye to future technological development. He does not get bogged down in some demonic attribute of modern technology (Jünger), or in technology as *the* danger, with a speculative hope of a turn for the better (Heidegger), or in acquiescence or flight before the threatening power of technology (Ellul). No, he tries to think normatively, in the hope of offering a perspective for the future.

Meyer wishes to understand and judge technology not on the basis of the history of technology as such but on the basis of the intellectual history from which technological development has come forth. This means that he would not try to resolve the dangers of technology with more technology; he proposes to reach above technology – or transcend technology, as he puts it – in order to reach its natural sources and so become its master once again. The great question that will engage us is whether he is successful. The answer to this question must be decisive for our evaluation of Meyer's views concerning the future.

My line of argument, accordingly, will be as follows. First I will investigate what the origin of modern technology is, according to Meyer. Then I will show that according to him, modern technology presupposes an entirely different relation to man and reality than did classical technology. Subsequently we will see that Meyer regards this new relation as being expressed in the positivistic philosophical conceptions that now function in the background of technology. The human perils that Meyer sees as following from these conceptions must also occupy our attention. Finally, before engaging in critical analysis I will devote a separate subsection to the perspective for the future which Meyer offers.

2.5.2. The origin of modern technology

Meyer is not content to understand modern technology and its significance for the future through a superficial periodization of the history of technology, such as may be found in Schmidt or Gehlen.[241] Both Schmidt and Gehlen, more or less in imitation of Kapp,[242] regard the content of technological development as a progressive objectivizing of human organs and functions whereby people are increasingly relieved of work. They emphasize the continuity in the development from tools to automatons by way of machines.

Meyer, on the contrary, regards the rise of modern technology as a break in technological development. He argues that Schmidt and Gehlen fail to do justice to the fact that modern technology arose only in western culture. They also misapprehend the character of modern technology, for it remains an open question whether modern people are unburdened by technology or entangled and hobbled by it.

Gehlen claims that technology is as old as humankind. He then proceeds to account for technological development from general anthropological propositions. However, "modern technology" is not to be accounted for from general anthropological propositions but is exclusively a product of European history" (TiS 750).[243] The intellectual history of the West has given modern technology a different character than classical technology. For Meyer, therefore, in opposition to Schmidt and Gehlen, the development from tools to machines or from material technology to energy technology[244] is a discontinuous development.

Meyer's objections, accordingly, are twofold and are closely related to each other. In the first place, the theory of the objectivization of human functions overlooks that modern technology is a threat to humankind. In the second place, such a theory forgets that modern technology presupposes a new worldview and that it implies a radically different relation between people and nature than the old technology did.[245] "It would be an expression of a lack of capacity to make distinctions to see in all the tools that man has brought forth objectivized mind and nothing more, to fail to take into account the altered conception of nature" (TdW 157). It is not true that machines take the place of human functions; rather, "people take the place of the mechanized way or working of machines in those places where hand movements are indispensable" (TiS 755). While it can be said of earlier tools that they were an extension and projection of human functions, this is no longer true of the machine. People are subordinate to the machine. The fact that machines relieve people of the burden of work does not mean that people escape their influences. People are enslaved by technology both in their work and in their leisure.[246] The connection between natural science, economics, and technology has brought humanity into the path of complete "technologization." The interplay of science, technology and economics has brought about a dynamic development. People are carried along by that development and become its victims.[247]

While modern technology is not to be accounted for apart from modern science, Meyer is not satisfied with the notion that modern science is the source of modern technology. A synoptic view of modern technology is only possible if attention is paid to the changes in the history of ideas, for that history is at the foundation of both science and technology.[248] Energy technology and information technology, which merge gradually,[249] arose from a new way of thinking. "The subject and object of the second and third phases [namely, energy and information technology] are different from those of the first phase [material technology]. That things can stand over against humankind, and that humankind can experience resistance, acquires a new significance here" (TdW 158).

Precisely what, according to Meyer, is the alteration in the history of the western mind that has been brought about by modern technology? Because of their worldview and their conception of nature, the Greeks could not hit upon modern technology. The way to modern technology was also closed in the medieval period because of the dominance in that period of Greek ontology, as the philosophy addressed to the whole of being, and of Greek metaphysical monotheism, both of which the medieval mind reinterpreted in the sense of the christian revelation. Moreover, "Where belief in the hereafter and the salvation of the soul were the ruling thoughts in life, interest in investigating the world could not be great" (TdW 17). When the old view of the world and humanity collapsed, modern machine technology became possible.[250]

According to Meyer, the new view of humanity and the world rests upon "the pretension of reason to be autonomous" (TdW 6). That is the source of modern technology. Meyer can accordingly state: "The breach between believing and knowing is also the beginning of modern technology" (TdW 6).[251]

Western science and technology can only be comprehended when this "pretended autonomy" of reason is kept in mind. The penetration of this autonomy in science and technology is the basis of Meyer's contention that there is a pause in the history of technology. Meyer believes that anyone who fails to perceive this will fail to understand the power of modern technology, together with the crisis it has caused in western culture.[252] Furthermore, he will also find it impossible to properly appreciate the impact of modern technology on nonwestern cultures. It is mistaken to maintain that the planetary power of modern technology does not impair the various cultures. Those cultures, too, must change. Modern technology does not tolerate a view of reality as being ruled by God or by divine powers. The integration of technology into nonwestern cultures signifies an alteration of the mind unprecedented in the history of humanity.[253]

I agree with Meyer that the integration of technology into nonwestern cultures will be accompanied by significant intellectual and spiritual upheavals. I also agree with his contention that the working of autonomous reason in science and technology has precipitated a crisis in western culture. (I will return to this matter later.)[254] To Meyer the crisis is inevitable, since he sees the idea of autonomous reason as the *absolute* origin of modern technology. The question that arises, however, is whether he really does justice here to the history of the West. Did the people of the Reformation not reach back to the fact that God, at the creation, gave man a mandate to subdue the earth and have dominion over it?[255] And is this not to be equally esteemed as a stimulus for scientific and technological development? Meyer refers only to the influence of the Renaissance. He believes that the Renaissance idea of the sovereign, free person who rules nature is *the* source and origin of modern technology.[256]

2.5.3   The altered relation between humanity and nature

The Greek conception of nature – particularly that of Aristotle and of medieval Christianity – was that it is one great, undivided whole. Every natural phenomenon had its assigned place. Meyer claims that philosophy, which was aimed at the whole of being, hindered the impulse to investigate things experimentally, which investigation is the hallmark of the modern age.

The investigation of nature with the help of exact methods in order to discover causes and functional coherences implies a new relation to nature. Meyer concurs with Heidegger that this new relation between humanity and nature rests in the human subject as the focal point of all being. Thinking in terms of cause and effect is "re-presenting" thought, in which nature becomes the object, or *Gegenstand,* of autonomous reason.[257] "In assuming a position independent of nature, man stepped out of his interwovenness with nature, and out of his hiddenness in it" (TiS 765).

Man used to be one with nature. He had a place of his own in it. Now he is no longer the child of nature, but the lord and master of nature. This means, in reference to technology, that man no longer uses what nature affords him but instead takes from nature whatever he desires. Thus the windmill *made use of* the forces of nature; in the new power machines, energy is *generated.* In short, according to Meyer, the organic conception of nature has vanished; in science and technology nature now stands at humanity's disposal. "We no longer have to do with a nature that is immediate to us, but with a second, isolated, abstract, artificial nature, the nature of the laboratory, of the machine" (TdW 112).[258]

Meyer regards Descartes and Galileo as the pioneers of the new view of nature. Descartes put the person – humankind – at the center as the subject to which all being is related. He thereby cleared the way for human intervention. "The conception of an unlimited comprehensibility, controllability, and availability of the world is founded in the subject of the modern age, which knows itself as the ground of all comprehensibility" (TdW 52).

Galileo was the first one to clearly employ a natural-scientific method. The method of quantification, by which things are reduced to quantities, is derived from Galileo. "Everything is reduced to quantitative relations" (TdW 121). Being thereby becomes a thing of mathematical computation, with the consequence that " ... all being is construed as mechanism" (TdW 49). Cartesian subjectivism and the natural-scientific method of Galileo produced a mechanization of the worldview and thereby heralded a new period in western culture.[259]

Nature was no longer regarded as having been given by God. Nature was de-apotheosized and de-sacralized. "Behind natural occurrences there stand no personal powers, neither a saving plan of God nor a damning plan of demons, but impersonal laws" (TdW 44).[260] The quantifying method kills the spirit of nature, making it a rather matter-of-fact affair, and leads ultimately to the destruction of nature by removing all obstacles to intervention.

The new relation of humanity in confrontation with nature implies a nature emptied of meaning, says Meyer.[261] This is the *basic problem* of the modern age. "The emptying of the world, the sobering of the world, is the correlate of the sovereignty with which the subject goes to meet the world, to arrogate to himself the world as object" (TdW 121).

When Meyer states later that through this new relation the humanistic tradition is imperiled[262] and western culture set loose from Christianity,[263] he means the humanism that is rooted in the ancients and in medieval Christianity respectively. He misses the fact that the Reformation – whose spokesmen acknowledged nature as created nature and who rejected any apotheosis of nature together with any instrinsically divine order of nature not to be tampered with – did not oppose the natural-scientific method in principle, provided that this method was seen as normed. In doing away with a deified nature, the Reformation did not cast out God as the Creator of man and nature. It is from the Creator that man receives the mandate to subdue nature – not to his own honor and glory but to the honor and service of God.

2.5.4 Philosophical backgrounds to technological development

At the outset the new relation of humanity to nature remained limited to natural-scientific thought. Moreover, that thought was still directed only toward inanimate nature. Yet, Meyer names Descartes as the founder of modern technology for having perceived that the theoretical-technological thought of natural science[264] would in time work its way into practice. Bacon, too, perceived that the new knowledge relation was a power relation. The practical effect of this knowledge-power would be the disappearance of necessity, anxiety, and suffering, and the ushering in of general welfare for the human race.[265]

The development of modern technology got underway at about the beginning of the nineteenth century. At the same time, natural science developed vigorously and expanded its terrain to include organic nature. The *idea* of universal progress was derived from the progress of science and technology. This idea took shape in philosophy first, and then it came to rule the popular mind. "Progress became the driving force in history; from then on, belief in progress ruled the idealistic and positivistic interpretations of history" (TdW 56). More than ever before, the *future* was made central. The progress of science and technology makes it clear to people that the future is an open one, and that *they* are to shape it.

The driving force behind the idea of progress is materialism as the absolutization of natural-scientific, technological thought.[266] "No limits seem to exist for natural scientific modification-thought; thus it elevates itself above all ethical and religious ties. The consequence is a high technological social order in which, next to a maximal technological control of nature, there is realized a maximal planning and control of human life by the state" (TdW 54).[267] In materialism, technology is proposed as the ideal for every practice. "It is more correct to say that materialistic thought makes use of technological models, since people find embodied in it that mechanical orderliness which they would apply to the whole of human existence" (TdW 67).[268]

This development was further strengthened by the French and Industrial Revolutions. The former strengthened the intellectual background – the will to autonomy[269] and the struggle to use technological methods for still more purposes. The ideal of technology, "Perfektion,"[270] had to be realized first in interhuman affairs and relations and then by making people themselves the objects of technological control. From the positivistic conception of nature springs the positivistic conception of society, "which is based upon the radical withdrawal of human independence, upon the radical functionalizing of the person" (TdW 241).[271] The person must be made ready in this process to assume the place of a component part in culture-as-the-great-machine. First the person is reduced to "object" (*Sache*),[272] then manipulated, and thereafter made anew. "There can be little doubt that similar methods will before long give us power, within wide limits, to create new individuals differing in predetermined ways from the individuals produced by unaided nature."[273]

Materialism goes by various names. Positivism, pragmatism, and communism all stand for a complete technocracy, says Meyer. A common trait is the denial of ethical and religious norms. The result of these philosophies, however different their nuances, is the rise of a "this-worldly," "purely worldly" culture[274] in which material well-being controls everything. An artificial paradise arises as "ersatz transcendence,"[275] where humanity's knowledge of itself is independent of God or any higher power. More about the dangers this entails below.

2.5.5   Modern technology as peril and menace

From what has been said above it is clear that Meyer's judgment about modern technology is less than favorable. In fact, he regards technology as the greatest menace to humanity. Meyer wishes to demonstrate this on the basis of modern technology's origin and the new relation between humanity and nature.

The cause of the danger in modern technology is the autonomy of reason, which signifies a breach between faith and science, between transcendence and the world.[276] Ethical and religious considerations no longer play a role in natural science and modern technology: "...the tie to the highest values [is] surrendered" (TdW 260). Although humanity thought to gain freedom through it, technology founded upon autonomous reason brought constraint. "To found human existence entirely upon itself and to choose the artificial and the unfree are the same" (TdW 206).

At first the dangers remained concealed. Not until the connection between natural science and technology became apparent did the dangers show up. As technological development advanced, they became increasingly conspicuous.

The autonomy of reason implied a different relation to nature. As we have already seen, the organic relation between humanity and nature was broken. Things were reduced in their being to the measurable, the countable, the weighable. Then, given this quantification, things were calculated. Thereafter, on the basis of the results thus obtained, things were manipulated, and so controlled.[277]

As a result of this way of working, the natural relation was exchanged for an artificial one. There was an emptying of meaning, a dehumanizing, a leveling, a pragmatizing of the relation between humanity and nature. The life went out of it; death began to rule there.[278] That was true of the natural-scientific approach, and it was all the more so when theoretical-technological thought turned to practice.[279] "The way of technology is...everywhere marked by the victory of the calculable over the unpredictable, the abstract over the visible-concrete, the functional over the independence and intrinsic value of things, the repeatable over the singular and individual, the artificial over the natural, the controllable outcome over the free activity of things" (TdW 189).

That technological power, through this development, has become a power turned against man becomes apparent from a comparison between modern and classical technology. In the case of the older technology, people had technology in hand. Technology had fixed limits that were tied to the senses and the morphological constitution of people.[280] Modern technology broke through those limits. The natural relation with technology was exchanged for an artificial one. Through the method of quantification, humanity's power over nature was enlarged. However, the technology based upon this enlarged knowledge-power led to a power opposed to humanity. The modern power tool makes that clear. "The power tool is the purest form of harnessing nature rationally, the epitome of its autonomous control, the embodiment of the victory over the organic connections of all the culture that has ever existed" (TdW 102). Modern technology surpasses people in tempo, in precision, in achievement; unlike people, it works continuously. Given people and modern technology, people are the lesser of the two. They no longer have technology in hand; rather, they do the bidding of technology, according to Meyer.[281]

The autonomy of technology is still clearer if we note its planetary character: it breaks the barriers of space and time.[282] Furthermore, the autonomy of technology is proven by its dynamic character. "We cannot escape its dynamism and must adjust ourselves to the world as the technological mind has installed it for us" (TiS 770). This adjustment to technology means that humanity must be made "technology-like."[283]

The peril of technology as an independent power opposed to humanity becomes even greater when the so-called engineers' technique is used as a model for controlling interhuman relations, human activity, and humanity itself. In short, "there must also be technology for the mind, for society, and for history. . ." (TiS 773).[284] Via planning, humanity is taken up into a future *calculable* beforehand. Here communism is the most advanced. Yet, while it promises increasing freedom, communism brings forth the total leveling and functionalizing of the person in total socialization.[285] Because people often resist this leveling and do not easily allow themselves to be made into component parts in modern culture-as-the-great-machine, they, too, become objects for technological control. With the aid of biotechnology, human propagation is already being tampered with; inherited characteristics can be altered. By extension, technology will be used in an attempt to make a person who will be fully a technological mass person – a product on the same plane as other mass products.[286]

The result is a nihilistic culture. The emptying of meaning in that process gains force. "Meaning . . . is transformed, by the most extreme pursuit of technological perfection, into meaninglessness" (TdW 244).[287]

It is often said that the danger of modern technology finds expression in the atomic bomb, in noise, and in soil, water, and air pollution. Meyer, as is probably clear already, disagrees. He sees the great danger in a kind of technological imperialism through which people are robbed of their freedom, their individuality, their personhood. That is the cultural crisis of our time. "The real danger of the technological epoch consists in this, that the technological mind, absolutely confident of its own autonomy, acknowledges as real only the calculable and material, and thereby so reduces reality that only a manufacture of it remains" (TiS 772).[288]

Meyer agrees with Heidegger that it is the essence and the origin of technology that poses the greatest danger. "The essence of technology is not machines, production, automation, or the production line, but the human will to deliver the essence of life itself over to technological manufacture" (TdW 186).[289]

Because Meyer *fully* disqualifies technology on account of its great peril, it is interesting to inquire whether he still offers any perspective for the future of technology.

2.5.6 Perspective for the future?

Although his view of technology owes a great deal to Heidegger, Meyer does not follow him concerning the future. He does not surrender to a speculative hope of deliverance, as Heidegger finally did.[290] In contrast to Heidegger, and also to Jünger[291] and Ellul[292] – each in his way – Meyer refuses to let technological development run its course undisturbed. He also clearly distantiates himself from Gehlen, who believed there is a possibility of avoiding the meta-human development of modern technology through a pragmatic orientation and through the application of the method of "trial and error."[293] No, Meyer is persuaded that an orientation to fixed standards and principles is necessary. In other words, he favors a normative development of technology. The great question is what these norms are to be. "For the question concerning the limits that must be set for rationalization [for Meyer this is identical with technologization], two considerations are decisive: By *which standards* shall the penetration of the possibilities of theoretical, natural-scientific control into the coherence of life be judged? And are the standards in effect adequate for the resolution of these problems?" (TdW 290).[294]

Prerequisite to finding correct norms is a standpoint outside technology. It is not technology but the controlling ideas concerning it that determine the demands which people should make of technological development.[295] Until now, if norms were spoken of at all they were drawn from technology itself. What could be made, was made. Materialism, positivism, pragmatism, and communism did not transcend technology. Their "norms" were those of progress – economic progress, amelioration of social conditions, aggrandizement of power. These "norms" contributed enormously to technologization. Meyer warns against substituting for the "norms" of the positivists norms which apply only to a part of technology – for example, repudiation of weapons of mass destruction.[296] Such an approach would remain piecemeal and would at the same time conceal the fact that technology is a danger to the whole person. "The norms must be of such a character that they bring to a halt the uncontrolled technologization of all the relations of mankind to being" (TdW 292). Meyer seeks the criterion for correct norms in the protection of the free, individual *person*. "A real control of the power given into the hands of man can only issue from principles which guarantee the protection of human personality — otherwise the road to totalitarianism lies open" (TdW 292). The elaboration of this norm means *positively* that freedom, justice, human worthiness, freedom of conscience, and the person as a human being responsible to himself ought to be guaranteed and protected by technological development.

Negatively, the protection of the person vis-à-vis technology means that people and their interpersonal relations may not be made objects of technological control, since in technology the person is made inanimate, is reduced to a thing, and is converted into a (mass) construction component. Meyer would limit technology to engineering technology, which he calls technology in the narrow sense, "where the activity has the character of manufacturing and making, and being is from the very start handled as material and object" (TdW 221). While he originally had objections even to the reduction of inanimate nature by way of the natural-scientific method, he now believes it is justifiable for inorganic nature,[297] since the reduction there is only of secondary importance. This cannot be said of plants, animals, and people. It is true that these beings, too, have a quantifiable or "objectively comprehensible side," but being more than that, they are not exhausted by it.[298] Mechanical processes are subordinate to life; they can perhaps be regarded as borderline cases of the organic. If this subordination is abandoned, life is destroyed; the spirit or mind of the person vanishes in biotechnology. This, at least, is what Meyer maintains. To my way of thinking, however, it is entirely conceivable that there is a place for biotechnology, and that the lives and minds of people can be advanced or restored through it, as medical science already envisions. Of course, it must then be very clear what the mechanical-quantifiable "side" as "*side*" (*Seite*) means, and *how* it *relates* to other "sides." Finally, it must be clear what the *integral* unity of these sides is.

In short, Meyer believes that technology is justified if the order of beings (thing, plant, animal, and person) given by reality itself is respected. Moreover, he believes that technology in the narrower sense may not be utilized if the free person is thereby imperiled, even if the utilization in question might afford social and economic advantages.[299]

Meyer is of the opinion that our culture still contains ethical-religious forces adequate to limit technology.[300] The basic attitude people take must not be that of lord and master: "This claim is false, because it is not identifiable with the true end of man's being" (TdW 296).[301] People must learn to live again from the natural sources.[302] This means that they must no longer accommodate themselves to technology but transcend it. They will then be in a position to distinguish the good from the bad in technology. They can then say yes to the good applications of technology, and say no " ... insofar as we prevent them from making an exclusive claim upon us" (TiS 778-9). When people assume this attitude, they can again exist freely, and they choose for self-determination.[303]

I find it rather surprising that Meyer should offer such a perspective for the future. First he made it clear that autonomous technology and human responsibility are mutually exclusive. Later he wanted to limit technology by appealing to human freedom and human existence. At that point it seemed that he was forgetting that modern technology in the narrow sense, too, has issued from autonomous reason, and that this technology, even according to Meyer himself, displays autonomy vis-à-vis humanity. In this situation, is it still possible to maintain the human person? That Meyer finally comes out in favor of it should have made him doubt his original comments concerning the absolute origin of technology in the autonomy of reason and the unavoidability of the danger of technology.

   The autonomy of technology and the protection of the person are in a dichotomous tension in Meyer, a tension which he cannot overcome. This tension is not broken through; in fact it is heightened. Meyer claims on the one hand that man must abandon his position as lord and master, since it derails human freedom, and that he must therefore return to the basic substance of our humanist-christian culture,[304] thus going back behind the Reformation and the Renaissance,[305] while on the other hand he claims that human autonomy (the Renaissance) is *the* source of modern technology. I believe that this tension might be resolved if Meyer would take into account the influence and import of the Reformation for the rise and future of technology and not limit his attention in this regard to humanistic subjectivism.

2.5.7   Critical analysis

2.5.7.1   Introduction

I will not repeat here the points already made in connection with Jünger, Heidegger and Ellul. These points involved primarily questions concerning what is to be understood by the term *technology*. Like many other thinkers, Meyer takes an expansive view. In general, technology involves "artificially giving form to existence" (TdW 188)[306] on the basis of "theoretical objectivization of being" (TdW 135). The most essential element in technology – creativity, issuing especially in invention – is missing in such a view.

Naturally, an erroneous view of technology blocks a proper view of the positive significance of technology for culture – and that is the case with Meyer.[307] But this does not detract from the fact that we can learn a great deal from him. He points out the undervaluation of nature in modern technology. He shows that our culture is becoming technologized, and that technology becomes an autonomous power as a result of the absolutization of natural-scientific technological thought.[308] Culture is therefore in crisis, and deliverance is possible only if technological development is put to the test of norms.

Meyer dedicated his book *Die Technisierung der Welt* (The Technologization of the World) to his mentor, Eduard Spranger, whose personalism he shares. For Meyer the central question is whether human personality is imperiled or killed by the advance of modern technology. With Spranger he opposes materialism and positivism because such philosophies regard humanity as material, denying freedom and equating people with things. It is this background especially that must be taken into consideration in any evaluation of Meyer.

We must first inquire whether Meyer is correct in identifying modern technology with the "power claim of autonomous thought" (TdW 299). Next we must ask whether the autonomy of technological power with respect to people is unavoidable. Finally, we must explore the perspective that Meyer's normative view of technology may offer for the future.

2.5.7.2 The origin of modern technology

Whereas Heidegger went back to Plato in his effort to understand the character of modern technology, Meyer goes back no farther than Descartes. For Meyer, cartesian subjectivism and the natural-scientific method of Galileo together form the source of the machanistic world – picture of which modern technology is a visible expression: " ... the power claim of autonomous thought [is] identical with the power of technology" (TdW 299). Meyer refuses to understand technology from the history of technology and certainly rejects the notion that modern technology is an extrapolation of classical technology. Modern technology derives from and is determined by intellectual history, or the history of ideas. "The question concerning the descent of technology is

... not in the first instance a question concerning the history of technology but a question of the history of ideas (*Geistesgeschichte*), which must take into consideration all factors that contribute to the advancement of culture" (TdW 69).

This is correct. Only, difficulties arise when it becomes apparent that Meyer views the rise of modern technology as marked by an interruption or caesura in the history of technology.

Meyer wishes to emphasize the unique character of modern technology, and he gives attention to the way in which the development of modern technology is attended by a "this-worldly" (*diesseitige*) culture in which loss of meaning is the leading feature. He concludes from this that there is a break in the history of technology, and that this break is to be understood only on the basis of the intellectual history of modern man. The source of modern technology, according to Meyer, is the autonomy of reason, the breach between faith and science, the severing of all connections with the transcendent.[309]

I agree with Meyer when he claims that the proposition that technology is as old as humanity and that modern technology can be understood from general anthropology does too little justice to the phenomenon of modern technology, which differs qualitatively from classical technology. Yet, in opposition to Meyer I would note that he fails to do justice to the continuity of the history of technology. There is continuity, notwithstanding the qualitative difference between earlier and modern technology. The history of technology is a continuous history, throughout which the discontinuities are intimately interwoven.

Viewed historically, modern technology develops from the technology of artisanship. It is characterized by its scientific basis, with reference to which its differences with the older technology are to be understood. The continuity of technological development is given in the specific *meaning* of technology as the free giving of form to the nature side of the creation. Aside from biotechnology, in which the physical substratum of life can be influenced by people in one way or another, the forming that has been characteristic of technology has always involved the tool. With the scientific basis of technology, the character of the tool is greatly altered. The tool becomes increasingly independent in technological forming, even if people still determine the conditions under which the tool will be utilized. They decide the goal and command the tool to start working. This relative independence of the modern tool can all too easily foster the notion – especially among those who do not know technology from the inside – that modern technology forms an independent power over against people. It then becomes less difficult to point out an interruption in the history of technology with the rise of modern technology.

This view gains greater emphasis as Meyer attempts to understand modern technology on the basis of intellectual history. He claims that the autonomous person has brought forth technology as an autonomous power. Indeed, that has often been the case. The question, however, is whether this information has special relevance with respect to the phenomenon of modern technology. Has Meyer not absolutized the autonomy of technology as such and the crisis of meaning which issues from it?

Meyer argues that people were originally inserted into nature and that they are the children of nature – not lords and masters. The question is whether Meyer does justice to humanity's original position. Did people not stand above nature in the old technology too? Is that not demonstrable from the many edifices erected before the rise of modern technology?

Indeed, the idea of the autonomous position of man as the ruler of nature has set a tremendous stamp on technological development. However, the Renaissance was not the only intellectual movement at work at the time of the rise of modern technology. Should it not be considered necessary to take these other movements into account as well when judging the history of technology?

Meyer has shown – correctly – that the Greeks and the medieval Christians could not arrive at modern technology because they either regarded nature as divine or construed it as being governed by an established divine order with which it was impermissible to tamper technologically. People shrank from any dynamic technological development because they foresaw that it would have many social consequences and because they accepted the existing social order as untouchable. When the synthesis between Greek thought and the biblical revelation was broken at the time of the Renaissance, the outset of the modern age, humanity was declared free and sovereign. Man ruled nature. The Reformation, however, sought restoration through a return to God's Word. The calling to work assumed a place of central significance. In the Reformation, people became conscious that they were called to serve God, and thereby also their neighbors, by subduing and controlling nature. In this way conditions were prepared for the development of the natural sciences and, concomitantly, the development of modern technology.

At the outset, the difference of principle between the Renaissance and the Reformation was not clearly visible. While it is difficult to determine whether it was the Renaissance or the Reformation that originally had the greatest influence on the development of modern science, the representatives of the Renaissance eventually gained the upper hand in the natural sciences. Since the eighteenth century, humanism has dominated the field of science.[310] The Reformation directed attention more to trade and to the technology of artisanship.

Because Meyer takes into account only the influence of the Renaissance and of later humanism and therefore does not investigate the interrelations between intellectual currents, he fails to do justice to the intellectual history of the West. The fact that humanism was later dominant in scientific circles puts him rather easily on the wrong track. It eludes Meyer that Christianity and humanism combat each other as cultural powers. The fact that humanism has gradually gained the upper hand causes Meyer to forget that it made its way through Christianity, and that humanism remains connected with Christianity as its secularization.

Because Meyer fails to take the intellectual history of the West fully into account, he views autonomous reason as the *absolute origin* of modern technology. This sets a stamp on his view of autonomous technological power and on the "perspective" he offers for the future. Modern technology, according to Meyer, leads inevitably to a crisis of *meaning*.

2.5.7.3 The autonomy of technological power

According to Meyer, then, modern technology's hour of birth coincides with the breach between faith and science.[311] This entails, for him, that the pervading influence of growing scientific knowledge in modern technology leads to a "purely worldly," "otherworldly" culture.[312] "When God vanished from human consciousness, culture sprang into the divine opening," (TdW 262) says Meyer, following Ortega y Gasset.[313] This culture is led by the idea of a material prosperity that is led in turn by science and technology. The life of humanity is controlled by an artificial paradise as ersatz transcendence.[314]

The main tendency of western culture is to technologize nature first, then human relations, and finally man himself. The positivistic view of society and of man follows naturally from the materialistic, positivistic view of nature. This technological development, striving toward perfection, borrows its norms from technological progress itself: " ... the ties to the highest values [are] abandoned." Step by step this has brought about a reduction of nature, society, and man to the quantifiable, calculable, functional, and makable.

Autonomous reason has created a technological power of which humanity is the victim: Meyer speaks of "the complete integration of man into the technological collective" (TdW 205).[315] In short and in summary: " ... *to found human existence entirely upon itself and to choose the artificial and the unfree are the same*" (TdW 206 – italics added).

To illustrate his view, Meyer uses the French Revolution's slogans of freedom and equality, which he believes were technological ones. The freedom of theoretical-technological reason brought about social unfreedom as social equality. The stimulus behind technological development is the idea of freedom, but the result is total technocracy, in which people become calculable and manipulable humanoids. "The process of equalization is unavoidably connected with the centralization of state power" (TdW 280). It was forseeable from the outset that equality would win the day over freedom.[316] The freedom and equality of the French Revolution are incompatible with each other because the freedom idea expressed in that slogan is not "the western idea of freedom"[317] but the freedom of autonomous reason. Man's normlessness brings forth a technological power that turns against him.

In my judgment, Meyer has understood the origin of the western cultural crisis better than others. When people acknowledge no norms for technology but instead absolutize that which can be made and manipulated, they wind up as slaves of their own manufacture – technological power.[318] Meyer has a sound view of the ruling dialectic between freedom and equality, freedom and power. However, he regards this development as inevitable. He absolutizes the autonomy of technology, for, as we saw, he sees autonomous reason as the absolute origin of modern technology.

Meyer verifies his view of the autonomy of technology by referring to a number of characteristics of modern technology. Modern machines work more rapidly and precisely than people. Moreover, they produce more and work longer. Furthermore, they work independently and continuously. According to Meyer, technology's planetary power and dynamic quality prove that man cannot escape its autonomy. Meyer forgets, meanwhile, that this autonomy is not absolute but relative. People are not subject to the machine, for they are the ones who determine the conditions under which the machine will function; they are the ones who turn the machine on. This *relative* autonomy is given with the scientific basis of technology. Meyer does not perceive that the autonomy of technology is a chimera;[319] he accepts it as such and therefore fails to grasp that which might make him moderate his opinion.

2.5.7.4   The normativity of technological development

Given Meyer's view of the origin of modern technology and his acceptance of technology as an autonomous power, it would be understandable if he either fled from technology or resigned himself to it. Yet, he opts for neither of these "solutions." He wants to put technological development to the test of norms. But is this not futile as long as he continues to accept the autonomy of technology? How can he square a normative development of technology with his judgment that technological development is controlled only by itself?

Meyer has summed up his view of the normative development of technology as follows: "Technology is not of itself human, however indubitable it may be that it is the work of people; its humanity depends upon human intellective decisions – for example, whether the utilization of its possibilities is to be tied to norms to protect the individual as person and to secure the safety of the conditions necessary for the unfolding of free human life" (TiS 782).

For Meyer, the substance of the normativity of technological development is first of all that technology be limited to technology in the narrow sense, to engineering technology. Thus he rejects technologies of human relations and of humanity itself. Secondly, engineering technology must leave the free person alone and must not tamper with him. The *use* of this technology, which "of itself" is autonomous and forms a neutral power, must be bound to norms.

After he has limited technology to technology in the narrow sense, Meyer must proceed to impose additional restrictions even on this technology, since it is autonomous. He must do so in order to once again make room for the free person. Given Meyer's judgment that engineering technology is autonomous technology, this must be quite difficult to accomplish. Has he himself not tried to demonstrate from modern machines and from technology's dynamic and planetary character that modern technology is autonomous? Afterwards, by limiting this autonomy, he tries to create room again for the free person – room within which this person must both take distance from his ruling position and tie the *use* of technology to norms. Technology as such, however, remains free from norms, and thus neutral.

The fact that this technological development is accompanied by the loss of the humanist tradition (by which he means the early humanists and the ancients) fills Meyer with concern: "What is more imperiled at present than the humanist traditions ..." (TdW 300). Elsewhere he claims that the christian conviction about the absolute value of the "individual soul" has inevitably been lost in and through technology.[320]

Meyer proposes a return to before the beginning of modern times.[321] Yet he cannot deny the significance and power of modern technology. Therefore he suggests a synthesis between modern technology and free personality.

Limiting the autonomy of technology to make room for the free person brings about the rise of a dichotomy between technological power and the free person. Meyer devalues technological reality in favor of free personality.[322] He allows no real place for a normative development of technology. Technology and freedom exclude each other, according to Meyer. Meanwhile, this "normativity" signifies that freedom and religion (by which he means Christianity) are continuously challenged by technology. Meyer believes that technology is the enemy of freedom and religion.[323]

Given his vantage point, Meyer is unable to reach an integral view of technology and freedom and of technology and religion. The "normativity" – as he conceives it – of technological development keeps him from making a fundamental breakthrough out of the dialectic between freedom and (technological) power. He remains stuck in that dichotomy. His personalism proves unfruitful for pointing out a meaningful direction for technology.

3
The Positivists

3.1 Introduction to the Positivists

In 2.0 we noted that the positivists choose their archimedean point in the reality given outside the person. They direct their attention to the facts at hand. This means that the orientation of a positivistic view of technology is derived from technological development and from the possibilities of technology.

In this chapter we shall deal with a number of thinkers, most of whom have played an important role in technological development themselves. The development of cybernetics, especially, has given rise to reflection on the possibilities and future of technology. The computer is by far the most fascinating and most promising instrument of modern technology. The results it attains, in contrast to the results yielded by other technological objects, are not predictable. This is one reason why it is important for experts in the computer field to engage in philosophical reflection on the computer. Another reason why such reflection is important is the remarkable fact that general philosophy is usually averse to technology (see Chapter 2) or indifferent to it, paying scant – and then incorrect – or insufficient attention to the phenomenon of modern technology. Marxism is a clear exception to this rule. When the technological specialists themselves begin to philosophize about technology, they at least provide us with information about the latest technological advances. Unfortunately, their philosophizing all too often turns out to be superficial.

The positivists derive their view from technological development, which they regard as the motor of cultural progress. The significance of this conviction for futurology is that the positivists place technology at the center of human thought. They give the advance of technology such a central, all-controlling place that technology becomes the model for controlling even human relations, society, and the future. The utilization of technological possibilities is made decisive for the development of society. The future as a technological future is regarded by the positivists as a meaningful future.

The positivists wish to exploit the method of modern technology and the principles of cybernetics. From this method and these principles they derive the idea of planning the future. Their futurology is inspired by the possibilities of modern technology and finds its stimulus in those possibilities. They are controlled by the thought that the technological-scientific method is the only correct one.

A feature common to the positivists is a rationalistic streak. They differ from the old rationalists, who tried to fit reality into systems of a priori judgments and of concepts that went beyond the scope of experience and observation. Science, to the positivists we are about to discuss, has become an instrument of technological control. To them, science is the knowledge that transforms praxis into technological practice. It follows naturally, then, that they would abolish the distinction between the natural sciences and the human sciences; at least, using the principles of cybernetics, they would like to resolve the differences of method between them. Moreover, they wish to reshape the sciences (e.g. the social sciences) into instruments for controlling society and its future. In their philosophy, technological-scientific thought is absolutized. In the area of praxis, this technicism finds its complement in a technocracy in which everything is controlled through technology.

The positivists are unanimous in their opposition to religion and philosophy, which they regard as being directed toward "another world." They reject "speculative" philosophy and Christianity, and they condemn the transcendentalists on the same grounds. They approve of secularization because it appears to remove every obstacle to increasingly more radical and integral technological control. They advocate a closed, technological-scientific worldview. Their philosophy is materialistic and is oriented to the concept of information.

Their general problem, as we will see, is that the future is a determined future. This determinism can be "softened" through the possibilities of cybernetics, but it cannot be eliminated. One of the positivists, Steinbuch, has been made aware of this; under the influence of contemporary revolutionary thinkers, he has wrestled with the problem.

In short, the positivists advocate a view of technology and of the future of society that leads to a technocratic society. They are proponents of the ideas held by many engineers. Their thinking is generally clearly opposed to that of the transcendentalists discussed in the preceding chapter, although one of them (Wiener) does resemble the transcendentalists in certain respects.

My method in approaching the positivists will be the same as the one I used in approaching the transcendentalists. After presenting a synopsis, I will look critically at the problems raised and at the way in which they have been handled. In this case, too, I will try to determine what motive unifies the thought of the positivists and how difficulties and dichotomies can arise in their views. The subjects that will occupy our attention are: what cybernetics is; what its origin and basis are; what information is; the relation between information and entropy; the categories and methods of cybernetics; the development of cybernetic machines; cybernetics as philosophy; the relation between the human being and the computer; the "thinking" of machines; the relation between schematic and creative work; the man/machine symbiosis; technological progress; cybernetics and the future; futurology; technocracy; and so forth.

I will deal respectively with Norbert Wiener, Karl Steinbuch, and Georg Klaus. Wiener and Steinbuch were trained as engineers themselves and earned their spurs in the development of cybernetics, especially in the development of the computer. Klaus is a marxist philosopher who takes a philosophical look at cybernetics from the perspective of marxism. The train of thought of the three can be summarized under the heading "instrumental positivism." Wiener and Steinbuch – to distinguish them further – are *analytical positivists* who regard the laws of technological development as all-controlling and all-dominating. Klaus, in contrast, is a *dialectical positivist* who examines the facts at hand in the light of their negation.

Steinbuch often invokes the neomarxists Marcuse and Habermas in his later publications. In my critical analysis, I will discuss the extent of his agreement and disagreement with these revolutionary thinkers. In discussing Klaus, I will observe his differences with Wiener and Steinbuch. There is good reason to do so, for Klaus bases his knowledge of cybernetics on the publications of Wiener and, especially, Steinbuch. In order to shed some light on Klaus's view of the possibilities of the computer, I will also refer to Turing's ideas about the "thinking" of machines.

3.2 Norbert Wiener: The Father of Cybernetics

3.2.1 Introduction

Norbert Wiener[1] was one of those rare men of modern science who are at home in many scientific fields. His great knowledge of modern mathematics and of the latest developments of modern technology, especially electrotechnology, formed the basis for his success in developing modern self-operating machines. Wiener can be called the father of cybernetics or steersmanship. He disposed over the far-ranging scientific knowledge that was prerequisite to initiating the development of cybernetics. His professional skill was indisputable.

In addition, Wiener was interested in philosophy. This interest was awakened early by William James, who was a personal friend of his father. Most importantly, contributions to his philosophical schooling were made by Bertrand Russell and John Dewey. From the former Wiener took over the doctrine of logical types, in connection with his view of the relation between people and the computer.[2] His conception of science has a dualistic character, since on the one hand he is oriented to logical positivism in the line of Russell, and on the other is under the influence of the pragmatism of Dewey. The latter influence is especially evident when Wiener reflects on praxis and on the *utilization* of cybernetic machines like the computer, and when he looks into the significance of cybernetics for the future.

Wiener's philosophical training enabled him to take a larger view of his work than was open to any straightforward scientific specialist. He regarded philosophy as important for reflection on the possibilities and the future of cybernetics. On the other hand, he believed that the principles of the method of cybernetics could shed new light on many philosophical problems.[3] Later I will inquire into the significance of this conviction and the extent to which it is an expression of the technicism that is opposed by the transcendentalists discussed in the preceding chapter.

Wiener's philosophical insights are inferior by far to his great mathematical knowledge and technological expertise. His philosophy often approaches the stage of fantasy. While he should therefore not always be taken seriously, his significance for philosophical reflection on the possibilities of modern technology is still considerable. When he launched ideas and grasped problems pertinent to the significance and the possibilities of the computer, he projected theses that are still the subjects of discussion today.

Wiener's dualistic conception of science, his brilliant conjectures, and his strongly developed practical intuition make it difficult to provide a clear total picture of his thought about cybernetics. I will attempt to distinguish the various lines that characterize his thought. This will make it possible to achieve a better insight into the various dichotomies that run through it.

In order to better summarize Wiener's ideas about the relation between people and the computer, about cybernetics as a philosophy, and about the significance of cybernetics for the future, it will be useful to consider aspects of the origin and foundation of cybernetics. Hence it will be necessary to explain briefly a number of frequently used terms, such as *feedback*, *information*, *entropy*, and *communication*. The tensions in Wiener's thought will be dealt with in the critical analysis. In a concluding subsection of that critique, I will inquire into the source of the difficulties in Wiener's thought and then into the possibility of finding a way to avoid them.[4]

3.2.2 The origin and foundation of cybernetics

During World War II, Wiener worked with the Operational Research Laboratory of Columbia University on air raid prediction mechanisms. At the same time he designed a system to solve the problems of aiming antiaircraft guns. The high speeds of modern aircraft made it necessary to have antiaircraft batteries operate extremely rapidly and therefore automatically, since the relatively low velocity of a bullet from a personnel-guided weapon and the "delayed" reaction time of the person firing it made it increasingly difficult to score a hit. A population not armed with automated antiaircraft batteries lay naked to the enemy. Accordingly, an investigation was made into the human functions utilized in firing, to determine how these functions could be built into a machine that would reduce the reaction time.

The designing of automatic antiaircraft batteries is the historical beginning of cybernetics. Thus cybernetics, or steersmanship, arose on the terrain of technology and technological science. But because the first cybernetic design was human marksmanship, there was from the outset a close relation between physiology and the independent guidance of the newest machines. Human physiological processes were used as an example in designing machines. Therefore it is not quite as remarkable as it is sometimes made out to be that cybernetics should be concerned with guidance and communication between machines and animals – the latter being a category in which Wiener includes humans as well.[5] Wiener says later that the principles of independently functioning machines appear to be the principles that govern life, the ones that are studied specifically in physiology. Naturally, the question then arises whether the human being is exclusively determined by these principles and can be entirely explained on the basis of them, or whether these principles control the human being only in some measure. We shall soon see that Wiener has no eye for humanity's transcendence – in its freedom – of cybernetic principles, and that he thereby places humanity on the same level (in principle) as cybernetic machines.

In 1947, in close cooperation with physiologists, psychologists, mathematicians, and electrotechnicians, it was decided to give the *whole field* of guidance and communications the name *cybernetics*. This new name is derived from the Greek word *kybernétés*, which means "steersman."[6] It is clear from the choice of the new name that cybernetics belongs to the field of technology and technological science. Wiener relates that this name was chosen first of all because steering machines for ships were the earliest and best developed forms of backcoupling mechanisms,[7] and secondly because it honors Maxwell, who wrote an important article about cybernetics as early as 1868.[8]

The scientific *foundation* of cybernetics, of course, is bound up with its *origin* on the terrain of modern technology. Technological science and physics, and especially modern mathematics,[9] constitute the basis on which cybernetics can develop. Of special importance to the development of cybernetics are the mathematical theory of probability, information theory, communication theory, "control engineering," and "communication engineering."

3.2.3 Key concepts of cybernetics

To understand cybernetics and its possibilities, and thereby its significance, it is necessary to take note of a number of important principles or components of cybernetics. With Wiener's publications in hand, we will look at what he says about feedback, information, and the connection between information and entropy, and also at how he uses the term *communication*. Discussing these terms in this order will shed some light on their mutual coherence.

3.2.3.1 The feedback principle

The principle of feedback (backcoupling) plays a large role in automatic guidance. Feedback goes into operation as soon as there is a deviation from a preset pattern. The discrepancy between the pattern and the machine's actual performance is transmitted to the head of the machine and introduced there in such a way that the machine corrects itself and brings the function involved closer to the preset pattern. Moreover, whenever the product of an independently operating machine shows deviations from the normed product, these deviations are carried to the head of the machine so that the starting conditions are appropriately altered to achieve the desired product. "Feedback mechanisms in general increase the uniformity of performance of a system whatever the load may be" (SP 440). The feedback mechanism aims, then, at an independently operating machine that deviates as little as possible from a preset pattern, or that delivers products that are as uniform as possible.

The feedback principle goes into operation on the basis of a deviation that occurs. Because feedback reduces the deviation, it has a negative influence on the system (countercoupling). This suppression of a deviation can be carried out hydraulically, mechanically, pneumatically, or electrically. Electrical backcoupling is the most prevalent kind because it works the fastest. In this case, too, the aid of mathematics makes a theoretical design for a feedback mechanism more easily attainable, leading again to better and faster guidance.

It must not be forgotten that the practical realization of backcoupling is no simple matter. The suppression of a deviation by means of the feedback principle can create many difficulties in cases of inadequate correction or of too strong a reduction of the deviation. Furthermore, a correction may come too late, so that the effect of the feedback is to defeat its intention. The deviation and the feedback are out of phase in such cases, and the deviation may be aggravated as a result, as negative feedback becomes positive feedback. Wiener theorizes at some length about these difficulties, but I will not pursue this matter further.[10]

The application of the feedback principle in modern technology has brought about a tremendous difference between the older machines and modern ones. "The older machines, and in particular the older attempts to produce automata, did in fact work on a closed clockwork basis. On the other hand, the machines of the present day possess sense organs; that is, receptors for messages coming from the outside. These may be as simple as photoelectric cells which change electrically when a light falls on them, and which can tell light from dark" (H 10).[11]

Feedback can mean correcting an existing deviation. Wiener calls this "informative feedback." He speaks of "anticipatory feedback"[12] when the feedback mechanism must be attuned to a future situation. Such a situation occurs, to cite one example, with automatic antiaircraft weapons. The future path of an aircraft is calculated with the help of mathematical theory. The feedback mechanism, which in this case is often called a servomechanism, tunes in on the future path of the aircraft in order to bring the weapon to bear on the correct target area.

The "learning" of machines[13] is also based on the feedback principle. "Learning is in its essence a form of feedback, in which the pattern of behaviour is modified by past experiences" (H 69). Also, "Learning is a most complicated form of feedback and influences not merely the individual action, but the pattern of action. It is also a mode of rendering behaviour less at the mercy of the demands of the environment" (H 69).

In summary, "informative feedback" is related to existing deviations from a given pattern, and "anticipatory feedback" is related to a future position. "Learning feedback," by way of contrast, involves altering a pattern of working, on the basis of experience. "Learning is, in fact, a most general sort of feedback, affecting the whole method of behaviour of the instrument" (H 73).

Because "learning" is concerned with improvements based on past performances, it seems to me that the term *retrocipatory feedback* should be recommended here.

3.2.3.2 Information

The feedback principle rests on information. Only when a deviation from a prescribed pattern, for example, is observed, and only when a technological instrument as a component of the machine is "informed," can the feedback principle come into play. In other words, *information is the beginning of the circular, feedback process*: observe a deviation and correct it.

Wiener is of the view that the heart of cybernetics lies in information theory, which is the foundation of steering or guidance: "It is this study of messages, and in particular of the effective messages of control, which constitutes the science of *Cybernetics* ..." (H 9).

The unit of information – called a bit – is a "choice" between two possibilities which theoretically each have a 50 percent chance of happening. Each "message" as a continuous or discrete course of measurable occurrences in time represents an *amount* of information. The *amount* of information is statistically determined. Wiener says of it: " ... we had to develop a statistical theory of the *amount of information*, in which the unit amount of information was that transmitted as a single decision between equally probable alternatives" (C 18).[14] This amount is the number of selective steps, expressed on the unit of information, necessary to arrive at a particular "message." "Amount of information is a measure of the degree of order which is peculiarly associated with those patterns which are distributed as messages in time" (H 21). The amount of information is the greater as the chance of a particular "message" is the smaller. The amount of information is the negative logarithm of the probability of information: " ... the usual measure of the degree of order to a set of patterns selected from a larger set is the negative logarithm of the probability of the smaller set, when the probability of the larger set is taken to be one" (H 21).[15]

3.2.3.3 The relation between information and entropy

In information theory, then, people work with the "order" of information. The more fully articulated this order is, the smaller the chance of an occurrence, and the greater the amount of information. In physics,

184

by contrast, people work with the concept of "disorder." The measure for this "disorder" is the positive logarithm of the probability of "disorder"; this is called *entropy*. The second law of thermodynamics states that an isolated physical system tends towards a maximum of "disorder." Entropy increases. In other words, a closed physical system strives for the greatest possible homogeneity, and thus loses "order" and "regularity."

Bok[16] says – correctly, in my view – that the term *order* might better be exchanged for *differentiation*. The word *articulation* might also be recommended. *Disorder* should be exchanged for *homogeneity*. What we call disorder is the order of the undifferentiated, which escapes our knowledge.[17]

Wiener says the following concerning the relation between information and entropy: "The notion of the amount of information attaches itself very naturally to a classical notion in statistical mechanics: that of *entropy*. Just as the amount of information in a system is a measure of its degree of organization, so the entropy of a system is a measure of its degree of disorganization; and the one is simply the negative of the other" (C 18).[18] Processes which lose information are thus the analogue of processes in which entropy increases: " ... a message can lose order spontaneously in the act of transmission, but cannot gain it" (H 7).[19] "This fact, that information may be dissipated but not gained, is the cybernetic form of the second law of thermodynamics" (H 88).[20]

I will not pursue the question of the precise relation between information and entropy any further.[21] It is clear, in any case, that there is a relation.[22] When we look at an "information system" as a physical system, the order of that system is observed to decline. The amount of information decreases; entropy increases. Wiener's observation that "the one is simply the negative of the other" (C 18), however, is a grave oversimplification and is at the same time incorrect. Information theory concerns a technological system, and physics concerns a physical system. The physical, particularly the energy structure, is the basis for that which, through technological forming, can become information.[23] Wiener fails to take the *boundary* between culture and nature into account when, the minus sign aside, he equates information and entropy. That oversight will prove to be of far-reaching significance when he comes to reflect on the future of cybernetics.[24]

3.2.3.4 Communication

The transmission of information is called *communication*. In addition to information theory, then, there is also communication theory. The former is the basis of the latter. The development of both theories is associated with the rise of telecommunications. Taken together, they make possible the theoretical control of the circular process of feedback, and so constitute the basis of cybernetics.

The transmission of information is always interrupted. Such disturbance is called "noise." During the conveyance of information, the information becomes mixed with noise, so that the information no longer comes through well. Thus noise determines the upper limit, as it were, of the possibility of information transmission.

Wiener says in summary: "In short, the newer study of automata... is a branch of communication engineering, and its cardinal notions are those of message, amount of disturbance or 'noise' – a term taken over from the telephone engineer – quantity of information, coding technique, and so on" (C 54).

For what follows it is again important to note that the transporting of information requires an energy basis. "The transfer of information cannot take place without a certain minimum transfer of energy, so that there is no sharp boundary between energetic coupling and informational coupling" (H 24). Yet Wiener errs in identifying energy with information. Physical energy is the basis for information conceived of as the analytical substratum of language. The transporting of information – communication – is a total event in which the lingual object function retrocipates, *via* the analytical function, on the energy function. The energy structure is not to be identified with information but is rather the basis for it. In communication man *structures* energy: the foundational function of communication is a *technological* function. When this is forgotten, every structure of matter and energy threatens to become identified with information.[25]

3.2.4 Computers and human beings

The most important fruit of cybernetics is the digital computer. Since I discussed its basic structure in the first chapter, there is no need to do so again here.[26]

The computer itself can be incorporated into the circular process of a feedback mechanism. This is the case not only when the feedback serves to maintain a preset pattern but also – and especially – when "anticipatory feedback" and "learning" are involved, which is to say, whenever a pattern alters itself on the basis of the experiences that are registered.

The use of the computer in technology has advanced automation enormously and banished people from various realms of technological forming. The introduction of computers into the production process requires the transposition of all data into the binary scale. The

186

calculating machine then calculates independently, according to the mathematical rules framed and incorporated into the memory of the computer, and corrects deviations from a preset pattern, or even alters the pattern itself on the basis of "experiences." These corrections and pattern alterations are automatically transferred to the head of the process.[27]

Because cybernetic machines have taken over various human functions, it is only natural that the relation between people and these machines should be a focal point of interest in cybernetics. That it is exactly that is clear from the title of Wiener's first major work, *Cybernetics, or Control and Communication in the Animal and the Machine*.

Cybernetics was initially interested in gaining a fuller understanding of various human functions. The goal was to become better able, on the basis of mathematical elaboration and designing, to substitute mechanical-electrical systems for them.[28] Wiener subsequently wanted to use the functioning of these automatic machines to cast light on the functioning of the human brain and the central nervous system. The neurophysiological investigation that issues from this approach in turn provides a stimulus for designing new machines. "The problems of interpreting the nature and varieties of memory in the animal has its parallel in the problem of constructing artificial memories for the machines" (C 22).

This development has led to the notion that human beings, too, are controlled by the feedback principle, and that machines into which simulated human functions have been built are like human beings. "To sum up: the many automata of the present age are coupled to the outside world both for the reception of impressions and for the performance of actions. They contain sense-organs, effectors, and the equivalent of a nervous system to integrate the transfer of information from the one to the other. They lend themselves very well to description in physiological terms. It is scarcely a miracle that they can be subsumed under one theory with the mechanisms of physiology"(C 55).

In cybernetics, the similarities between the human brain and the computer have come in for special study. It would appear that computers, too, are able to remember, associate, choose, and make decisions. It is clear that the computer can take over certain human tasks, such as measuring, weighing, counting, calculating, analyzing, retrieving data, and so forth. On the one hand cybernetics attempts to discover to what extent the human nervous system, which transmits its stimuli and signs to the brain, is a purely mechanical system. In particular, it has become evident that the nerve cells in the brain – the neurons – perform the same functions as the switching elements in the computer. "The all-or-none character of the discharge of the neurons is

187

precisely analogous to the single choice made in determining a digit on the binary scale, which more than one of us had already contemplated as a most satisfactory basis of computing machine design" (C 22). On the other hand it is also a question of importance whether the machine can become human or, more urgently, whether people can be surpassed and thus superseded by the machine.

Wiener's answers to these questions point in anything but one direction. Sometimes he regards the brain as a logical machine and compares it to a computer,[29] but at other times he still regards people as superior to the machine.[30] He seems to be torn between two views. We will see later that the dichotomy in Wiener's thought is attributable to his dualistic conception of science.[31]

3.2.5 Cybernetics as philosophy

From the beginning, cybernetics has overstepped the limits of the field of technology and technological science to ask whether a human being is a machine and whether a machine can surpass man. Moreover, cybernetics is not concerned with just a single special science but seeks a new view of a group of special sciences. Cybernetics aims at the integration of various sciences on the basis of the principles of steersmanship. Among these sciences (in addition to technological science, biology, psychology, and neurophysiology) are sociology, economics, and anthropology: "Sociology and anthropology are primarily sciences of communication, and therefore fall under the general head of cybernetics" (IAM 327).

Given this insight, the propensity of cybernetics to develop in the direction of a philosophy should not surprise us, for philosophy looks at the various special sciences in their mutual coherence and unity. From Wiener's autobiography it is clear that he regards the task of achieving a philosophical perspective as a task for cybernetics too: "It became clear to me almost at the very beginning that these new concepts of communication and control involve a new interpretation of man, of man's knowledge of the universe, and of society" (IAM 325). While Wiener has certain reservations about bringing sociology and economics within the compass of cybernetics, since in these sciences the application of mathematical probability theory (the method of cybernetics) to the future is extremely problematic,[32] he nevertheless believes that the social sciences are controlled by the feedback principle and that information and communication are pregnant with significance for these sciences. He writes: "It is certainly true that the social system is an organization like the individual, and that it is bound together by a system of communication, and that it has a dynamics, in which circular processes of a feedback nature play an important part. This is true, both in the general fields of anthropology and sociology, and in the more specific field of economics" (C 33).

For the moment I am content to establish that in Wiener's view, cybernetics brings disciplines that seem to be only remotely related closer together again. Wiener's reservations, where they exist, are not scientific or theoretical but practical in nature. Later we will have to take up the question whether cybernetics can really be accorded the significance of a philosophy. And that question will lead to another, namely, whether or not we are then confronted with an objectionable form of technologism.[33]

In his last book, *God and Golem, Inc.*, Wiener tells what cybernetics as a philosophy has to say about the newest technological possibilities, i.e. learning and self-replicating machines, and about their religious significance. He begins as follows: "It is here my intention to discuss not religion and science as a whole but certain points in those sciences in which I have been interested – the communication and control sciences – which seem to me to be near that frontier on which science impinges on religion" (GG 1). He also says: "It is hence germane to a revised study of the relations between science and religion that we should re-examine our ideas of these matters in terms of the latest developments of theory and practical technique" (GG 3). Instead of theological conviction, Wiener would place "intellectual forms" at the center.[34] In brief, the goal of his last book is as follows: "I wish to take certain situations which have been discussed in religious books, and have a religious aspect, but possess a close analogy to other situations which belong to science, and in particular to the new science of cybernetics, the science of communication and control, whether in machines or in living organisms. I propose to use the limited analogies of cybernetic situations to cast a little light on religious situations" (GG 8). Wiener realizes that objections will be raised, but he believes that the anatomist's blade is necessary to the discovery of new things by the grace of violence. "It is the part of the scientist ... to entertain heretical and forbidden opinions experimentally, even if he is finally to reject them" (GG 5). Moreover, "It is a serious exercise, and should be undertaken in all earnestness ..." (GG 6).

These quotations, which are drawn from the beginning of Wiener's last book, confirm the impression that he wished – if perhaps somewhat tentatively – to take cybernetics in the direction of philosophy and to offer cybernetics as a possibility for making clear in theoretical and scientific terms certain aspects of religious experience which had remained previously inexplicable.

Wiener absolutizes thinking in cybernetic categories; specifically, he reduces everything to information. He makes this clear when he claims that *theoretically* it is not impossible to transmit a human being via a telegraph line. "At present, and perhaps for the whole existence of the human race, the idea is impracticable, but it is *not* on that account *inconceivable*" (GG 36 – italics added). Wiener knows that there would be technological difficulties involved in preserving the organism during radical reconstruction. "It is not due to any impossibility of the idea. As to the problem of the radical reconstruction of the living organism, it would be hard to find any such reconstruction much more radical than the actual one of a butterfly during its period as a pupa" (H 111). Wiener does not set the matter up in this way in order to write a science fiction story but in order to make it clear, as he says,[35] that the fundamental characteristic of communication is the conveyance of information. A person is a quantity of information which, after conveyance, given the right materials, could be restructured into a complete person again.

With undisguised scientific presumption, Wiener goes on to suggest the importance of cybernetics for a number of religious questions. As he does so, he refers to learning and self-replicating machines, and preeminently to the computer.

According to Wiener, *machine learning* can be described as the alteration of an incoming message into an outgoing message according to one principle or another of transformation. This principle must embrace not only the most precise possible alteration of the incoming message but also the possibility of improving the alteration itself.[36] As an example, Wiener takes a game that is played according to fixed rules, where the criterion of success is winning the game according to those rules. Thus a chess machine might be able to play chess according to the rules of the game,[37] but an opponent would soon see through the machine's "tactics" and so prevail. A machine built to these specifications would repeat its blunders, given the same situation again. The machine would therefore have to be so constructed that it could learn from its own mistakes and its opponents' errors. It is only possible to speak of *learning* if a machine is able to alter its tactics on the basis both of the *rules of the game* and its *experiences* in past contests: "...it evaluates the scale of evaluation in terms of the games played" (CWF 22). The result will be a game-playing machine that constantly transforms itself into another machine. "In this, the experience and success, both of the machine and its human opponent, will play a role" (GG 21).[38]

Wiener's judgment that learning machines able to surpass humans in a particular function can indeed be built was strengthened by the checkers-playing machine of A. L. Samuel,[39] which ended up winning games from its maker. Prerequisite to such a device, however, is the possibility of establishing an objective scoring criterion for the function involved.[40]

While Wiener believes that establishing such sharp criteria for good achievements is extremely difficult, since the reduction of a criterion to formal rules is far from a simple matter, he does not doubt that the technology of the learning game will be used in the future in many fields of human activity where this has never occurred before, or never been possible before.[41]

This "learning" of machines casts new light, Wiener believes, on the relation between God and humankind: "This is the theme of the Book of Job, and of *Paradise Lost* as well" (GG 15-16). Just as people can lose to the machine they themselves have made, so, presumably, can God be the loser in the game between Creator and creation. Viewed in this way, God is less than *absolutely* almighty. God "is actually engaged in a conflict with his creature, in which he may very well lose the game" (GG 17).

According to Wiener, the "learning" of machines casts a new light on the relation between God and creation. This is no less the case, in his opinion, with respect to the "propagation" of machines. This propagation must be understood as analogous to biotic propagation. Machines can bring forth machines in their own image and likeness.[42] What Wiener means by "image" is as much *representation* as image in the sense of *operational likeness*. An operational likeness can perform the same functions as the original without having the appearance of the original. From this perspective, man is to Wiener an operational likeness of God, and the machine is an operational likeness of man.[43]

How does Wiener present the propagation of (information-processing) machines? He tries to clarify it by experimenting with thought as follows: "For us, a machine is a device for converting incoming messages..." or, to use engineering jargon, "...a machine is a multi-input, multi-output transducer" (GG 32). In general, the outgoing signal of a converter (the machine), which is set under tension by a given incoming signal, is a signal dependent upon both the incoming signal and the converter itself. If the incoming signal should contain a minimum of information, however, as in the case of a noise generator, then the information of the outgoing signal can be regarded as deriving from the converter alone. In that case the outgoing signal reproduces a function of the converter. According to Wiener, the characteristics of this outgoing signal, taken together, constitute the operational likeness of the converter. This operational likeness can subsequently be reconverted into a technological form – incorrectly identified by Wiener as a physical form. If this new machine again brings forth another machine with the same characteristics, their mutual relation is not operational but representational. Mindful of Samuel Butler's

191

remark that a chicken is the manner by which an egg produces a new egg, Wiener says: "Thus the machine may generate the message, and the message may generate another machine" (GG 36).

Although he does not doubt that machine propagation differs in details from the reproduction of living beings, he still says: "Furthermore, with all the differences between living systems and the usual mechanical ones, it is presumptuous to deny that systems of the one sort may shed some light upon systems of the other" (GG 46). "It will not do to state categorically that the processes of reproduction in the machine and in the living being have nothing in common" (GG 47).

In short, the conclusions Wiener draws from machine propagation are, among others, that God's acts of creation are to be understood from man's making machines, and that human reproduction is clarified by machine reproduction. He believes that old theological questions about God's omnipotence and His providence and intervention are rescued by cybernetics from a haze of mysteriousness and furnished with new, clear answers. He realizes that he is launching ideas that are even more difficult to accept that Darwin's theories concerning man's descent from the ape.

Although the significance and possibilities of learning, self-replicating machines need not be underestimated, Wiener has drawn from their development some extremely far-reaching conclusions which have only the remotest possible connection with those possibilities as such.

At the basis of all this speculation is the fact that Wiener has absolutized his own cybernetic-philosophical thought. He has reduced God and man to data-processing systems, to machines, in order to be able to liken them to be latest cybernetic devices. The result of this comparison is that while God and man are originally superior to these machines, they eventually lose out to them and become inferior to them in terms of power.

The comparison of God, man, and machine betrays Wiener's pre-theoretical standpoint. His whole experimental line of thought, therefore, has little to do with testing a hypothesis but is instead the application and elaboration of a presupposition. God, man, and the machine are placed – beforehand – on the same level as cybernetic systems.

The question is whether Wiener is justified in reducing man to quantities of information and whether he has grasped man's essence when he regards him as an information-processing system. These questions are legitimate subjects of discussion, but it must not be forgotten that what Wiener is engaged in here is speculation. Therefore he should at least be open to debate. But if it is affirmed that God is to be accepted first of all – and by faith alone – as the Creator, the possibility of making Him the subject of scientific discussion is precluded. If Wiener and others insist on dealing with God in scientific terms, the god that results will be no more than a pale figment of the speculative imagination.

Wiener's discussion of people as machines destined to become inferior to learning machines before long affords no genuine possibility of hope for a dignified future for humankind. It will become evident that Wiener himself is extremely pessimistic about the future. More about this in the next subsection.

3.2.6 Cybernetics and the future

It turns out that Wiener is fairly optimistic in his *thinking* about learning, self-propagating machines. Yet, when he reflects on putting his ideas into *practice*, his original optimism turns into profound pessimism about the future of humanity and technology. The reason for this is that he can give his *thoughts* about learning, self-replicating machine free rein, while *practice* involves the relation of people to the machine; it involves human beings with all their limitations and misguided intentions for the *use* of the machine. Wiener justifies his original optimism and his later reserve as follows: "As long as automata can be made, whether in the metal or merely in principle, the study of their making and their *theory is a legitimate phase of human curiosity*" (GG 52 – italics added). Yet, whenever the new machines are actually to be put to use, we must be aware of the unsalutary side effects they could have upon people. The tremendous possibilities of machines are in themselves no guarantee that man will not be burdened by their *application*. Therefore Wiener also assumes a dualistic standpoint with respect to automation. On the one hand he acknowledges that automation is a great step forward. With the invention of the steam engine, for example, people were relieved of some extremely heavy work, such as pumping water out of mines. But on the other hand, industrialization caused enormous distress. Women and children were forced to become factory laborers. These social evils were aggravated when many found it necessary to leave the countryside for the city.[44] Eventually the introduction of automatic machinery relieved some of these abuses. The positive side of all this is that "...we are now in a position to return to cottage industry..." (H 171).[45] People are relieved at once of routine and of very difficult or dangerous employment.[46] Moreover, the increased availability of leisure time affords people the opportunity to develop into many-faceted persons of culture. On the other side of the balance, however, is their reduced place in the production process, together with increasing unemployment: "... the

193

average human being of mediocre attainments or less has nothing to sell that is worth anyone's money to buy" (C 37).[47] The result of technological development, moreover, is that other human functions are called upon only too seldom, and the cultural products resulting from technological development, such as modern means of communication, have an even less positive influence on the unemployed masses.[48]

There is one area in which Wiener's optimism about technology is undivided. That area concerns the possibilities of cybernetics for the field of physiology. Wiener believes that the principles of cybernetics will lead to increasing success when applied to the making of prostheses for lost or lame parts of the body: " ... we are already in a position to construct artificial machines of almost any degree of elaborateness of performance" (C 36). Of lost sensory organs he says: " ... the problem of sensory prosthesis – the problem of replacing the information which is normally conveyed by a lost sense by one which is still available – is most important, and not necessarily insoluble" (C 166). Moreover, " ... the problem of sensory prosthesis is an extremely hopeful field of work" (C 167).[49]

Wiener summarizes his views concerning the development of artificial limbs and senses by saying: "The help of the machine may extend to the construction of better artificial limbs; to instruments to help the blind to read pages of ordinary text by translating the visual pattern into auditory terms; and to other similar aids to make them aware of approaching dangers and to give them freedom of locomotion" (H 195). Wiener toys in a similar way with the idea of letting the deaf "hear" again by bringing that which is to be heard into the brain by way of the organs of touch: " ... it appeared to us that it is perfectly possible to make the transition between sound in the outside air, and the semantic recognition of speech, by an artificial phonetic stage, making use of touch, and supplemented by adequate electrical tools" (SP 435).[50] Also, people can arrange to be served by machines that bear no resemblance to members or sense organs because their functions project beyond the functions of members or sense organs. Radar apparatus is an example.[51]

If Wiener is unmitigatedly positive about the development of prostheses, he is not at all positive when he considers the general use of so-called learning machines. He notes that the people who use these machines, whom he calls "gadgetworshippers,"[52] have an overweening desire to enlarge their power. "Power and the search for power are unfortunately realities that can assume many garbs" (GG 53). Regimes in both capitalist and communist countries are going to exploit these machines. The peril of worldwide fascism, "dependent on the *machine à gouverner*" (H 214), threatens. These machines will usher in a time of unhappiness: " ... we shall have to face a decade or more of ruin and despair" (H 189).[53] The danger that such a future will arrive is increased by the users' one-sided admiration for speed and precision. Moreover, they profoundly desire to be relieved of all responsibility for drastic, far-reaching decisions. "It is a desire to avoid the personal responsibility for a dangerous or disastrous decision by placing the responsibility elsewhere: ... on a mechanical device which one cannot fully understand but which has a presumed objectivity" (GG 54).[54]

The apparatus worshippers regard Wiener as a modern magician. Sooner or later, however, their illusion will avenge itself. The magic of automation supported by learning machines, Wiener points out, will produce to the letter what is asked of it. The result will not necessarily be what is best for humanity, or even what was intended.[55] Thus a machine set up to play a war game according to the criterion of victory may very well yield the conditions for victory while "forgetting" that the victorious party would like to survive the war. In such an instance people will have forgotten to include among the conditions the stipulation that victory is not to be purchased at the price of annihilation: "... if the rules for victory in a war game do not correspond to what we actually wish for our country, it is more than likely that such a machine may produce a policy which would win a nominal victory on points at the cost of every interest we have at heart, even that of national survival..." (SM 1357). The conditions for surviving a victory in war can sometimes not even be established beforehand, Wiener claims, citing the example of atomic war. The rules of the game may actually lag behind the facts. The results and effects of such a war cannot be predicted. "In short, when there is a war game to program such a campaign, there will be many to forget its consequences..." (GG 62).[56]

In general, it is Wiener's opinion that avoidance of the perils of learning machines is going to depend on increasingly higher demands being made of human ingenuity and technological training.[57] Instead of making a tentative prognosis which is valid up to a certain point in the development but must be worked out further as new difficulties present themselves, people must be prepared right at the start of the programming to predict all phases and stages of the process: "The penalties for errors of foresight, great as they are now, will be enormously increased as automatization comes into its full use" (GG 63). The speed at which machines work is so great that when an unanticipated danger arises during the process, people cannot intervene quickly enough to avert the danger. Thus the following situation could arise: "... if a bottle factory is programmed on the basis of maximum productivity, the owner may be made bankrupt by the enormous inventory of unsalable bottles manufactured before he learns he should have stopped production six months earlier" (SM 1358).

In addition to making higher demands of human ingenuity and technological training, Wiener seeks the solution to the growing perils in the "humanization" of technology: "As engineering technique becomes more and more able to achieve human purposes, it must become more and more accustomed to formulate human purposes" (GG 64). Wiener had warned even earlier that it is not only "know-how" that is important, but also "know-what," "...by which we determine not only how to accomplish our purposes, but what our purposes are to be" (H 210). In our use of robots we must beware of one-sidedness, and we must make enormous demands of our rectitude and understanding. "The world of the future will be an ever more demanding struggle against the limitations of our intelligence, not a comfortable hammock in which we can lie down to be waited upon by our robot slaves" (GG 69). "We must always exert the full strength of our imagination to examine where the full use of our new modalities may lead us" (SM 1358).

The solution Wiener advocates is that we ought to refrain from using learning machines independently of people. To this point in his analysis he has applied cybernetic thought to humans and machines separately. However, "Cybernetics is not only the study of control and communication in man and machine, but also *between* man and machine" (PBR 2). In other words, cybernetics itself must solve the problems it has created and destroy the dangers it poses for humanity. Concerning the interaction between people and the machine, Wiener offers this general advice: "Render unto man the things which are man's and unto the computer the things which are the computer's" (GG 73). So-called learning machines do not work effectively apart from people, for, as we saw, the success of an undertaking must be judged according to a criterion that involves people. Yet, because the problem of bringing this criterion down to formal rules is far from simple, there must be a permanent connection, a symbiosis between mankind and these learning machines. Thus it is impossible, for example, for translating machines to work perfectly without human intervention. The rules for translating will never be subject to perfect definition. In a symbiosis between people and machines, however, the machine can translate rapidly and people can make corrections, and these corrections can be stored in the machine's memory for later use: "It seems to me that the best hope of a reasonably satisfactory mechanical translation is to replace a pure mechanism, at least at first, by a mechanicohuman system, involving as critic an expert human translator, to teach it by exercises as a schoolteacher instructs human pupils" (GG 79).[58] After a long period of instruction, the translating machine should know the language well enough to be able to translate independently without making gross errors. Correction would still be necessary occasionally, for instance, with respect to new word uses.[59] Constant correction will be even more necessary and important in the case of machines set up to make medical diagnoses: "Such a closed, permanent policy in a medical machine is sooner or later likely to produce much ill health and many deaths" (GG 81). To take yet another case, the application of new inventions will proceed even less self-evidently and automatically, for in the context of a man/machine relationship, inquiry would first be made into how an invention can be used. "To be effective in warding off disastrous consequences, our understanding of our man-made machines should in general develop *pari passu* with the performance of the machine" (SM 1355). In this way, new elements can be absorbed into human history and petrifaction can be avoided.[60]

Thus Wiener warns against too facile a use of independently operating "learning machines." Precision and rapidity predominate all too easily at the expense of other human values. "The proper relation between man and machine is not that of competition, but in the development of systems utilizing both human and mechanical abilities" (PBR 3).[61]

In summary, Wiener states: "Not only is the problem of adapting the machine to the present conditions by the proper use of the intelligence of the translator or the doctor or the inventor one that must be faced now, but it is one that must be faced again and again. The growing state of the arts and sciences means that we cannot be content to assume the all-wisdom of any single epoch" (GG 82).

In looking back over all this, we can see that Wiener is clearly enthusiastic about the development of prostheses. It is not clear, however, that he regards the advantages of automation as outweighing its disadvantages. The independent utilization of learning machines will, he fears, produce a dictatorship and lead to a rigid social development unless man enters into symbiosis with the machine. Within that symbiosis, the machine may do only that which is beneath human dignity, and people must look after that which is worthy of themselves. This will help to prevent accidents.

All things considered, it is not clear whether Wiener takes a positive or a negative view of technological development.

He wants to have nothing to do with the American belief in progress, which anticipates a heaven on earth and expresses the expectation of perpetually advancing, of constantly moving forward to bigger and better things.[62] Yet, neither does he wish to be reckoned among the pessimists who interpret the development of the universe in agreement with the law of entropy. As we saw, entropy is the expression of the measure of disorder. In a closed physical system (for example, the universe), disorder will increase to a maximum of perfect homogeneity. Wiener says of this: "... it is necessary to keep these cosmic physical values well separated from any human system of valuation" (H 22). Wiener's reason for this is that in all of life, and thus with people too, entropy does not increase; rather, it decreases. Moreover, every cultural activity thwarts the increase of entropy. Nevertheless, he says: "The question of whether to interpret the second law of thermodynamics pessimistically or without any gloomy connotation depends on the importance we give to the universe at large, on the one hand, and to the islands of locally decreasing entropy which we find in it, on the other. Remember that we ourselves constitute such an island of decreasing entropy, and that we live among other such islands" (H 25).

If we gaze from our islands into a not too distant future, then there is still perspective for us, in Wiener's view. In the long run, however, he foresees a much less rosy future for humanity, partly because of the latest machines. "Yet we may succeed in framing our values so that this temporary accident of living existence and this much more temporary accident of human existence, may be taken as all-important positive values, notwithstanding their fugitive character" (H 26). But Wiener follows that up by saying later: "We shall go down, but let it be in a manner to which we may look forward as worthy of our dignity" (H 26).

In his autobiography Wiener has this to say about his view of the future of technological development: "I have not replaced the gloom of existence by a philosophy which is optimistic in any Polyanna sense, but have at least convinced myself of the compatibility of my premises, which are not far from those of existentialism, with a positive attitude toward the universe and toward our life in it" (IAM 328).

Wiener senses impending shipwreck on the premise that our culture has been made extremely unstable by the gains of technology: "... I am claiming... that a serious derangement and destruction... has become much more likely than in the past times" (H 42). World War II demonstrated that a conflict can no longer be localized. World communication has taken on such proportions that the transmission of a contagious disease has become easy, creating the possibility of some great future calamity. A similar disaster could befall mankind if there were a temporary interruption in the provision of food.[63] Then there is the additional unhealthy factor of man's increasing subjection to technology. The human is a communicative being, but "It is precisely this variability and this communicative integrity of man which I find to be violated and crippled by the present tendency to huddle together according to a comprehensive pre-arranged plan, which is handed to us from above" (H 217).

198

Wiener places the danger for the future not, for that matter, in a number of phenomena, however threatening they may be, but in what lies behind those phenomena: "I think that the over-all danger from the total situation is much greater than the danger from any of its particular manifestations, such as the atomic bombs or the learning machine" (CWF 27). The *deepest ground* of Wiener's negative attitude toward the gains of modern technology becomes evident when he states: "The sense of tragedy is the sense that the world is not a pleasant little nest made for our protection, but *a vast and largely hostile environment*, in which we can achieve great things only by defying the gods, and in which this defiance inevitably brings its own punishment" (H 211 – italics added). Prometheus is Wiener's prime example.

Yet, for Wiener there is no way back. It is as if he were ruled by fate when he says: "No, the way to survival does not lie backward. Our fathers have tasted of the tree of knowledge, and even though its fruit is bitter in our mouths, the angel with the flaming sword stands behind us. We *must* continue to invent and to earn bread, not merely by the sweat of our brows, but by the metabolism of our brains" (H 57 – italics added). Moreover, we have become heavily dependent upon machines: "...the machine habit is like the liquor habit. If we drink enough of it, and we already have, it is an awfully hard thing to stop. We have made ourselves dependent upon the machine, on the car, for instance." "We cannot go back to the old agrarian civilization" (CWF 28).

Summarizing his pessimistic view in brief, Wiener says: "Give us the freedom to face the facts as they are! We need not expect that the race will survive forever, any more than that we shall survive forever as individuals, but we may then hope that both as individuals and as a race we may live long enough to bring into the open those potentialities which lie in us" (H 58). As long as the human race endures, it must not put its trust in machines, or even in learning machines, unless it knows precisely what rules the machine follows and unless it is certain that those rules are not in conflict with human principles. "For the man who is not aware of this, to throw the problem of his responsibility on the machine, whether it can learn or not, is to cast his responsibility to the winds, and to find it coming back seated on the whirlwind" (H 212).

Thus, while Wiener sees no meaningful perspective for technological development in the long run, this does not prevent him from calling for the shouldering of responsibility and for the good use of cybernetic principles. If we refuse that call and place our trust in learning machines, we are shrugging off our responsibility, accepting a plan imposed upon us from above, and hastening the unavoidable disaster.

3.2.7   Critical analysis

3.2.7.1   Introduction

Through his knowledge of mathematics and modern technology and his readiness to work with many other scientists, especially psychologists and physiologists, Wiener gave a tremendous stimulus to cybernetics. At the same time, his interest in philosophy, which was awakened early by William James and strongly influenced by Russell and Dewey, brought him to reflect on the fruits and possibilities of cybernetics. Taking a broad view of culture, he launched a great many ideas that are still relevant today in philosophical discussions about the computer. A spirit rich in fantasy enabled him to sketch actual future developments; yet, it also led him into speculation. In general, Wiener's philosophical approach to problems is devoid of adequate theoretical underpinning; his view is weakly argued scientifically.

The weakness of his position stems from the fact that Wiener's thought is characterized by two incompatible lines. The one line is borrowed from the philosophy of Russell; the other displays the marks of the pragmatism of James and Dewey. The rationalism of Russell is not to be reconciled with the irrationalism of pragmatism.[64]

Along the line of rationalistic thought oriented to the *principles* of cybernetics, Wiener gives free rein to his ideas and comes forward with bold statements. These statements are subsequently corrected whenever Wiener goes on to construe information in a physical sense and to equate it with negative entropy and whenever he gives free rein to his practical intuition or bumps up against *practice* and develops an awareness of the dangers inherent in the latest cybernetic machines. Alongside this dichotomy intrinsic to Wiener's thought there is another one – a dichotomy between theory and practice. Because of the latter dichotomy, Wiener's science ideal disintegrates under the weight of the problems and demands of practice.

The dualism in Wiener's thought, added to the fact that his world is a closed one, finally prevents him from pointing out a meaningful perspective for the future. Wiener acknowledges no norms beyond the world of science and technology that might have any influence upon that world. Thus he is able to surrender to Russell's autonomous, self-normed, rationalistic conception of science. Since the practice of technology and the use of computers confront him with facts that do not correspond with this conception of science, he is forced to introduce corrections. And these corrections, since Wiener will hear nothing of supraarbitrary normative principles, must be subjective in nature. The question then arises whether such subjective ameliorations can be of any great significance, or whether their introduction is futile. In summary, Wiener's science idealism has another side – his culture pessimism.

In elaborating all this I will now show that Wiener's ideas about cybernetics as philosophy, about the relation between man and the machine, and about a future based on the possibilities of cybernetics all remain stuck in the mud of an unfruitful dichotomy. In a concluding subsection, the origin of these difficulties will once again be brought forward and an attempt will be made – while doing justice to the meaning of cybernetics – to suggest a way out of these difficulties.

3.2.7.2 Cybernetics as philosophy

As we have seen, cybernetics arose on the terrain of technology and technological science. With the help of the feedback principle, it is almost possible to eliminate unfamiliar, disruptive influences and to achieve the desired product or process. A thermostat functions according to this principle. A deviation from the desired room temperature is carried back to the heat source, which then readjusts itself to eliminate the deviation. Other forms of feedback have also been developed, as we noted; a machine can be attuned to a future situation or made to take account of the past in such a way that it can improve its own performance. In other words, it can learn. From James Watt's steam engine to the modern calculating machine, the feedback principle has found application in modern technology.

We saw, too, that cybernetics has significance beyond technology. Thus body posture, body temperature, and hormone regulation can be understood in the light of cybernetic principles.

Wiener foresaw this development. From the beginning of his involvement with cybernetics, Wiener suspected that cybernetics might be able to link many – or even all – of the sciences. Because sociology and economics are communication sciences, they fall under cybernetics, according to Wiener. Thus the idea crystallizes that cybernetics might well be able to play the role of a philosophy.[65] "Besides its function in these already existing sciences [i.e. sociology and economics], cybernetics is bound to affect the philosophy of science itself, particularly in the fields of scientific method and epistemology, or the theory of knowledge" (IAM 327).

According to Wiener, cybernetics as philosophy will be able to shed new light on a great number of philosophical problems that have either not been resolved or not been correctly posed. He contends that cybernetics is able to make short shrift of the point of contention between the mechanists and the vitalists: "... the whole mechanist-vitalist controversy has been relegated to the limbo of badly posed questions" (C 56). This problem is not to be solved through a one-sided orientation to matter or to energy as power; rather, it is to be clarified in terms of information. "The mechanical brain does not secrete thought 'as the liver does bile,' as the earlier materialists claimed, nor does it put it out in the form of energy, as the muscle puts out its activity. Information is information, not matter or energy. No materialism which does not admit this can survive at the present day" (C 155).

The substance and scope of Wiener's thought are determined by the far-reaching significance he attributes to the concept of information. He identifies information with structure; because every science explores structures, the sciences are to be subsumed under cybernetics. For technology, biology, psychology, physiology, sociology, economics, and anthropology, according to Wiener, information – or communication as the conveyance of information – is decisive. That is the reason why cybernetics integrates them. Mathematical information theory – which Wiener calls the heart of cybernetics[66] – is decisive for all these sciences. It is therefore understandable that Wiener should choose Leibniz as a "patron saint for cybernetics." "The philosophy of Leibniz centers about two closely related concepts – that of universal symbolism and that of a calculus of reasoning. From these are descended the mathematical notation and the symbolic logic of the present day" (C 20).

Cybernetic thought is imperialistic in nature not only because it oversteps the boundaries of technological science but also because Wiener absolutizes the concept of information. He identifies structures with information. The original meaning of the concept of information is thereby lost at the very outset. Information is always lingual in nature; hence it always derives from human cultural activity. In cybernetics, tools are designed for taking the analytical substratum of information as language and, with the help of signals, transmitting it (communications technology) or processing it (computer technology). These signals owe their structure to human formative activity. When cybernetics uses the term *information* – we have already noted the muddled use of the term[67] – it should not be forgotten first that it is not using it in the original sense, and secondly that *information* presupposes human form-giving labor. People reproduce in signals the analytic substratum of language.

Wiener errs, then, in identifying information with structure. This leads him into ill-founded speculation about the transmission of people themselves.[68] For such a feat to be accomplished, Wiener claims, a person's matter must first be transformed into energy. The energy has a structure, and the quantity of information depends on the structure of this energy. After transmission, the retransformation of this energy into matter must yield the complete person again. The notion remains a theoretical one for the time being, as far as Wiener is concerned, since one cannot conceive of a practical way of realizing the transmission. Nevertheless, "there is no fundamental absolute line between the types of transmission which we can use for sending a telegram from country to country and the types of transmission which at least are theoretically possible for a living organism such as a human being. Let us then admit that the old idea of the child, that in addition to travelling by train or airplane, one might conceivably travel by telegraph, is not intrinsically absurd, far as it may be from realization" (H 109).

All of this is to be understood on the basis of the rationalistic science ideal, which Wiener borrowed from Russell, and of logical positivism in general,[69] although as an ideal it is actually much older – and on that account more unyielding and unshakable.

The other line in Wiener's conception of science is pragmatism – specifically, the pragmatism of Dewey. Wiener claims, in Dewey's style, that the statements of science must be made operational. Only then are scientific statements true. "The science of today is operational; that is, it considers every statement as essentially concerned with possible experiments or observable processes" (C 147).[70]

The pragmatic conception of science competes with the rationalistic conception of science. Along the line of rationalistic thought, Wiener comes to bold pronouncements which must then be relativized and corrected along the line of pragmatic thought. This dualistic conception of science is the reason why his view of cybernetics as philosophy is far from a harmonious one. It will become increasingly evident that there is an irreconcilable dichotomy between these two conceptions and influences.

On the one hand, as we have noted, Wiener places sociology and economics in the field of cybernetics, but on the other hand he discerns from practice that cybernetic principles cannot be applied purely and simply in these fields. And Wiener is much less optimistic on this score than his friends are: "They are certain that our control over our material environment has far outgrown our control over our social environment and our understanding thereof. Therefore, they consider that the main task of the immediate future is to extend to the fields of anthropology, of sociology, of economics, the methods of natural sciences, in the hope of achieving a like measure of success in the social fields" (C 189).[71] The argument Wiener raises against them is that cybernetics has been concerned from the beginning to get a grip on the future. Thus a theory was developed, in connection with the designing of automatic antiaircraft guns, that was intended for establishing the future position of an enemy aircraft. This predicting of the future was based upon the past, upon the type of aircraft, velocity, acceleration, flight path, and so forth, and it was facilitated with the help of probability theory. Prediction was predicated on at least two conditons: first, the data of the past had to cover a long period, and secondly, there could be no discontinuity in the past. Now, these conditions are absent in a dynamic society where new technological discoveries are constantly occasioning discontinuities in the society's life. "The human sciences are very poor testing-grounds for a new mathematical technique..." (C 34). Elsewhere Wiener has given another summary of his objections: " ... in the social sciences we have to deal with short statistical runs, nor can we be sure that a considerable part of what we observe is not an artifact of our own creation." "We are too much in tune with the objects of our investigation to be good probes" (C 191).[72] Economists have developed the habit of dressing up their fairly imprecise thoughts in integral and differential calculations. Yet, before they get that far, Wiener argues, they use the mathematics and physics of a hundred years ago (without knowing it). Thus they forget that as *observers* they are themselves involved in their calculations. Furthermore, the relations in their field shift with the appearance of new technological discoveries, which necessitate a review of quantification. Through such oversights, a great deal of superficial, poor work is done. "Under the circumstances, it is hopeless to give too precise a measurement to the quantities occurring in it. To assign what purport to be precise values to such essentially vague quantities is neither useful nor honest, and any pretense of applying precise formulae to these loosely defined quantities is a sham and a waste of time" (GG 91). To all of this Wiener attaches the conclusion: "Thus the social sciences are bad proving ground for the ideas of cybernetics – far worse than the biological sciences, where the runs are made under conditions that are far more uniform on their own proper scale of time" (GG 92).

Wiener's reserve does not go so far that he would argue that cybernetic principles have no significance at all for the social sciences. It remains true that "... the society can only be understood through a study of messages and the communication facilities" (H 9). Only, if cybernetics is to be successful in these fields in the future, the past ought to be studied in an extremely precise way.[73] A knowledge of history over as long a period as possible makes the differences which arise in it relative, so that some grasp of it can be achieved with probability theory, and the future predicted and controlled. Wiener cites Russell in support of this view. Wiener believes that the various fields of cybernetics satisfy the various logical types as Russell developed them. There is a higher logical type for sociology than for biology. The higher the logical type for a particular field, the longer the period in the past of which knowledge is required before cybernetics can be applied to that field. "Each new level demands a study of a much larger past than the previous one" (SM 1356).

Knowledge of the past is indeed a prerequisite, but it is not in itself a condition sufficient for the application of cybernetics to the field of human society. To justify that, we must know humanity itself better: " ... the mechanical control of man cannot succeed unless we know man's built-in purposes, and why we want to control him" (H 209). The two conditions, taken together, afford perspective again.

Thus, while Wiener corrects his original stance, he retains the notion that cybernetics does afford a possibility of controlling people and their society. However finely nuanced he may ultimately make the cybernetic method, there is an implication of a technologized person and society in Wiener's quest, an implication that cannot be concealed. Through an absolutization of technological-scientific thought, he wishes to reconcile the rationalistic way of thought with the pragmatic one. He is ultimately concerned with "mechanical control." *His philosophy is a form of technologism.*

Wiener's absolutization of the concept of information and of cybernetics in general is to be rejected. This is not to deny, however, that the principles and technological fruits of cybernetics are of significance for the social sciences. Cybernetics can offer substantial *help* in understanding various phenomena in these fields. Cybernetic principles occur in these fields as analogies. Yet, people with their freedom play a decisive role in the fields of inquiry of the social sciences. And it is this freedom, which escapes science, that ought to be respected in the quest for scientific control. People do not satisfy mechanical laws alone, and they cannot be fully controlled by mathematically describable feedback mechanisms. People are free; they transcend cybernetic principles.

Wiener should have distinguished the various fields of inquiry. Then he might have avoided the dichotomies in his assessment of the significance of cybernetics for other sciences. Different laws and norms obtain for the different fields of inquiry. Also, when the cyberneticist proceeds to compare dissimilar objects, he should attempt to determine the aspect by which each of these objects is qualified. The ineradicable differences between the fields would then be taken into account right from the outset. A machine is technologically qualified, an animal is psychically qualified, and a human being transcends all modal functions. When these entities are regarded only in relation to the information-processing aspect, as in the case of Wiener, their qualifying functions are neglected – at least in the theory. In praxis, of course, all these functions come into play, so that Wiener is forced to retreat along the path of his original intention. As we observed, however, his retreat engenders dichotomies in his theoretical perspective. As long as the erroneous starting point is maintained, those dichotomies cannot be fundamentally resolved.

The danger of cybernetic technologism can be avoided by distinguishing theory from practice on the one hand, and on the other hand by taking into account the modalities of reality, which cannot be reduced to each other. In this way a safeguard is created against the irresponsible use of the technological-scientific method of cybernetics outside the field of technology. Cybernetics can render assistance in other fields, but then its use ought to be subject to norms from beyond the field of cybernetics itself. Only in this way are human freedom and responsibility respected, so that people are not reduced to machines. Wiener does not escape this peril, as we shall see in the following subsection.

3.2.7.3   The relation between man and machine

The dichotomy between science and practice noted in the preceding subsection becomes highly conspicuous when Wiener discusses human beings and machines from the standpoint of cybernetic principles.

Initially he believed that the difference between the human being and the animal consisted solely in man's being a speaking animal.[74] Accordingly he could say: "It is my thesis that the operation of the living individual and the operation of some of the newer communication machines are precisely parallel" (H 15). Modern machines and the human brain were ruled by cybernetic principles. Each was a "logical machine."[75] Wiener carried the parallel between the human being and the machine so far that in his last book he even wrote about learning, self-replicating machines. From this it might be concluded that according to Wiener, machines take on the likeness of a person. That, however, goes too far; he intended nothing more than a parallel. The human being is an extremely complicated machine, and modern machines can shed light on man as machine.[76] In Wiener's cybernetics, the human being and the machine are complementary, as it were: the machine can be elucidated through man, and man through the machine.

Even so, Wiener sometimes manifests reservations concerning this parallel and notes important differences. Successive computer runs have no bearing on each other – or only a limited bearing. The computer can be cleared after every use. This is not the case with the brain, where one processing influences the next. "Thus the brain, under normal circumstances, is not the complete analogue of the computing machine, but rather the analogue of a single run on such a machine" (C 143). This limitation does not, however, compel Wiener to be cautious. The consonance between man and the machine is too attractive to be surrendered. He remains of the opinion that learning processes, too, can be preempted by machines, and that machines will even learn to speak, for "... language is built in only as a possibility which must be made good by learning" (H 85).

Although the machine is more limited in its functioning than a person, it, too, according to Wiener, can possess the capacity for originality. It is even possible that in some single operation the machine will defeat man: "As machines learn, they may develop unforeseen strategies at rates that baffle their programmers" (SM 1355).[77] Therefore Wiener also states: "It is my thesis that machines can and do transcend some of the limitations of their designers, and that in doing so they may be both effective and dangerous" (SM 1355).

Still, in certain respects people remain superior to the machine. The human brain is better than a common, person-made machine: "The human brain is a far more efficient control apparatus than is the intelligent machine when we come to the higher areas of logic. It is a self-organizing system which depends on its capacity to modify itself into a new machine rather than on ironclad accuracy and speed in problem-solving" (SM 1357). It is probable that in the future the machine will take over more and more functions of the brain: "Yet for a long time at least there will always be some level at which the brain is better than the constructed machine, even though this level may shift upwards and upwards" (SM 1357). According to Wiener, the limit is connected to Russell's theory of logical types. The brain is best suited to the higher types, the machine to the lower ones.[78]

Wiener stands up for man especially when discussing the *use* of the newest machines. This seems to him to be necessary, since the users of these machines can be ruled by the wrong motives, such as the lust for power, the exploitation of people, or even a desire to pass off responsibility to the machine.[79] Wiener emphasizes that the machine may not be permitted to have power over man.[80] Not everyone, according to him, is similarly convinced of this necessity: "Precision and speed are valued by some engineers much more than the sum of all human qualities. This preference of machine over man displays a fundamental contempt for man and dislike for human values" (PBR 3).

Under the influence of *practice*, Wiener retreats somewhat – reluctantly – from his initial standpoint that the human being is a machine. He is concerned about the position that people would be in if modern machines should fall into the hands of power abusers who wanted to impose a plan upon humanity from above, thereby creating difficulties for the freedom of the individual.[81] And that danger is indeed very real. Wiener himself, however, is all too little aware that his own theory concerning the relation between man and the machine gives rise to this danger. Because human freedom is not the boundary of his thought, and because he fails at the same time to realize that freedom is a premise which must be taken into account positively *in* science, the free person escapes him. With Wiener, man is already the victim of the machine in the theory. Wiener has walked into the snare of the science ideal. In his reflections concerning praxis, this ideal conjures up human freedom as a counterpole; freedom intrudes, and it prevails. Because Wiener casts the free person overboard in his theory, that theory is eventually called to account before the bar of practice, and the resulting judgment also demands theoretical assimilation. As a result, Wiener's discussion about the relation between the human being and the machine is ambivalent.

Wiener would have been a better theoretician if he had maintained that the human being is free. Do people not think in freedom, judge in freedom, and decide in freedom? This freedom is normed, however, and this is precisely what makes it possible: *man is man-responsible-in-freedom*. This means that in contrast to machine behaviors, human behavior is not determined,[82] for in every process without exception, the free, responsible human being plays the decisive role. Through this very same human freedom, it is possible for people to objectivize certain processes and to allow them to be carried out by machines. The possibilities here have been increased tremendously with the help of cybernetic principles, even to the extent that within certain game-rules a machine can alter its tactics – something that Wiener calls "learning." Now, I have no objection to such terminology, as long as it is understood that it is being used to make reference to *determined learning*: given the game-rules (or learning rules), the machine *can* do no differently. In such machine learning, the rules of the game are not broken but fully exploited, which creates an *appearance* of originality. But in this context there is nothing that can be said about originality in the sense of creative freedom. The machine lacks it.

If Wiener had taken all this into account, he would have seen that the symbiosis between man and the machine is still there. The computer cannot be understood apart from people. People made the computer, and they determine the conditions under which the machine will work. It is therefore incorrect to place the computer on a level with people. A proper comparison might be made between humanity *with* the computer and humanity *without* it. When this comparison is worked out – Wiener's view offers a great deal of very fruitful material for it – dichotomies can be avoided and subsequent corrections rendered unnecessary.

It should be noted that in his later correction involving the symbiosis, Wiener introduces *no improvement of principle*. Under the pressure of the facts, he wishes to take a step backward along the chosen way. Nevertheless, it should not be forgotten that he does not abandon the way of rationalistic cybernetics. The symbiosis does not imply an acknowledgment of the human being as a free person; rather, Wiener continues to regard man as being in principle a machine, albeit a complex one. Cybernetics, having been concerned with circular processes as they occur in the machine and in man, would now simply extend its concern to the symbiosis between the human being and the machine. The circular process in this symbiosis of the human being and the machine is controlled, according to Wiener, by cybernetic principles.[83] In this circular process, moreover, the human remains ultimately a machine, as Wiener sees it. In order to preserve some semblance of human dignity, he offers the general and rather lame advice that in this symbiosis people should do the things that are worthy of themselves, while the machine does the things that are not worthy of people. In his view of man, Wiener hesitates between two conflicting views.

By now it should be clear that I cannot agree with Wiener when he contends that the circular interaction between the human being and the machine is a *closed* process subject to scientific control and therefore properly a part of the field of cybernetics. Yet this conclusion is unavoidable if it is not recognized that the human being is not a machine. Man is not exhausted in the symbiosis. Human freedom breaks through this process and goes on to create the possibility of controlling the machine meaningfully. However much the machine may be able to accomplish, freedom as people know it will never belong to it.

My main point of disagreement with Wiener ultimately stems from his view of man. It is my conviction that whenever man is not viewed integrally as being *free* and *responsible*, and that whenever it is not acknowledged that freedom and responsibility differentiate themselves into a variety of directions, then every view concerning the relation between man and technology and a future based on technology must lead literally to a dead end. How can people possibly expect a human future if humanity is already construed as a machine *to begin with*? Any meaningful perspective is then precluded.

3.2.7.4 Technology and the future

On the one hand, Wiener is somewhat optimistic about the future. The possibilities of cybernetics are unprecedented. Human limitations will be overcome by the new cybernetic machines. These machines will be able to learn and even propagate themselves. In "intelligence" and "originality" they will surpass people. With the help of these mighty machines, religious questions such as God's omnipotence and the creation of man will be brought out of the shroud of mystery and given clear, scientific answers through cybernetics.[84] In short, cybernetics as science will bring forth machines powerful beyond all conception.

On the other hand, this optimism gives way to profound pessimism. In practice, people will turn out to be victimized by these new machines. The machines will crush humanity. Furthermore, the utilization of these machines will eventually be accompanied by an unavoidable reduction of human labor; unemployment will increase. The rise of a global dictatorship is conceivable, for world culture is destabilized by the new technological possibilities.[85]

The ultimate basis of Wiener's pessimism, as the general discussion of various thinkers in Chapter 4 will make clear, is his conceiving of this world as a closed world. Such a conception casts the questions of meaning and responsibility into peril in the long run. Wiener does not appreciate the profound meaning and significance of human responsibility. Through the absolutization of his scientific thought, he has managed to exorcise it.

Wiener's thought is oriented, as he says himself, to modern physics and logical positivism.[86] He regards the universe as a closed physical system. Consequently, given the second law of thermodynamics, this system has a tendency towards increasing chaos. Yet Wiener is reluctant to draw this conclusion directly.[87] He claims that humanity is to be regarded as an island in this physical universe, and that people strive to resist entropy from their island base. They succeed in the short run, Wiener believes. Over the longer haul, however, a meaningful perspective is lacking. This does not prevent Wiener from advocating some appropriate utilization of such cybernetic principles as feedback, information and communication. Yet, on the basis of these same principles, the future looks anything but rosy, according to Wiener. Cybernetics facilitates the gathering, processing, and transmission of information. And because Wiener identifies information with structure, as we saw, it follows that cybernetics results in a reduction of the quantity of information, a leveling of structural differences, and thus an increase of entropy. Wiener finds that this result is clearest in the case of modern weapons of mass destruction. "The effect of these weapons must be to increase the entropy of this planet, until all distinctions of hot and cold, good and bad, man and matter have vanished in the formation of the white furnace of a new star" (H 142).

This argument must be regarded as anything but decisive, however, since with these weapons culture is dashed and turned into dead nature. It is precisely in their culture-forming activity that people have at their disposal a possibility of thwarting the increasing entropy. In fact, this is true for everything alive in nature. Moreover, even if Wiener is right, his short-term perspective cannot be reconciled with

210

his lack of ultimate perspective. Even in the short term, information loss occurs. Therefore entropy, in keeping with Wiener's theory, must increase. From all this it is again apparent just how perilous it is to identify information and structure.

Between information and structure there is (at most) an analogy. The difference can perhaps be made clear by noting that structures, whether given or formed by people, have a lingual *aspect* only, whereas information itself is lingually *qualified*. This means that information can constantly be supplemented through human freedom. Moreover, the fact that people form new structures in culture provides the possibility, in a physical sense, of resisting entropy.

Wiener's physicalism and his belief that cybernetic machines will develop tremendous power against humanity are the obvious basis for his claiming, with respect to the future, that he shares the premises of existentialism[88] – if only with a view to the short run. (How pragmatic!) We must consider the fact, Wiener argues,[89] that as individuals die, the human race will also die.

That same cybernetics that began with such great promise ends, according to Wiener, as humanity's fate – in two senses: "We *must* continue to invent..." (H 57 – italics added), while the end of the line signifies the annihilation of the human race. In Wiener, the marvelous possibilities of technology stand out in sharp, glaring contrast against the looming background of an ineluctable, foreboding future.

3.2.7.5 The source of Wiener's difficulties

We would do well to get the source of the contradictions in Wiener's fatalistic view into sharp focus.

Wiener's worldview is closed. He speaks of God as a data-processing system and thereby renders God an object of his cybernetic reflections. God no longer transcends this world but is contained within it. This implies a rejection of the transcendent. Wiener's thought aspires to autonomy; it seeks to fix its own laws in complete independence.

Meanwhile, we have already noted that this autonomous thought has an imperialistic character. Wiener begins by placing his confidence in humanity, and he arrives, by way of the absolutization of cybernetics, at humanity as a data-processing system. This deprives him of any possibility of offering meaningful perspectives on the relation between man and the machine or on the future of technological development. In principle he surrenders precisely that which is essential in humanity, namely, human freedom and responsibility. In the end this implies – such must be our conclusion, however much we may value his stimulus and contribution to the development of cybernetics – that in Wiener the problem of the meaning of humanity, the meaning of human work, the meaning of the result of human work, the meaning of technology, is conspicuous by its absence.

As I see it, a responsible look at the problems broached by Wiener requires first of all an acknowledgment that man was created by God. A human being is God's image-bearer. Man in his labor owes obedience to God. The fall into sin marred much in that image, but it is restored in Christ. In Christ it is possible for humanity to cooperate in the building and the coming of the Kingdom of God. Humanity's work in technology, too, is meant to contribute to this end. Yet this is not a matter of course. Technology can only be serviceable to that Kingdom if people acknowledge in faith that God calls them to labor in technology and that God requires them to be obedient to the laws – to the suprasubjective normative principles – that ought to both motivate humanity in technology and indicate a proper direction for technological development.

Coercive laws do not obtain for man as they do for the machine. In freedom people can obey or fail to obey the relevant normative principles, and people bear the responsibility for positivizing these normative principles in the form of norms suited to guiding technological development meaningfully. The use of modern machines, for example, ought to conform to the norm that the human being is responsible for the effects involved. Whenever that is not recognized, situations like those sketched by Wiener will arise.

The christian faith offers a meaningful perspective for technological development. It is erroneous to point to something incomprehensible in the rapid and precise calculation of machines and to believe that the machinery will get out of control. It can be said of tools and instruments generally that people can accomplish more *with* them than they can *without* them. Technology provides us with instruments that enable us to see and hear farther, go faster, apply more energy, and also calculate more rapidly than we were ever able to do without them. What is new about the computer is really this, that while the human being establishes the conditions under which it works and thereby unambiguously determines beforehand what the forthcoming result will be, he can nevertheless not foresee that result.[90] Whenever people fear unhappy results, they can simulate beforehand with a computer, and then decide whether to proceed with actualization. It would not be right to bar the computer just because we cannot achieve its results ourselves, or achieve them quickly. The very purpose for using the computer is to make it do what the human being himself cannot do – calculate rapidly and reliably in order to solve complex problems. To this end the computer is equipped with a memory that never fails it.

It is equally unnecessary to speak in negative terms about technological development on the grounds that such development will have dire consequences, which is what Wiener does. Rather, these consequences should be prevented through a normative determination of development.

Through automation, unemployment will indeed increase – or, to be more precise, work time put in on a wage basis will decrease. Wiener holds the development of modern technology responsible for this. I would dispute that. Moreover, I would add that one-sidedness can be a great problem here: a lessening of work time need not always be a danger. My own approach to this question is to maintain in the first place that humanity's calling to labor must always be taken into account. This calling is a calling to meaningful labor. Automation is therefore to be welcomed, for the repetition of routine procedures saps a person's attention and diminishes his freedom in labor, thus depriving human labor of meaning. By the same token, automation need not lead inevitably to unemployment: the calling to labor can be answered in the acceptance of new tasks that have not – or have not yet – been subjected to automation. Moreover, the increase in automation will require more education and study, through which an important condition will be met for guiding newer technological development meaningfully.

Nor should it be overlooked that a lessening of work time need not simply be disparaged as such. Rather, we ought to view it as delivery from certain consequences of sin. When we accept the calling to work and also gratefully welcome deliverance from the curse upon toil, we gain perspective for escape from the troubles that loom ahead. This calling and the gratitude we should manifest constitute the necessary and sufficient conditions. To the extent that they are not present for Wiener – and he makes no mention of them – he will indeed be proved correct: technological development will be a pernicious development. People allow themselves to be reduced to machinery and "know" no responsibility. With the acceptance of a closed worldview, the calling to work that comes from God vanishes, and then there is no more place for thanksgiving toward Him. Personal responsibility is "emptied" of meaning. Then the impending development can indeed be correctly characterized as utterly tragic.

3.3 Karl Steinbuch: Cybernetics and Futurology

3.3.1 Introduction

Steinbuch,[91] like Wiener, has high hopes for cybernetics. He even anticipates that cybernetics will be the universal science of the future.

In contrast to so many other sciences, cybernetics can make good what it says.

Wiener's view is clearly dualistic with respect to the future: the great promise of technological power ends up constituting a menace to mankind. Steinbuch, however, is thoroughly consistent in his optimism about the future. For him, the future will be a fine future if it is a technological future.

Steinbuch[92] continues the line of the Enlightenment. He agrees with Kant that humanity must dare to avail itself of its own understanding. "What has gone down in history as the Enlightenment is but a first, tentative forerunner of what still lies before us: the insight that man himself is a part of nature, an object of exact scientific inquiry" (P2 201).[93] This implies, as he says himself, that Steinbuch no longer needs to appeal to metaphysics to account for human intellectual functions. "In no case does it appear established, or even probable, that the explanation of intellectual functions depends upon the acceptance of presuppositions reaching beyond physics" (AuM 2). According to Steinbuch, people must free themselves of unscientific prejudices that start them on the road of speculation. He has therefore personally rejected metaphysical philosophy and faith, both of which point toward "another world." He wishes to remove all obstacles to technological progress. The philosophy put forward by F. G. Jünger[94] is a good academic example of the kind of thinking he opposes. Jünger's thought strikes him as so much pompous prattle and negativism.[95]

On the basis of natural-scientific explanations, it is possible to take the object of inquiry and simulate it in a machine. Given the principles of cybernetics, the possibilities in this field have been brought to such an advanced stage that it is now possible to duplicate even human intellectual functions in the new machines and, moreover, to improve and enhance them. This signifies to Steinbuch that in the future, man as an intellectual being will no longer be central: his place will be taken over by the computer. Steinbuch's thought is controlled by the method, categories and possibilities of cybernetics as a science. His perspective on technological development and his view of humanity are positivistic: "To the impartial observer, it is a real insight that that same physics which since the beginning of time has ordered all heavenly bodies also controls events in our brain" (AuM 4).

We will take a look at the obstacles which Steinbuch believes must be cleared out of the way in order to achieve a futurology based on the methods and possibilities of natural science and modern technology.

Steinbuch distinguishes the *study* of the future from the *control* of the future. He wants to achieve the latter democratically, by utilizing the potential of information technology. Steinbuch finds his view of technology and the future compatible with modern revolutionary movements. In our critical analysis we shall see whether his conviction in that regard is justified. We will also take up the other subjects mentioned, in order to show clearly what inner contradictions characterize Steinbuch's thought, what deepest motive controls him, and what image of the society of the future his philosophy implies.

3.3.2   What is cybernetics?

According to Steinbuch, cybernetics arises from three important sectors of modern technology: control technology, communications technology, and data-processing technology, also known as computer technology.[96] Information theory, which is the basis of these fields, creates the possibility of extending cybernetics into other, nontechnological areas. Thus cybernetics is on its way to becoming a universal science – "einer neuen Einheit der Wissenschaften."[97] It is true that we are only at the beginning of the development of cybernetics, but there is every reason for optimism. "Cybernetics is a happy illustration of the fact that scientific development does not run forever toward divergence. Indeed, cybernetics is a bridge between the sciences."[98] Cybernetics will increasingly play the role of a "mediator" between area specialists.[99]

According to Steinbuch, the penetration of the so-called human sciences by cybernetics has two important consequences. *In the first place*, the explanations available through cybernetics obviate appeals to metaphysics. The concept of information, which cybernetics adds to the concepts of matter and energy, makes it possible to give clear scientific answers to problems that have heretofore eluded resolution, answers that can subsequently be verified through technology. The concept of information has no metaphysical background, Steinbuch explains, because it has reference only to function or behavior.[100] "The cybernetic approach is typically suspicious of concepts that cannot be realized in constructible models" (AuM 326). "What is decisive, naturally, is that these technological models make it possible to say something about the *behavior* of the depicted system in interesting situations" (AuM 335 – italics added). Therefore the cybernetic approach opposes, for example, any thinking which, "like vitalism, postulates supra-physical law for the biological sciences" (AuM 325).[101] With its concept of entelechy, Steinbuch claims, vitalism resists the natural scientific explanation of life and the mind. The entelechy concept is rendered superfluous by Steinbuch's cybernetic thought. *In the second place*, the penetration of the human sciences by cybernetics has made an exact scientific approach to, and solution of, many problems possible. "Viewed from the standpoint of method, cybernetics can be typified as the penetration of mathematical aids into areas of science where they were hitherto inapplicable, for example, physiology and sociology" (AuM 323).[102]

Cybernetics contributes to the elimination of the boundaries between the various university faculties and to the explanation of the attributes of nontechnological systems with the help of technological models.

According to Steinbuch, the so-called intellectual functions of man can be explained by means of cybernetics and can be simulated in machines. To fully exploit these cybernetic possibilities, however, it is first necessary that human and social problems be addressed impartially and without ideological prejudice: "Undoubtedly our generation only stands at the beginning of the process. It is a matter for optimism, however, that this process has in any case begun" (AuM 378).

Because Steinbuch clearly agrees with Wiener concerning the principles of cybernetics, we need not review those principles here.[103] To the extent that Steinbuch's ideas do differ from Wiener's, they will be presented below. Like Wiener, Steinbuch includes under cybernetics both the science of steering and guidance, and the application of that science in technology. This is clear from his definition of cybernetics.

Steinbuch states: "By cybernetics is understood the *science of information structures* in technological and non-technological fields" (PuK 20).[104] Yet he also states: "By cybernetics is understood on the one hand a collection of certain thought models (controlling, information transference and data processing), and on the other hand the application of the same in technological and non-technological fields" (PuK 20).[105]

3.3.3 The impending technological development

In view of Steinbuch's conceptions of humanity and the future, it is essential to ask just what, according to him, is to be expected of technological development.

Steinbuch pays scant attention to developments in space flight and nuclear energy – not because he believes these sectors of modern technology to be unpromising for the future but because he views them as dependent upon the future possibilities of communications and information technology. Extended space flights, for example, will only be possible given the continuing development of machine intelligence. The development and application of nuclear energy, too, is dependent upon the future achievements of the computer.[106] Steinbuch is mainly interested in the relation between man and the computer. Therefore he concentrates his attention on the possibilities of communications and information technology.

I will summarize what Steinbuch, as a professional, expects from communications and information technology. I will leave out the technological details – Steinbuch himself offers extremely clear explanations for nonexperts. For example, he provides a very fine overview of the various features of a computer, such as "input," the "compiler," the memory, the calculating component, the program, and the "output."[107]

With respect to *communications technology*, Steinbuch expects the following. Telegraphy will advance, so that a great many people will be able to have at their disposal small, inexpensive, rapid, simple, and silent telex machines.[108] In telephonics, an apparatus is being developed in the form of an armband that can easily be carried anywhere. Through this apparatus, instant communication with anyone will be possible, as well as dialogue at any time with a computer to obtain any information desired.[109] Radio and television will increasingly inform the masses of what is going on elsewhere. Television's influence on human conduct will increase especially. Those who perceive in this only a power of "hidden persuaders" or a tendency towards an "Orwellian society," says Steinbuch, overlook the healthy influence television can have. Through this means of communication, people learn about the opinions of others. They relinquish claims to absoluteness and thus relativize their own views. The resulting abolition of ideological strictures will contribute to the improvement of East/West relations and the achievement of world peace.[110]

The future of *information technology* depends on computer development. New switch elements like the transistor require little space. They work rapidly, do not become "fatigued," and consume little energy. Therefore the computer's memory can be enlarged; it can store more information.

The computer spurs technological progress, leading to process management, numerical control of operating machinery, and the automation of various forms of transport. Its use will increase in the fields of economics, politics, administration, organization and command, scientific research, and technological designing for both military objectives and the regulation of traffic.[111] With so-called simulation technology, computerized attempts are made to simulate costly technological systems. "In simulation technology, the objective is to study the characteristics of systems which do not (or do not yet) exist by creating a computer mock-up" (P2 94). Simulation technology may be applied, by extension, in the fields of physiology, psychology, economics, and sociology. In the latter two cases, use is made of so-called society models, in which the consequences of certain decisions can be forecast quickly with the computer. The decisions to be taken can thereby be improved beforehand.[112]

The significance of the computer for education will be enormous,

Steinbuch believes. The so-called teaching automatons – many technological variations are possible – will relieve the teacher of much routine work and facilitate individualized application to each student of the material to be learned. An *advantage* of this is that rapid adjustment to constantly changing scientific and technological patterns is made possible. The great *disadvantage* is that it can lead to uniformity of thought and cause tremendous damage in cases of poor, yet universal, educational programs.[113]

Data banks with worldwide branches will put encyclopedic knowledge at the disposal of anyone seeking it.[114] The computer will also play an important role where a human overview of the great quantities of information that must be processed to achieve certain results is no longer possible. This includes, for example, long-range weather prediction, where a great deal of information has to be processed rapidly so that predictions are not issued after the fact. Other applications are the diagnosis of a disease on the basis of a number of symptoms, the identification of burglars or of the victims of a disaster, and the detection and repulsion of hostile attacks.[115]

Besides optimizing telegraph and telephone traffic and calculating the optimal orbits for space vehicles, the computer will increasingly support and take over various human functions. At the moment, machines cannot yet translate elegantly, because their store of words is still too limited. When that problem has been solved, however, a poor translation will not be caused by mechanical limitations but by many human departures from "normal" language, says Steinbuch. The same problem arises with respect to automatic symbol recognition. "Difficulties are formed above all by the enormous individual differences . ." (AuM 180).[116]

Computer development follows two lines – one toward the universal machine useful for many purposes, and the other toward the specific machine capable of performing only certain tasks. Both lines of development will probably contribute to the computer's being made available to large numbers of people in the future. A link between an inexpensive, personal computer and a costly, universal one affords a happy combination of the advantages of both types.

In short, the computer will liberate people in the future from routine tasks and heavy labor. Moreover, the computer will render services to people which they themselves, because of their inherent limitations, would never be able to perform. And the computer probably offers many additional possibilities too, possibilities of which people are not yet aware. "The present generation of computers certainly already makes possible many achievements which are not being realized simply because human fantasy has not come up with the necessary designs" (DiG 259). This state of affairs is new in technological development. Usually, new technological ideas precede realization. With computer development, the case is different; computers are capable of more than they are presently being used to achieve. In other words, "software technology" (programming and utilization) is running behind "hardware technology" (construction).[117] "In view of the almost unlimited functional possibilities of the computer, the future may very well be characterized by a great gap between the possibilities available for technological realizations on the one hand, and actual discoveries on the other, for the conceptual models prerequisite to full exploitation of these possibilities may be lacking" (PuK 150).

3.3.4 Learning machines

Computer applications are currently being improved by having machines "learn," which, according to Steinbuch, means that they are so programmed as to be able to adjust to altered external conditions and thereby improve their behavior.

Computers surpass human beings when it comes to rapidity, reliability, and endurance. With respect to learning, that is not yet the case. This is so because the human brain is more complex and because it contains "switches" which in both quantity and quality – they can repair damage to themselves – are superior to electronic ones. Humanity's advantage, however, is only temporary, according to Steinbuch. Machines will eventually be able to surpass people in learning. "Then learning automatons will have an influence on our intellectual and social lives which is at present beyond our imagination" (AuM 195).

Machine learning is presently being taken much too lightly; it is treated like some sort of game. It will be necessary to investigate human learning scientifically and to make use of the results in constructing learning machines, and for that purpose it will be necessary to ascertain the conditions under which learning is possible.

Human thought, according to Steinbuch, is characterized by model structure. "A system has model structure when a measure to be made operative in the external world can be tested against an internal model of the external world to ascertain the possible reactions of the external world to the intended measures, so that the only measure that is finally made externally operative is the one that will elicit the desired response from the external world" (AuM 195). In model structure, alterations can be introduced with a view to improving conduct. Steinbuch calls this "learning." "The learning of a system consists in this, that in keeping with earlier results or mistakes (experience), the internal model of the external is improved" (AuM 195). This form of learning is externally discernible since conduct, compared with the past in a specific sense to be further described, is improved.[118]

Computers, too, have model structure. If the program is not rigid and if alterations can be introduced, it is possible to speak of learning. It is obvious that various levels of machine learning must be distinguished. In the simplest case (for example, the guidance of a chemical process), the machine itself can take care of setting the optimal situation. "Of fundamental significance is the automaton's setting up an optimal situation, about which its fabricator knew nothing in advance and which in principle he cannot even calculate, since the physical-chemical laws of the process and its output are in principle unknown to him" (AuM 197).

The next step in learning is for the machine to be able to readjust itself automatically to a new optimal situation, given changes in the situational data. A further improvement is the machine's first simulating an optimal process on an internal model in order to ascertain that the optimal process will not be fraught with catastrophic consequences for the external world. In such instances the machine learns to keep inauspicious situations from arising, most of which are beyond the scope of human determination. All this learning is possible within, and on the basis of, the one model structure given to the machine by its fabricator. There can only be talk of real learning if the internal model can alter itself through "trial and error." It is true that the machine remains bound to mathematical-physical laws, but it can still accumulate and utilize experiences, that is to say, improve itself, without the fabricator having anticipated as much.

Steinbuch's conclusion is that "... objectively, learning processes do not represent a new category alongside normal processing processes. Their unique position, as it is subjectively experienced, stems from their similarity to human processes and from their description in terminology that has arisen in relation to the human being" (AuM 199-200).[119]

In the future, learning automatons will be most at home where mathematical, physical, social, and economic laws or parameters are not (fully) known or, if known, then very difficult to program. In those instances, the machine itself must have the capacity to improve its performance through communication with the external world.

Compared to organic systems, the learning of inorganic systems is more difficult because with the latter there is a separation of the functions of the remembering and calculating components. Steinbuch, however, has developed a so-called learning matrix – I will not attempt to describe its technical structure here – [120] by means of which it is possible, via linkage to similar learning matrices, to simulate organic systems almost perfectly. The impression arises that the differences

220

between the remembering and calculating components have disappeared. Such linkages have the characteristics of organic systems. They can correct themselves and locate short circuits. The machine can thereby learn better, and at the same time individualize learning.[121] What this means, with respect to automatic symbol recognition, for example, is that the symbols to be noted need not be known perfectly beforehand. The computer can seek out the most important characteristics of a symbol. "The technological problem consists in bringing all the various possible forms of the same symbol down to the same meaning, in other words, in realizing a great breadth of variety" (AuM 403). Once this is possible, machines will be able to read handwriting. It will then also be possible to speak of "understanding." "Learning matrices offer the possibility of adjusting to individual characteristics, for example, during the salutation 'Ladies and gentlemen'" (AuM 176).[122] Accordingly, it is also possible, Steinbuch claims, to deliver translations or short summaries of whatever is read or spoken. Consequently, Steinbuch believes that the computer, equipped with linkage to learning matrices, should easily be able to pick up new words. "Through linked learning matrices, words heretofore unknown to it should be capable of being added to the vocabulary of the automaton" (AuM 237).

In short, "the future development of automatons, like the logical analysis of intellectual processes, will...lead to the result that reading and understanding will be objectively describable processes subject to performance both by people and by automatons" (AuM 345). In addition to traditional information theory, the theory required for these machines is semantic information theory. "Future, non-traditional information theory will have to take into account the fact that the sender and receiver of information do not exist in a vacuum but have connections to the external world, and that both sender and receiver follow characteristic value systems" (FP 104).[123]

It would seem that Steinbuch wavers between two conflicting views when he writes about the development of learning automatons. On the one hand, the machines are to develop themselves independently into machines surpassing human beings, machines that increase in autonomy. The exploitation of the possibilities of learning matrices on the basis of semantic information theory points in that direction. On the other hand, people remain necessary in order to establish objectively the value or significance of particular information and to pass that on to the machine. "For logical reasons, every adaptive system must be furnished with a scale of values; otherwise the adaptation is undefined" (AuM 242, 243). Again, this last observation does not stop Steinbuch from going on to ask whether the automatons will continue to be restricted to the human intellectual niveau as the number of linkage elements increases and as these machines come to dispose over mechanical sensors for observing the external world. Steinbuch suggests that they will develop after the analogy of the evolution of human organisms. To shorten the development period, human intelligence must be imparted as a prior condition to the self-developing automatons. Starting from that level, the automatons, in communication with the external world, will be able to improve their intelligence and surpass human beings.[124] However, "As long as the automatons are forced to maintain human prejudices, their intelligence will remain limited" (AuM 210). In other words, whenever the machines are *not* restricted by human rules of conduct, they will surpass people. This is all the more so since they are unaffected by human biotic limitations, that is, by fatigue, aging, and death. Steinbuch therefore agrees with Wiener when the latter says: "As machines learn, they may develop unforeseen strategies at rates that baffle their programmers" (AuM 191).

The effects of the development of learning machines will thus be of great significance for humanity. "The insight that the human brain is not the only possible locus of intellectual processes will come as a shock to the human self-understanding, a shock no less disturbing than the insight that the earth, humanity's abode, is not the center of the universe. The future development of information processing will discredit all the philosophies that assume the uniqueness of the processes of the human brain" (AuM 191).

The changed relation between man and the machine, given machine learning, will engage our attention in the next section. We will see that Steinbuch's view of the development of learning machines has far-reaching consequences. It is important to discern whether his view contains speculative elements that put him on the wrong track when he discusses the future of mankind.

3.3.5 Man and the machine

When Steinbuch discusses man and the computer, he is well aware that at present any factual comparison turns out to man's advantage. In the future, however, that will be different. "Information-processing technology is at present in its initial, pioneering phase – about the same place mechanics was when a wheel first turned" (AuM 320). Nevertheless, as mechanical technology developed in marvelous fashion from the wheel, so data-processing machines will develop marvelously in the future. "Until that time, however, the embryonic technological systems of our era allow us only to make inferences about the systems that will be possible in the future" (AuM 11).

With the help of cybernetics, according to Steinbuch, a rational analysis of human intellectual capabilities is possible. *"What we observe in intellectual functions is the acquisition, processing, storage, and delivery of information"* (AuM 2).[125] And with this, the fundamental identity of human beings and information-processing machines is granted *in principle*.[126] He discounts the difference between the organic matter of the nervous systems and the materials of which machines are built. That difference in material is presently the reason why people surpass the machine, but in the future that will be otherwise.

Neurons are very tiny. They intermesh, and they are very numerous (15.10^9). Therefore they are difficult to analyze.[127] "Apparently purely quantitative factors render insight into human thought processes impossible" (AuM 17). Still, we can approach the human brain by seeking complexes of functions. And although neurons cannot be duplicated with elementary switches, these complex functions can be reproduced with switch blocks. Such switch blocks include, among other things, amplifiers, the calculating unit, the program instruction, the logical linkage, the memory, the learning matrix, and so forth. "The application of such switch blocks could make many discussions about intellectual functions more pregnant [*sic*]" (AuM 20).

Steinbuch's notion that there is a fundamental identity in principle between man and the machine arises from his conviction "that organic systems are subject to precisely the same laws as technological systems" (AuM 22).

These laws, according to Steinbuch, are *physical* laws. Nevertheless, he runs into a problem: Is perfect physical knowledge of the brain possible? Is this knowledge not always limited by the so-called uncertainty relation? "This effect, however, must not be allowed to alter our fundamental stance" (AuM 23). The uncertainty relation involves microphysics, and not macrophysics. Within total functions, it is true, uncertainties of detail are operative, but they have no effect on knowledge of the whole. Moreover, Steinbuch acknowledges that learning processes are not determined. This means, according to him, that these processes are not fixed. It is therefore possible to duplicate them in machines using random generators. Their operation can be explained theoretically with the help of probability theory.[128]

The basis of Steinbuch's view that the human being and the machine have fundamentally the same *cybernetic structure* and *behave* identically is to be found in the fact that one does not encounter an inexplicable "remainder" in explaining human intellectual functions. "One or another mysterious extra function is not necessary."[129]

Steinbuch disagrees with W. Wieser, who stated that an organic whole is more than the sum of its parts: " . . . a highly complex system such as an organism is not to be rationalized, then, and thus it also is not to be represented perfectly in a technological model" (AuM 19).[130] But in that case, argues Steinbuch, no theoretical model of a calculating machine is possible either. Nevertheless, at this point he overlooks Wieser's reference to a *given* organism, and not to a machine already brought into being on a scientific basis with the help of the technological-scientific method.[131]

The significance of Steinbuch's fundamental identification of man with the machine is that he believes it possible to bridge the cleft between subject and object, between the human sciences and the natural sciences. Subjective experiences like love, anxiety, desire, indeed, all types of consciousness, are susceptible of objective physical description and hence of duplication in machines. The problem of the coherence of mind and matter therefore makes no sense to him. Psychic concepts are justifiable on the grounds of man's complex structure. Ultimately, however, psychic concepts can be reduced to physical ones. "It is commonly accepted that the life process and the psychic processes can in principle be fully accounted for in terms of the arrangement and physical interaction of the organism" (AuM 9).[132] Since objectively describable traits obtain for both the human being and the machine, it is justifiable to refer to a function in a machine by the same name that we use for it in people – for example, thought, intelligence, learning, consciousness, and so forth.[133]

Steinbuch appreciates the fact that his view will not be received with equal enthusiasm by all. He attempts to answer the various arguments which he thinks may be raised against his position:

1. People say that machines can only do what they are programmed to do. However, Steinbuch claims that people are programmed too. Through birth (inherited characteristics), through learning (instruction), and through communication with the external world (experience), people acquire information. For the machine it is the same way – through the construction specifications, through the program instructions, and through the constant input of information. From these parallels the similarity between man and the machine is apparent: for both it is a matter of acquiring, processing, storing, and transmitting information.[134]

2. But is there not a difference inasmuch as the machine must receive a command? "Who participates in commanding the human learning system?" (AuM 267). Is a supraphysical source at work there? No, the information that people have derives from various sources. One special kind of information concerns the motive or mainspring behind human conduct. This *motive* is survival of the species, which is analogous to, and works in the same way as, the *command* of the

machine. "The view that survival of the species must be the primary motive may not be construed as a teleological notion, but is to my mind the only possible functional factuality that can explain the observable continuity of biological existence down through the generations" (AuM 268). Biological science teaches that this motive came into existence through coincidental organization, and that it guides and controls the natural process of growth.

3. In comparison with other organisms, it may be said of people that "the superiority of human beings is based on their capacity to foresee future situations in the external world and to influence them to their own advantage" (AuM 270).

People do indeed have "foresight," but because this trait is gained by learning, data-processing machines, too, can acquire it.

4. Steinbuch believes that there is no need to resort to supraphysical presuppositions in order to account for human intellectual functions. However, the other side of this question is: "Can consciousness be built into a cybernetic thought system?" (AuM 272). Yes, Steinbuch responds. Not all the physical relations of every subjective experience are known as yet. There is nevertheless no doubt that this will be made possible by continuing research. "A necessary consequence of the foregoing conjecture is the supposition that artificially fabricated technological systems can have consciousness" (AuM 273). A machine not inferior in structure and complexity to the human nervous system would declare for itself that it had consciousness. That we have not yet reached this stage is due to the present insufficiency of scientific research. Evidence to the contrary is lacking. "Where an idea cannot be convincingly refuted, it is admissible" (AuM viii).[135]

5. It is known that in man, the transmission of information is backcoupled to the acquisition of information. In that circular process, consciousness plays the decisive role. While this process can be duplicated in the machine, it remains a question whether the machine can possess *reflection* too. Human reflection is a circular process that neither begins nor ends in the external world. "In the case of reflection, information is not dependent upon the external world; it neither derives from it nor penetrates it" (AuM 279). This is also the reason why the human being posits the presence of a metaphysical substance behind reflection. Yet he is mistaken in this regard, Steinbuch claims, for reflection is nothing other than consciousness that has gotten detached from experience, from the external world, and thereby has become abstract and world-estranged. "Thought becomes world-alienated. It festers. At worst it becomes an end in itself, of no use to mankind" (AuM 281). Thus reflection, owing to its abstract character, is to be rejected as a corruption. In order to avoid the difficulties clustered around the notion of "reflection," Steinbuch proposes not to

talk about it any longer but to use the term *intelligence* instead, whenever the intention is still good. "Intelligence is the historical-developmental discovery of the human species, aimed at anticipating selection..." (AuM 281).

6. But can the machine, like the human being, do creative work? "Is this reaching beyond the boundaries of the already known to be accounted for by rational laws?" (AuM 283). As we saw, human thought is characterized, according to Steinbuch, by a model of the external world.[136] Creativity is "testing arbitrary opinions for their conformity to the established facts of areas already known" (AuM 284). Model structure is thereby improved. A machine is also capable of that. Just as a person who is fully enclosed within the information structure can be creative, cybernetic machines can be creative.

7. People speak of the human "spirit" when discussing the enjoyment of aesthetic information. But this "spirit" is not some metaphysical substance, says Steinbuch; there is a physical explanation for it. A machine with a stochastic generator can produce random signals, which accounts for the satisfaction of aesthetic experience, for " ... given a certain mixture of coincidence and law, we experience aesthetic gratification" (AuM 286). Steinbuch is accordingly convinced that machine art and machine art criticism are both possible, since an exact scientific theory of the production and observation of art works is possible.

8. From the above it follows naturally that Steinbuch believes that the newest machines will (eventually) make poetry: " ... the automaton will be able to select from among the tremendous number of possible coincidental letter combinations those that are formally correct, meaningful and attractive only after it has learned the language (and not just the dictionary!)" (AuM 314).[137] Poetic creativity by machine is possible through the combination of random letter selections, but it can also take place more systematically, for reasons of economy, with the help of a statistical analysis of the construction of the language.

If machines are actually to produce poetry, then it will be necessary to develop a so-called semantic information theory in which not the *form* but the *value* of information is analyzed and then quantified. Semantic information theory also offers the possibility of perfect machine translations, in which various significations of the same word might occur. "Words with several meanings can already be interpreted in many instances from the context of surrounding words. Here perfection is more a question of time and money than of possibilities in principle" (AuM 321).[138]

9. Whenever every subjective experience of an organism is describable in terms of objective physical laws, the question about the place of

personal freedom looms large. According to Steinbuch, the problem of human freedom is only a problem in appearance. "Nothing ... gives us occasion to believe that people can escape the laws of matter, energy and information" (AuM 406). That we still speak of human freedom is to be attributed to the fact that people are "programmed" in such different ways. "What freedom denotes in the language of ordinary intercourse is the possibility of carrying out activities answering to our inner structure without being hindered in doing so by coercive measures" (AuM 292).

Steinbuch, it would appear, takes a scientific view of humanity – and a natural-scientific view at that. The human being can be fully accounted for by physics, which includes information theory, and can be duplicated in machines. That we have made little progress along this road is to be attributed to our allowing technological development to be too strongly dominated by wrongheaded ideologies, preconceptions, and dogmas. To achieve progressive technological development, people must *free* themselves of all that.[139] Their possibilities of communication with the external world will be enhanced if they do so. "Freedom is thus a stimulus to further human development" (AuM 285). If machines have the capacity for free communication, there is no basis for believing that they will remain limited to the human niveau. This assures a beautiful future, but before that future can be made reality, many obstacles remain to be overcome. Society will first have to be converted from its "false programming."

3.3.6 Falsely programmed

Quite a number of objections have been raised against Steinbuch's view of technological development – especially his handling of the relation between man and the machine. Steinbuch claims that this is attributable to the fact that our society is "falsely programmed" by *philosophy* and *faith*. These two have never given technological development a fair chance.

Some believe that the impediments to technological development are to be sought in education, in the organization of our society – Steinbuch means German society primarily – and in poor financing of scientific research and technological development. These do indeed leave a great deal to be desired, but the societal malaise ultimately involves an intellectual problem.[140]

According to Steinbuch, our failure to tune in to the future and make full use of the possibilities of science and technology stems from an irrational, antitechnological, and antiscientific ideology which is "otherworldly" in its direction.[141] German philosophy inflicts tremendous damage on technological development: " . . . it seduces our people into wasting their power, intelligence and hope on the other side of this reality instead of using it here on this side to live successfully, peacefully and humanely" (FP 20). Nor is the revulsion for practical matters – for natural science and technology – regarded as an intellectual failing; on the contrary, the intellectual elite – the philosophers – believe that they would disqualify themselves if they were to orient themselves toward natural science and technology. According to Steinbuch, the world reaped the bitter fruits of this attitude in the rise of national socialism. Little or no resistance was offered, since people were thinking in an "otherworldly" way. Without an improvement in that way of thinking, changes in education, organization, and finance are of negligible value, unless the changes are aimed at eliminating falsely programmed thought from education. Only then can the perpetuation of errors from the past be prevented.[142]

To the extent that philosophy deals with technology, its critique is negative – and apparently ignorant and uninformed. This is evident, for example, in the arrogant pronouncements F. G. Jünger makes about technology in *Die Perfektion der Technik*.[143]

Another example is provided by K. Strunz when he criticizes the cybernetic way of regarding humanity: "Pure scientific thought is not competent here, for the simple reason that as long as it does not transgress the limits it has set for itself, problems of meaning and value are excluded. The intellectual and the moral are beyond the scope of its competence" (FP 27).[144] Steinbuch argues to the contrary that no more reliable science than cybernetics exists for making judgments about matters of principle. Information theory is not just classic information theory, which is concerned with *form*, but equally semantic information theory: " . . . the new information theory [investigates] the question of the value of information for a receiving system" (FP 27).[145]

Steinbuch's evaluation of (German) philosophy is: "The other world is focused entirely on the past; it opens no perspective for the future. . ." (FP 24). As for his own view, he often calls upon a certain philosophy that is not guilty of a *Hinterwelt*, of a "background world," namely, the philosophy of the Vienna Circle of logical positivism, of which Carnap and Reichenbach are important representatives for Steinbuch. This philosophy has moved beyond the speculative stage and finds itself in the logical phase.[146]

Steinbuch's appeal to this philosophy serves only to underline his view of cybernetics as *the* philosophy. He shares its repudiation of metaphysics, but he is not a loyal disciple of logical positivism. That becomes clear when he speaks about the relation between natural

228

science and faith. He does not reject faith as such; what he rejects is its "otherworldly" direction.

3.3.7 Science and faith

First of all, it is necessary to note an important change in Steinbuch's thinking. At the conclusion of an earlier subsection (3.3.5), I noted that according to Steinbuch, the machine will surpass man. He believes that the human being can be fully accounted for in a rational manner, that he can be perfectly described with cybernetic principles, and that no intellectual function can escape such explanation. This means that the machine could take over *any* human intellectual function. In short, Steinbuch needs no metaphysics in order to explain humanity.

In his book *Falsch Programmiert*, it appears that for Steinbuch man still has an a-rational core. Science cannot get a handle on its own *presupposition* (namely, the survival of the species) and its own *faith* (its directedness toward a world beyond or its acknowledgment of the possibilities of science and technology – in short, its directedness toward the future). This means that in the future, rational thought is reserved for the thought process subject to duplication in a computer. The computer will be able to surpass man only in rational thought. "The human is a being of weak intelligence, ruled by his desires" (FP 45). Yet, the end to which the computer is to be utilized will have to be determined by man. This is a-rational. "It is clear that the computer obviously offers us the best means of reaching our goals, but it does *not* take over valuation, nor does rational thinking take over this task" (FP 49). Here Steinbuch implicitly vindicates Strunz, although the latter is perhaps not sufficiently aware of the significance of semantic information theory.

Science and faith, says Steinbuch, have two things in common: the possession of information and the possession of an a-rational core.[147] With respect to the first point he declares: "The difference between faith and science consists in the nature of the information transmission, but especially in the criterion of truth" (FP 40).

Natural-scientific knowledge begins with observation, which people filter in an economical manner to obtain knowledge. This is to say that scientific knowledge is not free of biases that are beyond the capacity of science itself to establish and are thus a-rational.[148] Yet, there is considerable agreement in the natural sciences, since "the final criterion of truth in the natural sciences is the verification of the result predicted by the prognosis" (FP 42). This criterion of truth is not, however, value free but serves as the optimal strategy for preserving the human species.[149]

The a-rational core of science is not value free but is founded on an explicit or implicit value judgment.[150] The a-rational core of faith, which cannot be scientifically established, is not the same as that of science and is accepted without further foundation.[151] While Steinbuch states on the one hand that this a-rational core of faith is abstract, constant, and timeless, on the other hand it has a consoling function that science lacks.[152] The latter function is one which Steinbuch would not like to see missing in the growing power of science and technology.

The content of faith is derived concretely from the world we live in. Therefore this content must change now that the world is undergoing drastic change. "It is not God who is dead, but the God of the 'other world' who is dead" (FP 50). Faith cannot and may not be connected to otherworldly goings on, but must be focused on this world. At first God is phased out with the advancing development of science, and His role is reduced to that of a god of the gaps. "God as stopgap in natural events, God as saboteur of precise measurements, God as leader of the otherworldly masquerade – no, that God is really dead. If Christians should have no other God of whom to testify, then little would be lost in His death" (FP 51). Christian faith in this age must acquire a different content and must agree with the flexible attitude of physicists and engineers, who focus their attention on this world. The Christian must no longer say: "For what shall it profit a man, if he shall gain the whole world, and lose his own soul?" (Mark 8:36; FP 56). In short, if Christianity is to have a future, it must surrender its connections with the "other world" and not allow its content to be determined by it; it must focus rather on *this* world and attempt to provide *credible* answers to the questions of the future. Such a change for the better is discerned by Steinbuch in Dietrich Bonhoeffer and Karl Rahner.[153]

Faith as an abstract, timeless, a-rational core is filled, content-wise, by the world people live in. It becomes concrete. Thereafter, their patterns of thought and behavior are determined by that content of faith. "Faith always rests on an a-rational core, the truth of which is accepted without further verification. This a-rational core determines the value system and thereby the model of thought and behavior" (FP 54). Agreement between faith and science assures a meaningful future. Given a breach between faith and science, however, people are reduced to confusion in the face of perfected technology, scientific development, atomic technology, space flight, automation, the possibilities of artificial organs – in short, cybernetics.[154]

Given its consoling function, Steinbuch does not want to get rid of faith. Therefore he would be saddened by the loss of Christianity. And that loss is not necessary – at least, not if christian faith is willing to focus on the future of science and technology. A breach with the "other world" is called for in order to adapt the human system of values and moral behavior to technological development. In other words, scientific thought, led by the motive of the survival of the species, must be accepted by faith as all-controlling. Thereafter, faith can go on to stimulate science and technology. Should faith fail to fulfil this function, it would be a potential disrupter of human culture.[155] In other words, the content of faith should be in harmony with the content of science and technology. That will bring about the progress of science and technology.

3.3.8 Steinbuch's futurology: researching the future

In the place of "otherworldly thought," Steinbuch proposes future-oriented thought. The perfect lies of metaphysics must give way to the (not yet) perfect truth of science and technology.[156] Advancing technological development has distracted attention from the other world. The breach this has brought about between ideology and reality must be healed by thought focused on this world.[157] "The attitude brought to the center of thought by the rational sciences and technology affords a much better chance of survival than does the peevish irritation that Western intellectuals often exhibit towards technology" (AuM 3).

Technology and its development are the source of inspiration for Steinbuch's futurology: " ... it is precisely technology that liberates man from material need and furnishes him with food, raw materials, energy, hygiene, and communication, in short, a life free of want, thereby making life with human dignity possible" (DiG 32).

Technology as such works towards the abolition of all obstacles to a future of human dignity. We have already seen that Steinbuch believes that communications technology is accompanied by the surrender of people's claims to absoluteness and a relativizing of their convictions. The integrating tendency of technology is likewise the cause of the disappearance of the differences between cultures that constitute obstacles to technological progress. Distrust between the earth's peoples diminishes, and the chance for world peace, which is so vital in the nuclear age, is enhanced.[158]

On the one hand, technology is the most important cause of cultural changes and of peril for the future; on the other hand, it offers the opportunity to investigate and control the future. In the computer, cybernetics has designed a mighty instrument for conquering the future.[159]

Yet, technological progress is not accompanied by cultural progress as a simple matter of course. To achieve the latter, it is necessary to strive for the integration of technology and society. While not wishing to be designated a marxist, Steinbuch calls on Marx for the method of his futurology: "The philosophers have only given the world *different interpretations*; what is at stake is to change the world" (DiG 22).[160] Changing the world is possible – through technology. Thus a technological control model should exist for the control of the future. Natural scientists and engineers have the most important role to play here since they are the most progressive thinkers, because of their professions. Consider Galileo: "The work of natural scientists and engineers is incomprehensible apart from reflection on potential future improvements of reality. The basic attitude, really, of the natural scientist and the engineer is one of hope of discovering or inventing something that will assure a better future" (P2 198, 199). In futurology, the methods of engineers must therefore be made general. Steinbuch seeks "a natural and technological-scientific inquiry into the future" (P2 176).[161]

By "futurological research" (*Zukunftsforschung*), Steinbuch means a matter-of-fact, rational, ideology-free approach to future developments. "Futurological research is based on a careful analysis of development tendencies and on the prospect of returning probable verdicts about their future course" (DiG 297).[162] The methods of this type of research are extrapolation, inquiry, analogy, and the application of probability theory. The computer has an important role to play in all this. It serves to process enough information quickly and reliably enough to achieve results. Using communications technology, it is possible to connect information sources and users throughout the whole world, equipping them to receive, process, and return useful information. With wider application of the backcoupling principle in the methods mentioned, it becomes possible to gain more reliable results and to reduce the element of the unexpected. "Futurological research is meaningful when, with its help, the incidence of unexpected results is reduced and the incidence of anticipated results is increased" (DiG 310).[163]

Because technological development will have the greatest influence on the future, futurological research should start by examining such development. A great deal more can be learned about it. One should examine especially the already existing developments and the producers of new ideas. "The most important innovations of our time – nuclear, space, and computer technology – can already, with knowledge of scientific-technological development, be predicted years before their realization" (FP 128). To the extent that knowledge about particular developments is desired and the use of the computer is required, characteristic phenomena need to be quantified. However, certain limits to futurological research come to light. The conditions affecting the origin and advancement of certain developments are mostly unknown. Moreover, the information conveyance and storage capacity of the computer itself imposes limits on the method of inquiry, so that a totally reliable prediction is unattainable. These limits can be pushed back somewhat through technological improvements and a more effective application of probability theory: "...it is...certain that the share of questions that can be answered meaningfully will be greater in the future than it is today" (DiG 315).

Steinbuch wants futurological research to be kept neutral and objective; he thinks that such research should not be ideologically determined. Yet, this neutrality does not mean that futurological research must have no relation to *existing* value systems and their responses to future developments. Futurological research, according to Steinbuch, should predict even the psychic and social behavior of people. That is made possible by computer simulations of psychic and social behaviors. "I remind you once again of the possibility of so simulating human behavior on electronic calculating automatons that it is possible to make predictions about behavior in future situations" (AuM 359).[164]

The result of futurological research breaks down into projections, prognoses, and plans. The projections represent various future possibilities. The prognoses suggest the most probable of them. The various alternative plans supply us with instruments for realizing a particular projection.[165] These plans form the basis for controlling the future.

3.3.9   Futurology and democracy: controlling the future

By extension, researching the future leads to controlling the future. The various plans which are the final result of futurological investigations form the basis for making decisions to get the future under control.

Until now, futurological control has been exercised for the most part under the direction of economic and political power holders. They are the ones who determine which future is to become reality. The people get coercive schemes imposed upon them from above. Because the representatives of the people no longer have access to information sources and information systems, democratic control has become illusory. Society becomes one-dimensional: it is the image of the idea of the "technological state," Steinbuch comments. Society is not determined by scientific people and engineers but by economic and political power holders who reduce democracy to a chimera: "With Herbert Marcuse's 'one-dimensional man,' a great present and future danger is strikingly presented. However, this danger is not based on technological-immanent laws but on technological-estranged power structures" (FP 70).[166] Thus, if futurological research should pass into the hands of natural scientists and technicians, whose desire would be to perform a service and to carry on research free of any ideology, the technological state could be replaced by a cybernetic state, which would make direct democracy possible. Modern technology provides the opportunity for this. Technological development has often resulted in a certain "tyranny of events" (*Sachzwänge*). Information and communications technologies provide practical possibilities, in the form of data banks, for breaking through this tyranny and letting *the people* choose between a number of alternative plans.[167] "Democracy still has its chance; information technology is a powerful weapon against the tyranny of events, which ostensibly outgrows people's capacities to cope with it" (FP 152).[168]

According to Steinbuch, then, democracy gets new opportunities with the development of modern technology. The democracy of the future will be *direct* – not *representative* – democracy. It will have the task of seeing to it that all remaining necessary labor, leisure time, and other fruits of modern technology are distributed equitably. Democracy will remain feasible even if technological development should demand and make possible a *world state*. Anyone who fears the rise of an undemocratic world state underestimates the possibilities of cybernetics.[169]

In summary: "Over against the model of the technological state we posit the model of the cybernetic state, in which, it is true, the functions are rationalized to the greatest perfection but serve no other end than the realization of conscious human goals. By the same token, we do not allow ourselves to be shunted onto the wrong track by the tyranny of events. Humanity is the steersman, the *Kybernétés*, of all these political events, setting the standards and establishing the goals" (FP 152).

Direct democratic control has the advantage of furthering futurological research. With the help of the laws of large numbers, it is possible to reduce the element of surprise in the future; then, from the simulation of human psychic and social reactions to the latest developments, conclusions can be reached. Those conclusions can be incorporated into planning for the future.

According to Steinbuch, the informed society is the democratic society. Later I will return to the question whether this is perhaps an illusion. Steinbuch seems to forget, since he maintains that futurological inquiry can be neutral, that the plans available for the people to choose between must derive from an elite of futurological researchers, the specialists in the field of cybernetics.[170] They are the ones who draw up the various plans for the future. Do they not thereby have the power to guide the future in a particular direction – the one they desire? And is the people's freedom of choice not bound to their plans, and therefore extremely limited? If these questions are answered in the affirmative –

I will return to them in the critical analysis (see subsection 3.3.11.5) –
it follows that technological development and society's future are
considerably less dependent on the people than Steinbuch has allowed.

3.3.10 Futurology and revolution

Steinbuch claims that we have arrived at an important turning point
in history. We are moving from the irrational, prescientific period into
a rational, scientific one. The ideology of science and technology should
replace the ideology that is aimed at a background world.[171] As we saw,
Steinbuch is in favor of a world government led by a natural-scientific
and technological elite. At the moment, the "intellectual worker" is
still being repressed, just as the workers were in the last century, but
when he is liberated, the way will be open to a technological future.
Steinbuch does not have a rigid future in mind when he says this, but a
flexible one. He would prefer to speak of a cybernetic future.

Steinbuch applauds student revolutionary actions because they
resist undemocratic and authoritarian political and economic rulers.
As I read Steinbuch, however, his agreement with the revolutionary
student movements extends only to the *rejection* of existing power
structures and of "over-the-shoulder" vision. "In place of the prevailing
backward-looking thought, authoritarian thought, yes, in many cases
the renunciation of thought, there must come progressive and original
thinking" (P2 29).

Steinbuch argues that student resistance should not be satisfied
with negation, but that it should further technological development.
"He whose heart beats on the left side need not have a hollow head"
(P2 31). He disdains the revolutionary movement's manifestation of a
certain abhorrence of natural science and technology. He wants to
promote technology instead. "Any design devoid of rational connection
to science and technology is irrelevant to decisive future states of
affairs" (P2 200). The cultural revolutionaries, according to Steinbuch,
have insufficiently understood until now that the natural-scientific
and technological revolution is not their enemy but their ally. "They
will achieve success when – and only when – they bring their goals into
conformity with the demands of natural-scientific technological progress" (P2 201). This implies, among other things, that they will have
to recognize "that man himself is a piece of nature, an object of exact
scientific inquiry" (P2 201).

To explain the intellectual functions, Steinbuch repeats here –
apparently having forgotten what he said about the a-rational core of
faith – that cybernetics has shown that no appeal to metaphysics is
necessary. Therefore, "the human understanding must surrender its

hitherto often fairy-tale-like principles and strive for a more credible view of man" (P2 201). I might add that human freedom as pursued by the revolutionaries is a pseudoproblem for Steinbuch. Thus he would also establish ethics scientifically. "The preconception that moral problems are far too exalted and subtle to be subjected to such rational, yes, mechanical methods has led to the development, under the cover of an extremely lofty ethic, of an extremely immoral praxis" (P2 202).

Steinbuch supports the students who protest against present forms of education and against an antiquated university. Nevertheless, he proposes remedies that differ from theirs. Education must be reformed so that people get a feeling for a more flexible kind of planning. Moreover: "It is necessary that more logic, mathematics, physics, chemistry, and biology be taught at a more advanced level" (FP 160).[172] The university must look to the future and concentrate especially on the natural sciences and technology, so that students, "unhindered and engaged, can produce new scientific, technological, social, and political designs" (P2 37). The university must no longer be dependent on practice; instead, practice should be dependent on the university. The production of ideas at the university should benefit practice.

In short, Steinbuch welcomes the revolutionary struggle of our age. However, he does not wish to be simply negative. He seeks not to demolish social structures but to remodel them along the lines of scientific insights and technological possibilities. The new social structure will have to be a hierarchical one, Steinbuch thinks, with natural-scientific and technological people at the top. They are the ones who will have to see to it that everyone possesses something and shares in a fair and reasonable way in the fruits of technological development. But every individual must have this freedom: "... we must plan freedom" (FP 174).

Given this view of society, which is based on the ideology of science and technology, Steinbuch repudiates Christianity. Christianity, he says, seeks to mortify the flesh most unnaturally and is turned toward the past. He repudiates marxism because it rejects private property, and neomarxism – although he often appeals to Marcuse and Habermas in support of his own position – because it advocates an unrealizable power-free society.[173]

Although Steinbuch often appeals to the revolutionaries to clarify his own views, his differences with them remain substantial. The question to be raised here is whether Steinbuch's striving for a world government in which technicians and engineers are kings would not bring about a technocracy to which the revolutionaries would be even more vigorously opposed than they already are to a society ruled by political and economic power-holders. In such a society, would human freedom not be totally encapsulated, despite Steinbuch's notions about direct democracy? I will have something to say about this in the critical analysis (see subsection 3.3.11.6).

3.3.11 Critical analysis

3.3.11.1 Introduction

The significance of Steinbuch's writing and thinking about technology consists primarily in his using the present state of technology and its development to show what the possibilities are for the future. Certainly, Steinbuch is correct in stating that information and communications technologies are just at the beginning of their development, and that they promise many surprises for the future. That Steinbuch tenders information as a specialist in these fields and at the same time offers his own views concerning their cultural and social effects is to be applauded. The development of modern technology does entail tremendous changes. It can never be too early to begin reflecting upon them.

Yet, as an engineer Steinbuch has had little philosophical training. That becomes apparent whenever he steps out of his field of expertise to engage in philosophical reflection. His theoretical stance is that of positivism; he naively identifies his view of facts with the facts themselves. He looks at facts from a natural-scientific viewpoint exclusively. He ignores the distinction between concrete experience and natural-scientific theory about it. Actually, this is still not the proper way to put it, for it is above all cybernetics, with information theory as its basis, that Steinbuch selects as the point of departure for his philosophy. That he fails to distinguish between cybernetics, physics, and information theory, speaking now of cybernetics and then again of physics or of information theory while always meaning the same thing, is not only confusing but also incorrect. In my critique I will pass over these errors. This confusion of usage and terminology has been noted often enough in the analyses of other thinkers.

In contrast to other thinkers, Steinbuch—like Klaus (see section 3.4) – is extremely clear about his starting point. He accepts natural-scientific thought as the only correct manner of thinking. In his hands this approach has an imperialistic character, and the matter clearly comes to a head in cybernetics as the science of guidance and as the application of this science in technology.

After seeing that Steinbuch has absolutized cybernetics on the basis of a concept of information, we shall look at the significance of this step for his view of the relation between man and the machine and for his view of technological progress. After that we will compare the dominant streams in futurology with Steinbuch's futurology in order to explicate the content and significance of the all-controlling motive that makes Steinbuch's thought a unity. Finally, in a concluding subsection we will seek to determine to what extent Steinbuch is justified in appealing to the neomarxists Marcuse and Habermas for ideas about technological development and the future, and also whether an unbridgeable gulf separates these revolutionary thinkers and the "cyberneticist" Steinbuch. We will see that the freedom ideology of Marcuse and Habermas cannot be reconciled with the science and technology ideology of Steinbuch.

3.3.11.2 Informationism, physicalism, and cyberneticism

Like Wiener, Steinbuch grafts the category of information to the existing ones of matter and energy. "Information is understandable in terms neither of matter nor of energy" (AuM 22).

Matter and energy are carriers of information. Until that insight was achieved, it remained impossible to account for human thought scientifically, Steinbuch argues. To explain the inexplicable, people appealed to metaphysics, to the notion of substance, or to the notion of entelechy. Steinbuch rejects these solutions, which he considers speculative. "I believe we can analyze our thought functions just as rationally as we can our metabolism or our muscular system" (AuM 4).

Steinbuch clearly distinguishes signals from information. Signals are information carriers. Steinbuch proposes to analyze information scientifically and then quantify it. Information, to be "information," must be defined in a natural-scientific manner.[174] Quantification concerns both the form of information, in connection with which he speaks of classic information theory, and the value of information, where he speaks of semantic information theory. An example will serve to make this distinction clear: the *quantity* of information, as far as its weight is concerned, is the same for a pound of sand as for a pound of gold. But the quantity of information in the sense of its *value* differs. Semantic information theory concerns not the *quantity* of information which must be transported as such – thus, not the *form* – but the value which the conveyed information has for a receiver. In semantic information theory, the contribution which the conveyed information makes to the solution of a receiver's problem is taken into account quantitatively. Classic information theory was developed in abstraction from the sender and the receiver. Semantic information theory aims to embrace both the sender and the receiver. Accordingly it must entertain certain

238

assumptions about the structure and position of the sender and the receiver.

I am persuaded that Steinbuch is correct in introducing additional distinctions into the concept of information. I do not believe, however, that the distinctions he makes are adequate. I would rather follow a number of others in distinguishing between syntactic, semantic, and pragmatic information theory. These are concerned respectively with the *form* of information, its *content* or *significance*, and its *value* for the receiver.[175] What Steinbuch calls semantic information theory is designated pragmatic information theory by others. For real semantic information theory Steinbuch has no definite place. He gives the impression that semantic information theory can be identified with pragmatic information theory,[176] which is consistent with his positivistic point of departure. Steinbuch absolutizes the element of quantification in information theory. However, he seems to forget that information can be *new* – which is true of information in the sense either of *content* or *significance* – and thus not subject to theoretical determination with the help of quantification. This is the very reason for his expecting too much of translation machines.

Steinbuch believes, in fact, that with information theory – whatever further distinctions there are to be made – he is able to comprehend all human activities, including the intellectual ones, and to objectify them in machines. An *assumption* he makes in this regard is that with his scientific *theory*, which is oriented to the concept of information, he will be able to completely explain the human being in his activities and to duplicate all that in machines. "What we observe in intellectual functions is the reception, processing, storage, and sending of information" (AuM 2).[177]

Since the concept of information is a scientific concept, it is clear that Steinbuch identifies human beings with whatever information theory has to say about them. The basis for this identification, however, is not information theory itself; rather, it is Steinbuch's *reductionistic, positivistic* point of departure. Steinbuch reduces observable facts to what information theory – which he incorrectly identifies with physics – has to say about these facts. "The oft presumed supraphysical means are pure figments of the imagination, the existence of which is made probable by no *natural observation* whatsoever. There is *no discernable reason* why any supra-physical components should have to be accepted in connection with events in the higher organisms (up to and including human beings)" (AuM 394 – italics added).

Physics as matter and energy theory could not fully account for humankind, but Steinbuch claims that physics as information theory can do so. Whatever does not fit, such as self-consciousness or reflection, he takes to be abstract or decomposed thought.[178] However, this *disallowance* of reflection occurs in the *theory*; only then is humanity able to be fully accounted for by the theory. This, then, is Steinbuch's internal contradiction.

Steinbuch has absolutized the concept of information and has gathered everything under this appellation. His philosophy might therefore be designated *informationism*. Because he fails to distinguish between physics and information theory, his philosophy might also be called *physicalism*. For the present I will follow Steinbuch in his terminology. It is an incorrect terminology, and it obviously creates confusion. It does so because Steinbuch does not distinguish information as language from "information" as the analytical substratum of information as language, and because he also speaks of the theory of this "information" as physics, thereby giving the impression that the signals are identifiable with the "information" itself – even though we saw earlier that he still regards the signals as carriers of "information."

This all becomes a bit clearer when Steinbuch discusses the relation between psychology and physics. "It is accepted that the life process and the psychic processes can in principle be fully accounted for from the arrangement and physical interplay of the components of the organism" (AuM 9). On the basis of his positivistic point of departure, Steinbuch identifies the observation of behavior with the subjective feeling that accompanies behavior: "To the information-processing system itself (whether it be a person or an automaton), certain processes appear as subjective, psychic experiences. Precisely the same processes appear to an outside observer to be processes that can be described in the language of physics. Yet, the two phenomena are identical. Accordingly, one must speak not of psychophysical parallelism but of psychophysical identity" (AuM 12).[179]

Why we speak of the psychic in distinction from the physical is a question to which Steinbuch gives no adequate answer. He does not account for pretheoretical experience, in which feeling, for example, is distinguished from motion. In theory he reduces the psychic to the physical, but he acknowledges that psychic concepts are justified, since humans have a complex structure. But does the nonidentity of the psychic and physical not confront us precisely in that difference of terminology? And is Steinbuch not intrinsically contradictory in his theorizing when he assumes from the outset that a person's "inside" and "outside" are identical, which is the very point that his theory is supposed to demonstrate?

To carry his identity theory through consistently, Steinbuch would have to conclude that all talk about feeling, anxiety, willing, and so forth is based on mere *appearance*. This he does not do. In this way he obscures the fact – however clearly he absolutizes information theory –

that the substance of his *prescientific* point of departure is that science (as information theory) legislates for ordinary experience.

Information theory is the basis for the technological application of cybernetics. In other words, what information theory has analyzed can be duplicated in machines.[180] Inversely, it is also true that the principles of the machines shed light on the various sciences. "It would appear that 'controlling' is an intellectual tool shared by scientific disciplines that people have thus far construed extremely differently" (AuM 152).[181] From this it follows that Steinbuch uses cybernetics as the science of control to unite the various sciences. *His informationism finds its complement in his cyberneticism*: "It is less than reassuring when the terrain of experience is divided into layers and when various kinds of laws are attributed to these layers" (PuK 145).[182]

At bottom, Steinbuch's absolutization of information theory and cybernetics rests on his *belief* that only they do justice to the unity of reality. He thinks it possible to both explain and control everything through cybernetics. It is understandable, then, that he should think it possible to shed some new light on people and their activities, such as learning, knowing, problem-solving, and so forth, through cybernetics. Cybernetics will clear up a great deal of confusion in the so-called human or cultural sciences. "I am convinced that in the next decades, a cybernetic anthropology will arise that will reduce human thought and behavior to the function of information structures" (P2 96).

Thus, cybernetics does not render a service to the human sciences (a consummation devoutly to be desired). No, according to Steinbuch cybernetics absorbs them. For Steinbuch, true philosophy is not only informationism but also cyberneticism: " ... where philosophy and cybernetics contain no errors of thought, they cannot lead to different statements" (PuK 184).

The consequences of this philosophy for Steinbuch's views of the relation between man and the machine and of technological development will engage our attention in the following subsection.

3.3.11.3 The relation between man and the machine

Steinbuch's view that people are controlled by the cybernetic principles of receiving, processing, storing, and sending information is obviously of far-reaching significance for the relation between man and the machine. The computer, that most modern of machines, functions in conformity with these principles too. Only, in today's computers these principles have not yet been fully applied. Their operations are therefore somewhat less complicated than those of a person, especially those of a person's brain. However, advancing technological development will alter this situation rapidly in the future.

To realize this future, states Steinbuch very dogmatically, it is necessary to proceed on the basis that the brain and the computer are in principle the same. "Otherworldly" philosophy denies this similarity, as does the christian faith.[183] But the force of their argumentation is negligible, according to Steinbuch, since it is based on what the computer cannot do yet, to the neglect of what will be possible eventually. Yet, even where the development and possibilities of the computer are favorably regarded, the consequences for humanity are not being drawn. "It seems clear to me that our philosophy has not yet entirely succeeded in making the transition from alchemy to chemistry; some chunks of the 'other world' still swim about undigested" (FP 98).[184]

Steinbuch begins by "humanizing" the machine, and he ends up "thingifying" man: the person is a machine. He speaks anthropomorphically about the machine and goes on to regard people in the light of the possibilities of cybernetic machines. That we nevertheless continue to speak of the person in distinction from the machine is not accounted for by Steinbuch, for he sits imprisoned in a cyberneticism that denies all given, pretheoretical distinctions in advance. The consequences for his anthropology are enormous.

Steinbuch maintains that the difference between the person and the machine is quantitative in nature – not qualitative. He believes that every (intellectual) human function can be described logically and can, in principle, be transferred to the machine. Steinbuch uncritically identifies the analysis of the intellectual functions with the quantification of these functions. "Cybernetics also makes it possible to subject intellectual functions to exact scientific analysis" (AuM 333). From this it follows that with the transference of such a function to a computer, the function can be improved and even enhanced, so that the computer is superior to the human being in that respect. The computer will increasingly surpass man as more functions are transferred to it. The machine will not have to remain limited to humanity's intellectual level. "There is certainly no solid ground for the notion that computers have to halt precisely at man's intellectual niveau" (FP 116).

Steinbuch acknowledges that machine intelligence derives from the designer. He argues, however, that this does not put the designer in an exceptional position with respect to the machine. The designer himself is programmed too – through heredity, nurture, education, and experience. In fact, it is here that the fundamental similarity between man and the machine is apparent, according to Steinbuch. It is often believed, he says, that intelligence is connected with generating new information, and that this is beyond the machine's capacity. Not so. New information appears when people, through reduction from *exist-*

ing information, select information. Here learning leads people to greater success. In principle, the machine does nothing different.[185] Steinbuch prejudges the question in that he slights human innovation, or equates it with coincidence, in order to get a cybernetic grasp on it.

Should machines begin to develop in agreement with cybernetic principles, their possibilities will be legion.[186] They will be able, for example, to replace people in outer space. As yet this is not possible. "There is no technological device as broadly capable of motor actions as the human hand" (AuM 302). However, *"Within a few decades* it must be possible to realize all essentially intellectual functions for space flight through automatic systems" (AuM 310).

Steinbuch is persuaded that the possibilities of machines will be increased – and rightly so. Only, he forgets that the possibilities involved will be opened up by scientific analysis and synthesis, and that, given the nature of science, they will have a general character.[187] The human hand will remain superior to the machine in that respect, since it can adjust to new situations, which may in turn be subject to permanent change. Should Steinbuch aspire to achieve an equal machine, equipped with a learning matrix, it will have to be designed in a scientific manner. And that implies abandoning that which is unforeseeable – the new as the unexpected. It is only after the fact that the new can be "captured" again through the feedback principle, though it is not possible to determine this perfectly in theory.[188]

That Steinbuch is insufficiently aware of the element of novelty in human creativity is confirmed by his claim that in the future computers will be able to translate perfectly. A certain difficulty does present itself, he acknowledges, in that language has not been systematically constructed and therefore displays tremendous variety. A good translation machine will be possible only when the following conditions are met: "Actually, the languages would first have to be rendered systematic" (AuM 123).[189] Steinbuch slights human creativity. Furthermore, he reflects neither upon language nor upon the structure of science. Scientific knowledge, as general knowledge, is always abstract and therefore is not to be identified with the concrete and the particular in ordinary experience – in this case, with the great diversity in ordinary language.

Compared with man, the machine presents still another problem, in a different context. The human being is able to arrive at scientific – thus, general – pronouncements. Steinbuch acknowledges that the machine cannot (yet) do so: "For the induction process proper, there presently exist no effectual algorithms" (AuM 333). Steinbuch believes, however, that these possibilities are not excluded as machine learning advances with the help of the "trial and error" method. Yet, he has here implicitly suggested the difference between man and the machine – even if he feels compelled to conclude that they are identical. What it all boils down to is that whatever success the (trial and error) method may have in the machine can ultimately be attributed to people's having analyzed that method in a scientific manner and having allowed it to function in a technological manner in the machine, with the aid, for example, of random generators. Thus the machine does not develop this method independently but remains dependent in it upon man, the designer. Moreover, the "trial and error" method is not arbitrary in people's hands – an impression Steinbuch creates – but is directed. It is on the basis of this direction that errors are discernible.

Steinbuch is not entirely oblivious to the dependence of machines upon people. When machines can eventually read, speak, learn, translate, and so forth, they will still have the same limitations as human beings: "Automatons possible in the future naturally have the same limitations to which man is subject. They are not clairvoyant, for example" (AuM 342). On the other hand, Steinbuch claims that machines will be able to evolve independently and reach out beyond humanity's grasp. The designers will have to provide the right conditions for this machine evolution: "The achievement of the engineer consists in this, that he so felicitously gives form to the conditions for mutations and procreative selection that the system rapidly converges in the direction of intelligent behavior" (AuM 346). The result of this evolution may even be that machines will say of themselves that they have consciousness. He thinks this "pronouncement" of the machine is to be anticipated at the time when the machine is so programmed as to be free of human biases and able to communicate freely with its environment. "Until the contrary has been proved, it is my guess that such a system would assert of itself that it has consciousness (and feelings too)" (AuM 351). By the same token, the machine would know the difference between "I" and the world "out there," says Steinbuch. He arrives at this *speculative* thought because he believes that people are exhausted in, and can be fully explained by, the principles of cybernetics, and because he imagines that a machine in which these principles are fully applied will be "anthropomorphic." Given this assumption, he goes on to say that "the only possible proof consists in this, that it cannot be falsified" (FP 89). This unscientific speculation, which can operate like a myth and which Steinbuch cannot make concrete, is attributable to the fact that he sees neither humanity nor the machine properly. The machine is a "thing," and its last subject-function is the physical function. The technological function as a qualifying function is an object-function, which means – even in the case of the most complicated mechanism – that it must be actualized by human beings. Every machine remains subordinate to man.

Meanwhile, Steinbuch's a-rational assumption[190] that man and the machine are ruled by identical principles has far-reaching implications for human freedom and responsibility.

Steinbuch talks about human freedom in a double sense. *First*, freedom for Steinbuch is a *being free from* antiscientific and antitechnological biases. Freedom fosters technological development, the ultimate result of which will be machines that surpass human beings. For people, too, this freedom means progressive intellectual development through learning, and thus through the *acquisition* of new information. That one person is more successful at this than others gives rise to great diversity, which is incorrectly attributed to human freedom. *Secondly*, Steinbuch denies human freedom in the sense of a being free from natural laws. That, to him, is an illusion. This is consistent with his point of view: "...it would appear that the problem of freedom is a phantom problem, caused by perspectival re-formations of the subjective experience. On what ground does the human being contend he is free? On the ground, for example, of the laws established by the Creator for all that comes to pass? Would that assumption not be a sort of nihilism?" (AuM 292).

In itself, there is no objection to acknowledging that human freedom must be obedient to laws – but then to laws which people must obey in responsibility, laws which they can, in unfreedom, evade. Steinbuch, however, speaks only of natural laws that man "obeys" as a matter of course, in that he is coercively subject to them. Therefore freedom and responsibility are concepts with which Steinbuch cannot work: "they are science-alien elements within our science" (FP 95). According to Steinbuch, the human being is not free but determined.

The question that might be posed here is whether Steinbuch's judgment is determined too. If it is not, he is in conflict with himself. If it is, then the same determination applies to the "background" philosophers and to the Christians, in which case his appeal to these people to yield to science and technology is out of order.

Steinbuch does not take up this problem, for he has not reflected upon the transcendental presuppositions of his philosophy. He posits dogmatically that the human being is a cybernetic system and proceeds from there to construct his philosophy.

The denial of freedom entails many consequences. For one thing, guilt is abolished. There can be no such thing as guilt when we are speaking of something that goes wrong with a person-designed machine. "In the case of regulating, judgment will result in the fault being attributed to various causes with various likelihoods. The content of the notion of 'responsibility' is thereby sharply altered" (AuM 371). Since the human being, according to Steinbuch, is a machine – albeit an extraordinarily complicated one – it follows that "a person, too, can bear no responsibility in the sense of guilt" (AuM 371).

The inquiry into the origins of a mistake can be pursued indefinitely. That someone errs may be the "fault" of a teacher, for example, but the teacher's "guilt" may go back to his own milieu and experience, which were formed, in turn, by still other people. The quest for a "guilty" party is therefore impossible. Remarkably enough, Steinbuch does not go on to draw the conclusion that silence about freedom, responsibility, and guilt would be appropriate. Accepting that conclusion would mean simply accepting whatever mistakes are made. That, of course, he does not wish to do. Instead he suggests "punishing" mistakes with feelings of discomfort or loss, so that a person (or the machine, as the case may be) will alter his learning program, which will prevent repetition. In this manner guilt and responsibility are so diluted that Steinbuch thinks himself able to get a grasp on them with cybernetics. "The postulate 'Particular faults are punished with particular penalties' is in my view decidedly applicable to the regulation of the behavior of both the person and the automaton" (AuM 372).

Steinbuch never even approaches the question why we still continue to speak of freedom and responsibility. He leaves it utterly untouched. But does not the very thing he denies continue to obtain here? And are freedom and responsibility not the cause of the inner contradiction in Steinbuch's philosophy as described above?

The denial of freedom inflates Steinbuch's expectations of the machine. Not only will the computer be able to translate, learn, hear, and speak, it will be able to play the poet and composer too. I shall return to these matters in my discussion of Klaus.

Steinbuch's anthropology is *totally* determined by science, that is, by cybernetics. He explains human thinking by referring to the cybernetic model he has borrowed from cybernetic machines. Whatever fails to fit that model, such as reflection, he labels "corrupted" thought.

From the fact that Steinbuch derives his definition of thought from the model of a learning machine, it follows as a matter of course that the machine can think, and that the thinking human being is a machine.[191] Steinbuch is guilty at this point of a *petitio principii*, for he starts out by abstracting from thought proper, only to end up taking the insights gained through reference to learning machines and applying them to human thought. This is the dichotomy inherent in Steinbuch's positivistic prejudice.

3.3.11.4 Technological progress

Before critically analyzing Steinbuch's futurology and his appeal to Habermas and Marcuse, I will probe the dominating motive in Steinbuch's view of technological progress.

Steinbuch notes that technology, in contrast to other human activities, has advanced. "While people can squabble, using weighty arguments, about the reality of human progress, technological progress consists of an indisputable increase of registerable capacity."[192] According to Steinbuch, the advance of technology means progress, health and peace for humanity.[193]

In short, cybernetics based on information theory is the prerequisite for a "humane" future.

This future entails a total alteration of our conception of the world, an alteration that can only be compared with the copernican revolution: "Copernicus demonstrated that the earth, humanity's habitat, is not the center of world events. Cybernetics will teach humanity that the brain is not the only possible locus of 'intellectual' events" (AuM 387).[194] The advance of modern technology will reveal that man is not alone in having an "intellectual" capacity. Machines will make up for all of humanity's shortcomings and deficiencies and usher in prosperity.

Steinbuch has sold his soul to technological development. The progress *of* technology will become progress for humanity and society *through* technology if the methods of cybernetics are universally applied. Steinbuch advocates an integration of technology and society abetted by cybernetics. This is possible because cybernetics abolishes the distinction between the natural sciences and the human sciences: "A framework of thought applicable to the fields of technology, biology, and the social sciences presents itself, to which the distinction between the natural sciences and the human sciences is alien" (AuM 13). Cybernetics will become a bridge between all existing sciences. Steinbuch agrees with the statement made by Helmar Frank: "One can conjecture that by the end of this century ... the human sciences, modernized by then, will be of such a nature that they will no longer speak of the mind and its derivatives but will have broken it down into components and thus have 'de-spiritualized' it into a systematics of information and information processes" (FP 92).[195]

Although Steinbuch's optimism with respect to the future as a future of technology has the upper hand in his thinking, and although he initially creates the impression that the progress of humanity and society occurs as a matter of course since cybernetic machines will develop nicely in an evolutionary manner, he also expresses concern: "Technological development is an inevitable influence upon our destiny, an influence against which we can offer no effective resistance" (AuM 346). The cultural pessimism of Spengler and Jünger and of the neomarxist philosophy has made him aware that technology can become an autonomous power.[196] It will do so especially if cybernetic principles are not applied universally. Therefore machines for governing must not be technological machines but cybernetic ones, so that the people can exercise direct influence on the decisions to be made. That way the "tyranny of events" can be broken through.[197]

This wending of the machines toward humanity is a happy inconsistency in Steinbuch. Does it not mean that man still surpasses the machine, even though Steinbuch initially argued that the machine had taken over man's intellectual position? This adjustment enables Steinbuch to deny the autonomy of technology. "This notion must be opposed. It is not the perfection of technology but the reasonableness of its use that is decisive" (P2 78). Technology does not exist independently of humanity: "Technology knows no mysterious 'tyranny of events'" (P2 79).

For the rest, this turn in his thought does no injury to his expectations concerning the potential of technology for the future. Information technology does not coerce people to employ the most awful possibilities of science and technology to their hurt. Steinbuch contends that all ways are open for *people* to use science and technology to their benefit.[198] The first thing required for this outcome is a positive attitude toward natural science and technology. The connection with the "other world" in philosophy and in christian belief will have to be abandoned. Then it may be anticipated that people will have the insight to make the correct decisions concerning technological development – and none too soon, for "We saw that scientific-technological progress is a vehicle racing ahead with accelerating speed, for the time being without any effective control."[199]

We must attempt to keep the process of technology, which is running wild, on the right track. Together with the development of a futurology (see the following subsection), social structures must be adapted to technological development. "What political and social structures are suited to neutralize the impetus toward unlimited technological progress?"[200]

For the time being, unhampered technological progress will remain an inevitability, since politics and economics require it. Yet, "Are we not perhaps sitting on a volcano which sooner or later will erupt, consuming all these political and economic structures of our age in the holocaust?"[201] When things have gone that far, we will need "dynamic, stable social structures"[202] upon which man with his critical consciousness will have to have such an influence and over which he will have to have such political control that the "tyranny of events" will be broken through. If we move in this direction, estrangement, dehumanization, and the disturbance of psychic harmony can be avoided.

In the long run, even more serious alterations will have to be made. Neither a socialistic nor a capitalistic society will be able to satisfy the requirements. "It may be presumed that neither of these models is suited to deal with the problems of the future. We really need to develop new social models; we require new social inventions for dealing with the problems of future reality" (PuK 150). Steinbuch introduces the idea of a world state in which technological development can at last be limited and stabilized. At that point, everything should be aimed at protecting humanity from technology. "To realize the conditions for a desirable human existence, we must strive for political relations of such a nature as to neutralize the impetus toward unlimited technological progress. The notion – for the moment, no doubt, utopian – of a world government is probably the only principle that can lead to this" (FP 170).[203] "Then perhaps an industrial organization is possible free of the dictates of competition and progress, following rather an anthropological orientation. Yet, these considerations are at the moment extremely utopian, and meaningful only as the long-range goal of a global, peaceful, post-industrial society" (P2 82).

The contradiction running through his view is that Steinbuch absolutizes technological-scientific thought on the one hand, hoping to extend it to all the sciences and to society, while on the other hand he is aware of the danger of the autonomy of technology, through which humanity becomes a victim of technology. Thus he suggests that human freedom be "planned."[204] Yet, because he perceives that this planned freedom, too, will have to obey the "clock as the instrument of repression," he believes that the increase in leisure time offers a good counterbalance for this unhappy situation.[205] With this, Steinbuch implicitly acknowledges – I will take this matter up again in the following subsection – that his idea of direct democracy is an illusion. Such a notion is inadequate in the face of the continuity tendency of absolutized scientific thought. The deepest reason for this is that Steinbuch, although he speaks of human freedom – but then of a freedom prescribed by natural law[206] – continues, at all costs, to view the human being as a cybernetic system. "I am convinced that in the next decades a cybernetic anthropology will arise that will trace human thought and behavior back to the function of information structures" (P2 96).

The deepest motive operative behind Steinbuch's absolutized, technological-scientific thought is the *passion for control*, his *belief in* and *hope in* technology. He believes it possible, with the help of cybernetics, to satisfy humanity's desires. This belief is so strong that Steinbuch simply does not see that what will be attained will be the opposite of what he intends.

Actual practice already shows that in the future, people will find themselves in an ever greater bind as a result of the development of science and technology. Although Steinbuch would like to avoid this outcome, he is not willing to do so at the expense of his belief in the

249

science of cybernetics. Ultimately, then, nothing is changed. The future of society as a cybernetic society is a *determined future*.

The only real possibility for avoiding this outcome is for Steinbuch to reflect upon who man is. He would then have to perceive that human motives lead technological development. He would realize that when human freedom is denied in principle, the *advance* of technological development brings *no progress* for humanity but ushers in the abolition of freedom instead.

3.3.11.5 Futurology

The problems concealed within Steinbuch's futurology can best be clarified through a discussion of the major tendencies of the dominant streams in futurology. A rough distinction is made between the *futurology of order* and the *futurology of conflict*, which is also called critical or revolutionary futurology.[207]

Steinbuch belongs to the futurology of order.[208] Via projections and prognoses, he attempts to draw lines to the unknown future. The past and present serve as models for the future – not in the sense that Steinbuch would see the future as an enlargement of the present, but in the sense that the future must be built upon the past and present, with the deficiences of the past and present eliminated.[209] Nor does Steinbuch turn a blind eye to the element of novelty that the future will bring. He is of the opinion, however, that the new, as encountered in discoveries and inventions, for example, must be foreseen by futurology and incorporated into models of the future.

Steinbuch refuses to be satisfied with a single model of the future to be realized by scientists and engineers. He contends that futurologists must construct a number of models of the future from which the people, by means of direct democracy – modern communications technologies make it possible[210] – can make a choice. This choice must then be translated into reality, under the leadership of the scientific and technological elite. With the help of the cybernetic principle of back-coupling, it will be possible to incorporate in the planning the element of the unexpected in humanity's future, so that the "surprising" will not be disruptive and the future will turn out to look as much as possible the way the democratic referendum indicated it should.

Steinbuch believes that the people decide. But do the people not decide from among the alternatives laid before them by the futurologists? And is the choice then not already forced in a certain direction? Is this not all the more so if the communications media serve up one choice as self-evident?[211] When Steinbuch goes on to say that the decisions of the direct democracy can be predicted according to the law of large numbers and allowed for in drafting plans, and that in these plans, furthermore, the psychological and social reactions of the people can be taken into account since these reactions can be simulated in a computer, then this democracy is an illusion. Steinbuch's "direct democracy" affords merely another, and still better, possibility of determining the future.

The methods of the futurologists of order can be summed up in the term *planning*. Planning can become differentiated in several directions: technological planning, scientific planning, economic planning, social planning, the planning of education and of research, and so forth. The possibility exists of fine-tuning these types of planning to each other and of integrating them. Planning is fully determined by science, with the help of the computer. After a choice is made (whether democratically or not) from among the alternative futures, the plan is carried out with the aid of scientific and technological possibilities.

The futurologists of order tend to lean toward a collectivist society in which human freedom suffers under the pressure of the power of science and technology. The *critical futurologists* oppose the futurologists of order because they see in that futurology a real threat to human freedom and a perpetuation of existing social injustice. They fear the power of an establishment that employs science and technology as instruments of domination. Often they also fear science and technology as such, since these two threaten to become autonomous, anonymous powers to which humanity will be utterly subservient. According to the critical futurologists, the future must not be construed as a continuation of the past and present, with the deficiencies rectified; no, the future must be a *radical* correction of the past and present. Therefore they advocate methods of a bold, revolutionary character. The issue is a *free* future. Following the line of Herbert Marcuse and A. Waskow, they echo the call for protest, conflict, and sometimes even revolutionary violence.[212] Their analyses of the past and present are aimed at disclosing precisely where freedom encounters difficulties. Changes must then be made accordingly. The structures within which culture has developed up to the present must be altered – and altered permanently, to the maximum extent possible, since in social structures petrifaction always sets in sometime, which is pleasant for a minority (the elite) but oppressive to the greatest portion of humanity (the masses). Nothing is really changed if through scientific and technological development the numbers of the elite increase and those of the masses decline. This is all the more true, according to the critical futurologists, since both categories ultimately come to be ruled by science and technology as autonomous powers.

The extreme revolutionary futurologists – let us limit our attention to them to make the tendency clear – desire to smash the existing powers of the establishment and of science and technology in order to make room for *creative disorder,*[213] as Waskow puts it.

In this creative disorder, heralded by the "creative passion of destruction" as total revolution, the people attain an opportunity to work on a free future. *Prophecy* as concrete action or protest and *utopian thought* as a liberating glimpse of the today that ought to be made real this instant are terms that have a meaning entirely their own within the context of the futurology of conflict.

In summary, the methods of the critical futurologists derive from an integrated critical attitude toward the past and the present. In contrast to the futurologists of order, they advocate not quantitative but qualitative change for the future. This approach to the future is a leap from the present into the "kingdom of possibilities." Consistent with Marx, the ideas of the futurologists of conflict must be "abolished" by becoming reality – through violence, if need be.

I believe that a better understanding of these two approaches to futurology is to be gained with the help of an analogy from technological development. This development has two characteristic traits – continuity and discontinuity.

Relying on the so-called *technological-scientific method*, the designer arrives at a design that can be carried out. After the elementary building blocks have been discovered by way of an analysis of a technological problem into its component problems, the synthesis of these elementary building blocks leads to the establishment of a complete plan, which serves as the blueprint for the execution, the actual making.[214] Technological development itself is characterized by this technological-scientific method. Left to this method alone, technological development, remaining under the control of theory, would become petrified. *Invention*, as the expression of human freedom, creativity, and ingenuity, prevents such petrifaction. It is invention that endows technological development with an ever fresh perspective. Given the possibility of an invention, technological development is not determined; it does not stand forever fixed if theoretical control leaves room for invention. By means of an invention – think of the computer – technological development can become differentiated in various directions. This development is not continuous only; it is also interwoven with discontinuous development. In invention, continuity is broken through in a surprising way. But even continuous development is not perfectly continuous, since in theoretical control, which is led by the productive, free person, permanent changes, however slight, which no one would characterize as inventions, can be, and are, introduced.[215]

The analogy here with the methods of the futurologists of order and conflict is that they are geared exclusively either to technological-scientific control of the future or to the character of invention as the new. Continuous and discontinuous development are sundered. The one makes technological-scientific control of the future so central that the future comes to look determined. The other gives such tremendous emphasis to freedom and the new that all continuity of order and structures is repudiated. Neither the technological-scientific method nor the method of permanent revolution can exist by itself. Whenever one of the two methods is absolutized, it evokes the other as a reaction. Any technological-scientific control of the future that shows no concern for human freedom evokes, in protest, a formidable opponent. Any absolutized freedom results in powerlessness and chaos and thus evokes technological-scientific control as a reaction. We may conclude that the method of futurology must not be allowed to become oriented to just one of the characteristics of technological development. The method should take into account the fact that technological development is both continuous and discontinuous, and that human freedom ought therefore to be acknowledged as the boundary of the scientific method.

Now, the condition I have stated here is a necessary one for a responsible futurological method, but it is not by itself a sufficient condition. Where the method of futurology is concerned, the analogy of continuity and discontinuity in technological development must remain just that – an analogy. Technological-scientific control is only possible in the case of inanimate matter. It is not suited (though Steinbuch, as we have seen, advocates it) for controlling the whole of society. Such control of society, of humanity, and of history would lead to a technocracy in which people were trapped like flies in a bottle.[216] This is not to deny that for the various facets of human society, it is possible to analyze in a scientific way the possibilities for the future. This is already being done in the fields of sociology, economics, and politics. However, such inquiry into the future is limited in principle. It cannot simply pass over into actual control of the future, for the human being as a free and responsible person appears within the stated fields of inquiry. Precisely because freedom and responsibility are the fundamental boundaries of futurological inquiry, such inquiry cannot automatically yield to control. The results of futurology – of inquiry into the future – must be placed, as advice, in the hands of the responsible people of practice. They, in their freedom and the exercise of their responsibility, must then make the leadership decisions required for the future.[217]

What I have just outlined is, of course, not an exhaustive treatment of the problems pertaining to futurology. I have only tried to make it clear that a futurological method oriented to technological-scientific control of technology stands in the way of human freedom, and that this is the cause for the rise of the method of revolutionary futurology. Yet, even if an attempt were made to harmonize these two methods (as it ought to be made in modern technology), the result should not be permitted to lead to technological-scientific control of the future and of humanity without due regard for people, since people, after all, are the ones responsible for the future in actual practice. They would be deprived of their freedom to make responsible decisions – and not they alone: those who were busy with praxis even later would be deprived of their freedom too.

Steinbuch's futurology does not satisfy that last-mentioned requirement. After the people have made their choice from the various models of the future, the scientists and engineers will set about bringing the "chosen" future to pass. The consent of the people – a good thing in itself – can be chimerical, as we saw: it has a determined future as its reverse side, no matter how much more supple that determination may be rendered by the cybernetic principle of backcoupling. Steinbuch sees no room for the *responsible* human being in future practice. His futurological method binds humanity too tightly.

3.3.11.6 Habermas, Marcuse, and Steinbuch

Steinbuch often appeals to Marcuse and likes to associate himself with the thought of Habermas. In doing so, he is somewhat eclectic. He introduces the ideas of these thinkers without doing justice to their original framework, thereby creating the mistaken impression that he should be numbered among the revolutionary thinkers of our time. For all the corrections Steinbuch introduces into revolutionary thought, the deepest intention of his "cyberneticism" is never altered.

In this subsection I will begin with Habermas's interpretation of Marcuse's view of technology. Then I will take a very brief look at Habermas's own conception. In conclusion, I will add up the balance to show where Steinbuch, Marcuse, and Habermas are in agreement, and where there are differences of fundamental importance.

Habermas begins his treatise on technology with this important quotation from Marcuse: "The liberating power of technology – the instrumentalization of things – is perverted into a fetter of liberation: it becomes the instrumentalization of man."[218]

To Marcuse, the innovation of our time is the elevation of technology and science to the rank of ideologies. Our age no longer fits the older ideologies, such as Marx's. Economic realities have now become joined to modern science and technology in such a way as to render human freedom an impossibility, whereas Marx viewed freedom as an inevitable consequence of economic development. On the contrary, technological progress has become so all-dominating that the economic factors of capital and labor have been taken up into a technological collectivism.[219] Technology *as* technology has become a power, to the detriment of freedom. In other words, technology and science legitimize the institutional power, which oppresses humanity. Habermas says in summary: "For this unfreedom appears neither irrational nor political, but is rather subjection to the technological apparatus, which increases life's ease and raises labor productivity" (TuW 53). Scientific technology as production power has become an ideology. It is a "historical totality" (*geschichtliche Totalität*).

Habermas claims that Marcuse wavers between two views on the issue of achieving a breakthrough against technological power. *On the one hand*, Marcuse regards science and technology as a prioris of our society leading inevitably to an absence of freedom. A revolution would be required to *radically* alter the nature of science and technology. This means that the technological-scientific method as a method of oppression will have to be exchanged for a liberating method, namely, that of tenderly nurturing and cultivating.[220] *On the other hand*, Marcuse creates the impression that a revolution would alter only the institutional framework of technological power. Accordingly he resists the political and economic power-holders, the establishment, those who employ science and technology as instruments of dominion. Technological development must be given a different direction: "...the new must be the direction of this progress, but the standard of rationality itself must remain unaltered" (TuW 58).

Habermas rejects the first "solution." He agrees with Gehlen that technology belongs with humanity and cannot be some passing historical phenomenon. "Marcuse has an alternative *attitude* towards nature in mind, but there issues from it no idea of a new *technology*" (TuW 57). Habermas believes that Marcuse states his second "solution" with insufficient nuance when he lays the blame for the loss of freedom not on technology as such but on the institutional framework within which technology is developing. Yet, apart from that Habermas agrees with Marcuse's basic thesis: "For the analysis of the altered constellation, Marcuse's basic contention that technology and science presently hold the function of legitimizing power is in my opinion the key" (TuW 74). Technology and science have become ideologies themselves.[221] The reason for this is that the state increasingly intervenes to assure industrial stability. Moreover, science and technology, in combination with research, have become "production forces," replacing the institutions of an earlier period. The result is that solving technological problems takes priority over realizing practical goals. Politics, led by the ideology of science and technology, excludes practical questions and allows no room for establishing guidelines for technological development through expressions of the democratic will. "The solution of technological problems is not dependent upon public discussion" (TuW 78). The masses are gagged by the ideology of science and technology. "Thus a perspective presents itself in which it would *appear* that the development of the social system is determined by the logic of scientific-technological progress" (TuW 81). The "tyranny of events" of technological progress seems inescapable. This "technocracy thesis" has been articulated scientifically by various philosophers. Habermas refers, for instance, to J. Ellul and A. Gehlen.[222]

According to Habermas, the autonomy of technology is a "quasi-autonomy,"[223] but it has established itself as a common illusion. "It strikes me as more important that it can, as an ideology in the background, penetrate the consciousness of the politically innocent masses of the people and operate with legitimizing power" (TuW 81). Society comes to be built up according to some technological model. It becomes a self-regulating cybernetic system. "Industrially advanced societies seem to approach the model of behavior control guided by external stimuli rather than by norms" (TuW 81). Also: "...they [the technocrats] desire to bring society under control *just like* nature, by reconstructing it after the model of self-regulating systems of goal-oriented doing and adaptive behavior" (TuW 96).[224] Freedom of choice, of consumption, and of leisure time is an illusion. This is true for both "capitalistic" and "communistic" society.

In contrast to an earlier day, ideology is no longer just ideology but a driving power. In other words, the ideology of technology and science has reflected itself in practice and in institutions and has thereby sundered people from communication and interaction as possibilities of human freedom. Marx failed to foresee that technological progress would acquire the function of "legitimizing power." Technology does not inevitably liberate humanity.[225] Futurologists such as Kahn and Wiener will eventually be able to control both personality changes and conduct through technology.[226]

Habermas sees through the quasi-autonomous character of technological power, which enables him to regard that autonomy as something less than a matter of course. Compared with other views,[227] this is an undeniable gain. Habermas states: "I proceed from the fundamental distinction between *labor and interaction*" (TuW 62). Next to – and not over against – the consideration given to technology as instrumental doing, Habermas calls attention to communication and social interaction. He advocates – alongside the development and expansion of technological power – the emancipation of the individual. "Dominion-free communication" must break through the old institutions and make possible new forms in which science and technology can be of service to humanity. "Therefore *political* dominion can henceforth be legitimized 'from below' instead of 'from above' (through an appeal to cultural tradition)" (TuW 69 – italics added).

Habermas opposes the ideology of science and technology on the grounds that "dominion-free" interaction ought to take precedence over the autonomy and self-evident character of science and technology. Interaction and communicative doing will have to take the place of rational and goal-oriented doing. Only then can the society constructed on the model of cybernetic machines be broken through. Then technology will be able to fulfil its *liberating* function too. To avoid misunderstandings, Habermas says that communication can and should be rational, but not in the sense that technological control is rational. Rational communication "does not lead, as does the rationalizing of goal-oriented systems, to aggrandizement of technological control of objectivized processes in nature and in society; it does not lead *per se* to improved functioning of social systems, but it will give the members of the society chances for a more far-reaching emancipation and advancing individualization" (TuW 99). What is decisive is not that we develop technology but that we choose what satisfies human existence, Habermas maintains. And what this is is not up to the specialists to decide, but is to be decided by every person for himself.

Yet, at this point it is not yet clear *how* the present development, in which science and technology have dominion as ideology, according to Habermas, can be broken through. Marcuse believes it will take a revolution, a revolution brought about by those who exist on the *periphery* of the technological system and are therefore unable to pluck its fruits. Habermas thinks Marcuse is mistaken. He wishes rather to return to the *core* of the system in order to appeal to future leaders not to allow themselves to be swept along by the ideology of science and technology, but instead to seek answers to the questions that are important for life. According to Habermas, some categories of students are sensitive to this appeal, namely, those coming from better social milieus, who have had no authoritarian upbringing, are high academic achievers, and therefore do not have to be busy making their own careers. They are the ones who can take distance from technocratic basic principles; they do not have to work for high income and leisure time. "Their protest is [therefore] directed much more against the category of 'compensation' itself" (TuW 102).

The condition for Habermas's "revolution" is that technology must first usher in wealth and a high level of material prosperity. In this he differs fundamentally from Marx, for whom the exigencies and suffering of the workers were the condition for revolution. Given prosperity, says Habermas, people are independent and are in a position to relativize the fruits of technology. "In the long run, student protest will therefore be able permanently to destroy this weakening achievement-ideology and thereby bring about the collapse of the already brittle basis of legitimacy, concealed only by de-politicization, of late capitalism" (TuW 103). What humanity *can* do must be subordinated to what humanity *chooses* to do. That will guarantee freedom.[228]

A remarkable difficulty in Habermas is that while he seeks to promote a "dominion-free" technological development, he begins by isolating and accepting the present technological-scientific system as being *only* technological. Within that system he appeals to a student elite to bring about the revolution.[229]

From all this it is clear that Steinbuch's appeal to Marcuse and Habermas can at best be only partially justified. It is even a question whether he understands them correctly. Steinbuch appeals to Habermas to make it clear that there is a gulf between the human sciences and the natural sciences. Steinbuch believes that he is following Habermas when he says that "praxis is unaffected by theory" (P2 33). Steinbuch draws the conclusion that the natural sciences and technology ought to be granted control of the terrain of the human sciences too, while Habermas means to say that the human sciences themselves have little to say on their own terrain precisely because praxis is ruled by technology.[230] The great difference between Steinbuch and Habermas is clear when the former speaks of primitive technocracy, and the latter of total technocratic consciousness.[231]

From Marcuse and Habermas, Steinbuch has indeed picked up the idea that the autonomy of technology is not a mere matter of course and that the "tyranny of events" can be broken through by means of democratic control. Only, Steinbuch does not want a "dominion-free" democracy but a hierarchical one in which the technicians enjoy preeminence.[232] *They* are the ones who must work out the plans for the future, lest technological development get out of control. Marcuse and Habermas, by contrast—despite the differences between them—want to establish the guidelines for technological development in a democratic manner, since they perceive in the current development of technology a threat to human freedom. It is for this reason that Marcuse on the one hand wants to return to the preindustrial epoch, while wishing on the other hand to stabilize technological development in a *revolutionary* way, or even to restrict it to the niveau of such elemental necessities as food, clothing, and shelter. To him, work is play and play is work. To the extent that this requires central planning, Marcuse is not opposed to it.[233] Here he differs clearly with Habermas. Steinbuch, too, aims to stabilize technological development, but he would like to do so in an *evolutionary* way. This utopia becomes possible only if existing institutions are exchanged for technologically controlled ones and integrated into a cybernetic world state.

Marcuse has objections to the existing social structures because they are unduly controlled by technological thought. This is what he calls one-dimensionality. Steinbuch, however, is of the judgment that those structures are simply still too much stamped by the "otherworldly" "background thought" of economists and political theorists. He wants to exchange these structures for technology – especially cybernetics.[234] Steinbuch's conception of the cybernetic world state would evoke objections from Habermas too, for as we saw, he perceives that people will be manipulated in a cybernetic state just as if they were part of inanimate nature.

We can conclude that Steinbuch has borrowed a number of ideas from these revolutionary thinkers. Actually, he is opposed to them. Their ideas have lent "suppleness" to his initially rigid anthropology. (He initially regarded freedom and responsibility as totally illusory.) He proceeds to acknowledge freedom – a freedom that must again be planned in a technological-scientific way. As a rationalist – because he absolutizes cybernetics – Steinbuch often appeals to logical positivism,[235] while Marcuse and Habermas, as irrationalists – for them, human freedom is all-controlling – oppose philosophy. Steinbuch does not regard technology as a threat to humanity; according to him, it brings deliverance from existing tribulation.[236] Habermas and Marcuse, however, are one in their judgment that technological power as ideology abolishes freedom.[237]

Although Steinbuch seeks association with revolutionary movements, wishing to make it clear to them that the technological revolution is not their enemy but their ally,[238] an unbridgeable chasm continues to divide them. Steinbuch fails to plumb the depths of the problems of contemporary revolutionaries,[239] and they, in turn, are well aware of the dangers of his philosophy. Häusserman even calls Steinbuch a priest of technology and says that he is driven by "a very devout faith in science." According to him, Steinbuch *believes* in the mountain-moving power of modern technology. His cybernetic conceptions cannot conceal his advocacy of a technocracy to which critical students and their teachers are justifiably opposed.[240]

Steinbuch bases his view on technological-scientific possibilities and *power*, at the expense of freedom. The revolutionary thinkers champion individual *freedom* as absolutized freedom. They recognize no "dominion" (*Herrschaft*) – or perhaps better, no authority – although in the period of revolutionary transition they do require a certain elite.

This need for an elite, however, does not detract from the fact that opposed to the ideology of science and technology there stands the ideology of freedom. Nor are these ideologies reconcilable. The first view either denies freedom, making it only an apparent problem, as Steinbuch initially does, or reduces it to a planned freedom; the second view resists every power and every authority as a threat to freedom. The harmonious coexistence of these two views is impossible. They exclude each other and evoke one another as dialectically opposite poles.

Habermas[241] comes closest to the idea that people ought to control technology in freedom. He advocates an appropriate place for technology and unmasks its pretended autonomy. However, I disagree with his assessment that "dominion-free communication" can assure harmonious agreement of technological power and human freedom. Even the kind of "dominion-free" decision about technological development that Habermas advocates would place heavy restraints upon the man of the future, seriously impairing his freedom, even from Habermas's standpoint.

In the concluding chapter I will show that this problem can be solved when suprasubjective normative principles are taken into account and the human relations that contribute to technological development are governed by the norm of authority and freedom. I will argue that this problem is ultimately traceable to the search for meaning. Meaning finds its substance neither in absolutized power (Steinbuch) nor in absolutized freedom (Marcuse and Habermas), but is given in Jesus Christ. In Him, freedom and power are made one. They are no longer mutual adversaries, as we have found them to be in all the philosophers we have examined. This affords a liberating perspective for technological development.[242]

3.4 Georg Klaus: Marxism and Cybernetics

3.4.1 Introduction

Any study of technology and the future would be incomplete without a consideration of technological development from the perspective of marxist philosophy.

Georg Klaus[243] promulgated his views widely in the 1960s and early 1970s and included cybernetics in the scope of his reflections. Marxism has always been interested in technology. As a means of production, technology constituted the most important element in Marx's notion of "production power." Nevertheless, marxism initially greeted cybernetics with suspicion, perceiving in it a capitalistic instrument for suppressing the workers, for exploiting the proletariat.[244] When Klaus posited a connection between dialectical and historical materialism on the one hand and cybernetics on the other, this was not well received in the communist world, he reports. "This intended marriage of historical materialism and cybernetics would – so it was alleged – mean a desertion of the foundations laid by Marx, Engels and Lenin" (KuG ix). Klaus would deny this charge. He seeks to demonstrate that dialectical materialism is the only legitimate basis for the existence of cybernetics, that it is the only philosophy "which emerges confirmed and enriched from the confrontation with cybernetics and information theory, to which all philosophies are presently exposed" (KuE 206).

According to Klaus, both Wiener and Steinbuch have often inadvertently given expression to dialectical-materialist trains of thought while developing cybernetics. Klaus can therefore appeal to them. At the same time, he passes up no opportunity to show that in principle they are inadequately equipped, as servants of capitalism, to solve the problems pertaining to cybernetics. Their lack of a dialectical-materialist starting point prevents them from giving a correct interpretation of the development of cybernetics and thereby of modern technology. This deficiency obviously entails in addition a misconception of the significance of modern technology for society and its future.

Marxism-leninism is taken for granted in Klaus's publications. He embraces this philosophy as an incontestable dogma.

Before presenting Klaus's view, I must say something about the laws and categories of dialectical materialism. Once this is taken care of, it will be easier to understand why Klaus believes that dialectical materialism and cybernetics belong together and require each other – a belief that determines his answer to the question what cybernetics is.

After this general orientation, it will be possible to address particular subjects. Concerning the question whether machines can think, I will look first, in a kind of intermezzo, at Turing's answer. Then I will go on to Klaus's view. Turing's positive answer is denied by dialectical materialism on the basis of its own main question, namely, the inquiry concerning the relation between matter and consciousness. Matter is primary and consciousness secondary, since the latter derives from the former by way of a dialectical process.

Klaus elaborates on his negative answer to the question whether machines can think by referring to the categories and method of cybernetics. Only on the basis of such reference, he believes, can the real significance of the development of cybernetic machines be appreciated. These new machines will have consequences for working people, for production relations, and for society. Society will have to be – or become – communist, since only then will the fruits of technological development be fully realized and the opposite of what technological development really contains be averted. Klaus sees in the possibilities of cybernetics a confirmation of the laws of dialectical materialism. The workers will be liberated, and classes will be abolished.

Klaus expounds this expectation by taking up the following subjects: the change in the relation between schematic and creative work; the altered man/machine symbiosis; and the relation between consciousness of self and societal consciousness. Klaus persistently deals with these subjects against the background of views such as those implied by behaviorism, pragmatism, and physicalism, with which Turing, Wiener, and Steinbuch respectively present capitalism, according to Klaus.

I will conclude this section with some observations on Klaus's view of technology and the future.

It should be noted, finally, that neither in the synopsis nor in the critical analysis of Klaus will I be able to engage in an extensive discussion of the development of dialectical materialism and the divergences within it. I shall consciously assume this restriction, trusting that in doing so I am not short-changing the fascinating view of technology and its development presented by this philosophy.

3.4.2   The laws of dialectical materialism and certain of its categories

Dialectical materialism goes beyond the old materialism. The old materialism – especially French materialism – was static and unhistorical; it did no justice to development. Dialectical materialism desires to avoid being abstract; it wants to be an instrument to change reality. Marx was not concerned to understand or interpret societal life but to *change* it. Stalin put it tersely: "Marxism-leninism teaches that theory severed from praxis is dead and that praxis without the light of revolutionary theory is blind."[245]

Marx borrowed the dialectical method from Hegel. He filled it, however, with a content diametrically opposed to that of the idealist method. He spun it around 180 degrees, thereby placing it – so he claimed – upon its feet. For Marx, matter is not a product of (transcendental) spirit; rather, spirit is itself matter in the sense of the form of motion of matter. Marx recognizes no transcendence whatever. Matter, to him, is the only reality. The thinking consciousness is the active processor of material reality.[246]

In his dynamic dialectic, Marx often uses the (German) term *aufhe-ben*, which has a twofold meaning. In the first place it means that the contradictions of the dialectic are overcome in some subsequent stage, but in the second place it means that the contradictions are merged and preserved on a *higher* niveau in a new unity in which the contradiction no longer operates.

One refers to *historical* materialism in the case of Marx, since for him the dialectical method concerns history – specifically, the development of the societal process. But in Engels, nature, too, is governed by the dialectic and is regarded as the starting point of the dialectic. Historical materialism is broadened and then designated *dialectical* materialism.

According to dialectical materialism, the development of societal life is controlled by the production of material goods. Two factors work together here – production forces and production relations. The production forces include raw materials, tools, instruments, and machines (collectively called the means of production) and the human labor force, labor experience and labor skill. By production relations Marx understands the relations between the people involved in production. The existing production relations are a reflection of property; they are property relations.[247]

There is a dialectical connection between the production forces, which include the workers, and the production relations, which are established by the capitalists. Production forces and production relations presuppose each other and penetrate and cannot exist apart from each other, but at the same time they exclude each other. This means conflict – the conflict between the working class and the propertied class, the capitalists.[248]

The class struggle is historical and dialectical, which is to say that it generates an uninterrupted process of becoming and decay. In his main work, *Das Kapital*, Marx describes the laws that govern the course of the conflict between the proletariat and the capitalists. He shows that the class struggle must result inevitably, through the development of the forces of production, in a process that is to be characterized by the "abolition" (to be gained with the help of the revolution) of the class struggle – in short, by the rise of a communist society. In this society the means of production will be held in common and the exploitation of workers by capitalists will be at an end. Workers and capitalists will be absorbed into a new society, the communist one, in which there will be no exploiters and no expoited. The relations between them will have made way for communal freedom. In a communist society, however, the development of the means of production does not come to a standstill. From then on development is continuous and progressive.

The laws of societal development find general expression in the laws of dialectical materialism. Matter moves permanently in an ascending line with ever new qualitative forms. These new forms surpass the earlier ones, although the earlier ones constitute their necessary presuppositions. As far as societal life is concerned, the dialectic attains completion and is abolished in the communist society.

Klaus presupposes the laws posited by dialectical materialism and the categories that flow from them. To follow Klaus's argument, it will be helpful to state these laws and categories briefly.

In dialectical materialism, three laws are credited with the control of matter. The *first* is the law of quantity changing into quality, which is also called the law of transition from a continuous to a discontinuous development (for example, rising temperature changes ice into water). The *second* is the law of the unity and conflict of opposites, which is also called the law of the whole and the parts (for example, a magnet with positive and negative poles). The *third* is the law of the negation of negation as an expression of dialectical development which presupposes the first two laws (for example: ice, water, vapor).[249]

These three laws concern respectively (1) the form, (2) the origin and essence of the self-sufficient (self-)movement of matter, and (3) the coherence and progress, that is, the direction of dialectical development.[250]

While the laws of development are a means for predicting the future scientifically, doing so is possible only along general lines. For specific components, these laws need to be complemented by the so-called categories, even if these categories themselves also express the significance of these general laws in a specific way. "By categories are to be understood the logical basic concepts which reflect the most general and most essential attributes, aspects and relations of the things and phenomena of reality."[251]

Among the categories to which Klaus refers in connection with cybernetics are the following:

1. The category of *essence and phenomenon* is the expression of a dialectical unity: " ... the phenomenon is identical with the essence, and yet it is not; it expresses it adequately – and not adequately."[252] Klaus identifies this category with that of *structure and behavior*. Sometimes he speaks of the category of the *whole and the parts*.[253] This category is then barely distinguishable from the second law.

2. The category of *causality and finality*, sometimes called that of *interaction*, makes it clear that the mechanical materialists' classic notion of causality fails to do justice to dialectical material reality. Cause and effect or end reciprocally determine each other.[254]

3. The preceding category is complemented by the category of

necessity and contingency, which is also called the category of *necessity and freedom*. For society this category is rendered more precise in the category of *the social and the individual* or *the societal and the personal*. Occasionally it is also called the category of *the universal and the individual*. The dialectical unity is present in all these categories: "... in the consciousness of the acting subject, necessity coincides with freedom and freedom with necessity...."[255]

4. The category of *possibility and reality* seeks to express the fact that dialectical materialism has no room for a fatalistically understood historical determinism or historical idealism. "Possibility is the necessary prerequisite for the rise of a new reality."[256] This category is decisive for the relation between theory and praxis – the abolition of philosophy through its realization.[257]

5. The category of *content and form*, which is closely related to the first category above (that of essence and phenomenon), is given more precise definition in the relation between production forces (content) and production relations (form). The "content" determines the essence, the ground, of an object; the "form," in contradiction, provides for the binding of the elements of the content into a unity: "The dialectical unity of form and content is clear in its contradiction only when form and content are viewed not as static but as dynamic, in the development process of the object."[258]

From the main laws of historical materialism and from the thesis that labor constitutes the real essence of humanity it is apparent how much marxist philosophy is interested in the means of production, the technological panoply of tools and instruments. The development of the means of production is the basis for societal development. It is the basis *par excellence*, especially now that people, under the influence of the possibilities of cybernetics, are gradually being weaned from their tools. The machine takes over more and more human functions.

In more than one recent study of Marx, it has been emphasized that Marx has been given less attention as a philosophizing technologist than as a philosophizing economist. The renewed interest in Marx's view of technology is not unjustified on the basis of its merit; yet, it should also be attributed to the extremely fascinating, surprising development of technology itself, whereby the interest in what Marx had to say about it naturally gains more emphasis.[259]

From the following we will see how Klaus, too, seeks to make it clear that a responsible view of technological development is intrinsic to dialectical-materialist philosophy. According to him, this view leads to new perspectives, perspectives that could never occur within a capitalistic outlook.

3.4.3 Dialectical materialism and cybernetics

Before Klaus undertakes an explication of cybernetics, he must make it clear that dialectical materialism provides a basis for doing so, the more so since he will have to appeal to Steinbuch and Ashby for many ideas and trains of thought about cybernetics.[260] "I could do this because both of them, whether they will admit it or not, and in spite of serious philosophical shortcomings appearing in their works, constantly and clearly present dialectical and materialistic trains of thought" (KiS 9).[261]

According to Klaus, cybernetics is not opposed to dialectical materialism, as was originally thought; rather, it will only come to full development on the basis of this philosophy. Cybernetics is to be regarded, then, as one of the most impressive corroborations of dialectical materialism. "Cybernetics is ... in essence dialectical and materialistic" (KiS 484).[262]

According to Klaus, Karl Marx used the *method* of cybernetics in describing the capitalistic system, although he was not aware of it himself. Moreover, Marx believed that the laws of economic crises are subject to mathematical description, and that mathematics is the heart of cybernetics.[263] Therefore it is correct "to point to Karl Marx as the first cyberneticist" (KiS 219). "Karl Marx can be pointed to with far more justification than James Watt as the forerunner of cybernetics, as the first cyberneticist" (KiS 467).[264]

Cybernetics deepens dialectical materialism, and vice versa. Whenever cybernetics is developed on the basis of dialectical materialism, it is freed from a great deal of speculation and fantasy, such as appears in the West. For example, there is no reason, given marxist philosophy as basis, to speak of cybernetic machines as a "Satanic power."[265] "In any case, we believe that this new science will raze every wall thrown up against it, and that much that is presently declared impossible for theological reasons, or emotional objections, or reactionary political considerations of the most various kinds, will actually be demonstrated by cybernetics in five, ten, or twenty years, through technological realization" (KiS 8).

With the development of cybernetics, dialectical materialism has been afforded an instrument or a method of making known the value of dialectical materialism for all the sciences.[266] Cybernetics constitutes a kind of mediator between philosophy and the many special sciences[267] and plays a kind of synthesizing role in the system of the sciences. "Philosophy alone cannot solve this problem. It requires as a mediator between the sciences just such a discipline as cybernetics ..." (KuG 335). On the one hand cybernetics belongs to the field of philosophy,

266

and on the other hand it has a special significance for every special science: "...it has a dual countenance – as both philosophy and special science" (KiS 121).

Thus cybernetics creates unity among the sciences, which are constantly growing in number: "It counteracts the tendency toward more far-reaching preoccupation, splintering and isolation of the special sciences" (KiS 32).[268]

The method[269] of cybernetics is universal, thus bridging the gap between the natural sciences and the social sciences. Both for nature and for society, cybernetics is an instrument of control. "It can be regarded as an effective instrument of socialistic planning and organization of the economy and of societal relations" (KuG xi).[270] Cybernetics provides for the confirmation and strengthening of the unity of human culture. The methods and categories of cybernetics are even applicable to history, since history progresses on a historical-materialist basis in accordance with objective laws. "History as the history of actions stands theoretically in the vicinity of cybernetic systems theory and game theory" (SGF 79).

In conclusion, cybernetics involves not only human relations but even the essence of humanness. With cybernetics it is possible to make models of the logical and mathematical thinking of the human being. For this reason especially, cybernetics is much more important to the development of modern technology than atomic energy and space flight.[271]

3.4.4 What is cybernetics?

Cybernetics, for Klaus, is first of all a theory about certain machines. "Cybernetics as machine theory is not in the first instance concerned, in contrast with all earlier machine theories, with ways of processing materials by machine, or with the transformation of various sorts of energy by machines; it is interested instead in the control processes which the functioning of both these types of machines requires. It has – and this is what is essential – gone beyond this and introduced a new type of machine into the circle of comprehensive and general theoretical investigations" (KiS 373). The electronic machines seem ever more similar to human beings in their behavior or functioning. They begin to perform tasks which were previously carried out by people working together. Cybernetics thereby assumes significance for the individual, for people cooperating, and for the relation between human beings and the machine. Stated very generally, we could say: "Cybernetics is the science of the possible modes of conduct of possible structures, and then not of arbitrary structures but of dynamic structures, that is to say, of structures that are imbedded in time-dependent processes" (KiS 21). Cybernetics investigates the coherence of dynamic systems with their component systems – with an eye, thus, to the whole/part relation – and it attempts to duplicate the result of that inquiry in machines, so that cybernetic machines can simulate the working of the investigated system.

Cybernetics possesses four essential aspects: the system aspect, the control aspect, the information aspect, and the game aspect. The system aspect is foundational. Regulation and control occur within or at a system. Information conveyance and processing are important for cybernetics insofar as they occur between or within dynamic self-regulating systems. In cybernetics, game theory concerns the conflict between the goal and the disrupting factors of the system, the conflict between various systems, and the conflict between various subsidiary systems and a greater system.[272]

Klaus concurs with Wiener that mathematics – especially probability theory and information theory – is the heart of cybernetics. The various aspects of cybernetics can be treated mathematically. The possibility of mathematizing is given with the fact that systems undergo a "measure" of influence and that there is a "measure" of resistance to this influence.

Although Klaus recognizes the significance of cybernetics for many sciences, this does not mean he wishes to level the differences between the sciences. The qualitative diversity is not abolished by mathematics. The category of quantity and quality assures this.

Meanwhile, cybernetics does imply that machine models can be made of all systems to be studied. The question to be answered is the question of the relation between a model and reality. Reality may not be identified with a model. That would lead to faulty conclusions – for example, that the computer, which simulates the brain, can think. It would then be forgotten that cybernetics as a science makes use of abstraction and that the influence of this abstraction extends to cybernetic machines. "What, now, is cybernetics as science? What is unique about its scientific abstractions?" (KiS 12). Only when this question is answered correctly – and according to Klaus, it can be answered correctly only with the help of dialectical materialism – do the conduct and significance of cybernetic machines come to appear in the correct light.

3.4.5   Can machines think?

One of the first questions evoked by the possibilities of cybernetics is the question whether cybernetic machines can think. This question brings to the foregound the content of the basic theme of dialectical materialism, namely, that of the relation between matter and consciousness.

Klaus handles this question against a background of answers that are often given in the West. The one that is proposed by Turing plays an especially important role.

3.4.5.1 Intermezzo: Turing's answer to the question

In 1960 Turing[273] wrote an important article about the question whether machines can think. He is of the opinion that definitions of "thinking" and of "a machine" provide no adequate answer. "Instead of attempting such a definition I shall replace the question by another, which is closely related to it and is expressed in relatively unambiguous words" (CT 11). A clear answer can be attained via a thought experiment. In an imitation game, interrogator C asks questions of person A and machine B, without, however, having any way of learning which is which except by posing questions and evaluating the written answer (yes or no). This experiment is thus substituted for the question whether a machine can think.

According to Turing, the thought experiment has the advantage of providing for the elimination of human qualities which he believes are not decisive for thought, such as finding something attractive. Giving a written answer to questions eliminates the fact that machines work much more rapidly than people. "The conditions of our game make these disabilities irrelevant" (CT 13). Turing's prediction is that " ... a machine can be constructed to play the imitation game satisfactorily ..." (CT 13), and that the question posed is thus to be answered in the affirmative. "The question is to be decided by an unprejudiced comparison of the alleged 'thinking behavior' of the machine with normal 'thinking behavior' in human beings" (CT 9).

The machine Turing has in mind is a digital computer: " ... they can in fact mimic the actions of a *human computer* very closely!" (CT 15 – italics added). That is also true, and especially so, when these machines are fitted out with a randomizer. Looked at theoretically, a computer with an infinite memory (the so-called "infinite capacity computer") would also suffice. This universal machine would be in a position to imitate the behavior of every other digital machine, among which Turing reckons the human brain. Turing believes that by the year 2000, computers will be built with a capacity of about 10^9 bits (the human memory is estimated at 10^{10} to 10^{15} bits). Turing believes that if such a machine were to be employed in the test mentioned, the interrogator would have only about a thirty percent chance, after five minutes, of making the correct identification. "I believe that at the end of the century the use of words and general educated opinion will be altered so much that one will be able to speak of machines thinking without expecting to be contradicted" (CT 19).

As machines with a large memory require rather a good deal of difficult programming work, the computer in the experiment might be replaced by a learning machine. The program of such a machine would be similar to the content of the memory of a child. Thereafter, the learning process might go on in roughly the same way as the human learning process. Given these learning machines, the question at issue here is answered even more easily in the affirmative.

Initially, according to Turing, the similarity between the human being and the machine will only be purely intellective. However, he anticipates that machines in the future will be able to hear, speak, buy, and sell.[274] Turing's assumption is that the human mind is wholly mechanical. Expounding, he compares the human mind to an onion. Peeling an onion layer by layer is similar to the progressive formaliz-ing of the function of the human brain. The question is whether there exists some remainder not subject to formalizing and hence to me-chanization. Does the onion have some irreducible pit, or is the last layer empty of all content? "In the latter case the whole mind is mechanical" (CT 30).

On the basis of this assumption and of his behavioristic point of departure, as evident in this example, Turing refutes many objections raised against his view that machines can think. These I shall not pursue. Turing admits two objections, however. The first is that tele-pathy and clairvoyance might render his experiment questionable. The human being has more possibilities in this regard than the computer does. In order to be able to carry on his experiment in spite of this objection, Turing proposes a "telepathy-proof room," without, how-ever, suggesting how such a thing might be possible.[275] The second objection is based upon the theorem of Gödel, who states that within a consistent logical system, propositions can be formulated which can be neither demonstrated nor refuted within the system.[276] The computer cannot formulate such propositions itself. Even in the case of a com-puter with an infinite memory, " ... it is argued that it proves a disability of machines to which the human intellect is not subject" (CT 22). According to Turing, however, undue weight must not be attached to this conclusion, since the human being has more deficiencies in comparison with the universal digital computer. The human being makes more mistakes, and while on the basis of Gödel's theorem he may in some particular respect surpass one machine, he will never be superior to all existing machines simultaneously.

3.4.5.2 Klaus's answer

According to Klaus, the question whether machines can think points to the basic question of dialectical materialism, namely, the relation between matter and consciousness. "Cybernetics resurrects for discussion certain aspects of the basic question of philosophy, which of course does not mean that this question can or must be answered anew or at variance with dialectical materialism" (KiS 35). Cybernetics can contribute to a clarification of the relation between matter and consciousness.[277]

Dialectical materialism posits that matter is primary and consciousness secondary. In other words, matter ultimately determines all forms of consciousness. Lenin emphasized that consciousness is an inner state of matter, and that consciousness does not allow itself to be separated from matter. The vulgar materialists are mistaken, of course, in believing that consciousness is something material, to be regarded as matter. Consciousness is even not an attribute of matter; it is an attribute of the form of motion of matter. In other words, it is an attribute of an attribute of matter.[278]

This fundamental relation between matter and consciousness, which Klaus assumes *a priori*, is decisive for (a) the question whether machines can think and (b) the question whether concepts of information and communication belong to the field of matter or to the field of consciousness.

Re a. A clear answer to the first question comes down to clarifying the boundary between matter and consciousness. Klaus attempts to do this through confrontation with various answers that are given, answers that he cannot share himself. In general it can be said that in the West, the mere *performance* of cybernetic machines has been sufficient for people to conclude that they think. It is clear to Klaus that in that case, the boundary between man and the machine fades, ultimately to disappear, to humanity's detriment. This is especially so when the future possibilities of technology are taken into account in the comparison. "Just because the behavior and the results of two systems are the same, it does not follow that we are to identify them" (KiS 44).

Klaus's opposition to the mechanical materialists and behaviorists does not imply that he wishes to join the ranks of the idealists. Human thought is not something supernatural. Therefore " ... the 'no' of dialectical materialism indicates the proper place of that relative truth which is present even in the grossly exaggerated claims of many a cyberneticist concerning machine thinking" (KiS 45). A good answer to the question whether machines can think is possible with an ontological approach.

It is striking at the outset, says Klaus, that whenever there is talk of machines thinking, it is always mechanical and schematic thought that is meant – not creative thought. But schematic thought presupposes creative thought. The former always requires an algorithm, and the algorithm, according to Klaus, must have been discovered by human beings.[279] "By an 'algorithm' is meant an acquirable and reproducible exact formula, with the help of which it is possible to solve problems of a specific type step by step" (KiS 40). Klaus agrees with Turing's contention that with the help of algorithms, thought processes can be built into machines.[280]

However, even though every algorithm is theoretically capable of realization with the universal Turing-machine, it must not be forgotten that such a machine, theoretically regarded, is a borderline case. Given its infinite memory alone, this machine is impossible to manufacture. Yet, in principle the human being's superiority to the machine involves precisely this: human beings must find the algorithm, and therefore it is impossible for the machine to surpass the human being as long as there is no algorithm. The following argument is also valid: "It is mathematically proven that there are various problems – for example, the general decision problem of mathematical logic – which it is impossible in principle for the Turing-machine to solve" (KiS 41, 42). Moreover, the machines are made of inorganic matter. Human beings have projected thought structures and thought processes into this matter. This projection into inorganic, physical matter has been made possible via abstraction from organic matter. "Having built such a physical model, however, we are in a position to translate particular thought problems into physical terms, and vice versa" (KiS 50). Even when the machine functions more rapidly and precisely than humans do, we must not forget that the machine knows only physical processes, says Klaus. If this point is overlooked, then of course it will be alleged that machines are in a position to solve logical problems and to think. The physical processes of the machine yield a physical result, one that human beings must translate into the sphere of consciousness. Therefore Klaus says: "Strictly speaking, the machine cannot perform even the tiniest operation – for example, the sum $1 + 1 = 2$" (KiS 53).

Klaus's hardheaded realism is attractive in comparison with the many fanciful and scientifically irresponsible expectations associated with the latest machines. Nevertheless, his ultimate conclusion goes too far. The last subject-function of the computer is indeed the physical function. The computer cannot count subjectively, let alone think. Counting processes can, however, be objectivized in the computer. The counting function as an object-function is actualized in the process of the machine. In this sense it is possible to speak of an analogy with human counting. Klaus speaks of it too. He correctly perceives that this analogy does not constitute an adequate basis for concluding that there is an identification. The mistake that is often made – Turing makes it too, says Klaus – is to reject the matter/consciousness problematic, and thus human thought, only to conclude subsequently on the basis of their performance that machines can think.[281]

Talk of analogies and cybernetic equivalencies[282] is legitimate, since what is accented there is simply the agreement between qualitatively different areas. "In this case scientific accuracy requires that one proceed to carefully ascertain if it is admissible to transfer concepts that are meaningful there to the lower niveau, and whether or not essential qualities connected with the higher niveau are lost in the transfer" (KiS 54). An analogy passes over into an identification not when either the behavior or the results alone of two systems agree but when the material, structure, and functions of the two systems are all the same, says Klaus.

Klaus's answer to the question whether machines can think is implicit in all this. Machines cannot reflect; they have no self-knowledge, and they are unable to hold sway over themselves. "However complicated machines may be, they cannot think; they can have functions and achieve results which equal our brain – yes, and even surpass it. Yet, what they do is not thinking in the sense of epistemology" (KiS 65).

Re b. The answer to the preceding question is decisive for the question: Do information and communication belong to matter or to consciousness?

This question has been raised legitimately by the development of cybernetics. Wiener stated that information is neither matter nor energy.[283] Information and communication (as the conveyance of information) represent, alongside matter and energy, a third aspect of matter.[284] Klaus, however, does not see the need for reserving, alongside matter and consciousness, a third "zone of being" for information.[285] Information, according to Klaus, is linguistically determined and therefore belongs to consciousness; information is a derivative of consciousness. Information is "a whole consisting of a physical bearer and a semantic" (KiS 81).[286] That we nevertheless speak of "information technology" is the result of our being able to draw an analogy with information as a derivative of consciousness. Klaus is not unwilling to use the term *information technology*, provided it is not forgotten that information technology involves abstraction from the semantic. "Purely physical operations, which occur in a closed control system, have no semantic" (KiS 82). Only when the above-mentioned process of abstraction as the basis for making cybernetic machines, which project logical and mathematical thought into physical matter, is taken into account is the mathematical kinship between information theory and thermodynamics clear. Entropy affects the physical bearers of information, "while the semantic of information is itself a fact of consciousness" (KiS 92). In cybernetic machines the structure or organizational aspect has assumed importance alongside the matter and energy aspect. Information, which belongs to consciousness, is imitated in these machines with the help of physical signals.[287] "Viewed purely physically, information is a row of signals lined up in a certain order" (KiS 80).[288]

From Klaus's answers to both questions, it is apparent that he rejects the informationism and behaviorism of Wiener and Turing respectively, together with the physicalism of Steinbuch.[289] He considers the machine to belong to the field of matter. The human being, however, consists of matter and consciousness. This is precisely what enables him to make machines which, in their conduct and results, bear a *likeness* to man.

It may not be concluded from Klaus's answers that he posits the human being as free person over against the machine. As he says himself, he rejects both the vulgar materalists and the idealists. However severely he differs with Steinbuch on the computer's capabilities, he does agree with him that consciousness is objective nature, with nothing "supernatural"[290] about it that might put it beyond the reach of science. That man might be free seems to be suggested by the phenomenon of reflection, because reflection "is a movement of thought with a lawfulness of its own" (KuE 40). "This aspect undoubtedly arouses in the individual a feeling of freedom and of the possibility of spontaneous decision" (KuE 46). This suggestion cannot be justified, however, on the basis of dialectical materialism.

Klaus's view that consciousness can be fully investigated scientifically means that he cannot exclude the possibility of an artificial consciousness. "It is in principle not impossible to imitate technologically the operations of nerve cells, the internal processes of which are essentially determined by the interaction of ribonucleic acids" (KuE 253). In that case it would also be preferable not to speak of machines any longer. A transitional phase is displayed by the machines which, equipped with the "trial and error method" or with Steinbuch's learning matrices, are no longer limited to schematic thinking but also partially take over creative thinking. In the case of these machines, one can speak of a relative consciousness.[291]

As I see it, this last conclusion of Klaus is in line with his materialism, even if he seems to contradict himself. "It should be remembered above all that cybernetics finds itself at the very threshold of its development, and that what it has achieved in such a short time only allows us to surmise what great scientific advances undoubtedly can be anticipated in this extremely complicated field of scientific inquiry in the future" (KuE 255).

3.4.6 The categories and method of cybernetics

3.4.6.1 The categories

We learned above that Klaus believes cybernetics to be justifiable only on the basis of dialectical materialism: "Cybernetics confirms dialectical materialism, cybernetics is based upon dialectical materialism, cybernetics needs dialectical materialism" (KiS 97).[292]

The various categories of dialectical materialism crop up again in cybernetics. In the definition of cybernetics, one finds the category of the dialectical unity of possibility and reality. Cybernetics is "not a theory of actually existing dynamic self-regulating systems but of the *possible dynamic self-regulating systems*" (KiS 99 – italics added). A variant of this category occurs in the dialectic of theory and practice. Every scientific abstraction, as possibility, has a reverse side in technological reality, in the application. This shows that cybernetics has a legitimate place within marxism as the philosophy of action. "Cybernetics...has, as it were, made Marx's eleventh thesis on Feuerbach its leading motive, the one that says that what really matters is changing the world" (KuG 4).[293]

The category of quantity and quality also occurs in cybernetics. "Cybernetics has shown that with the increase of the number of elements, more and more new qualities appear in the material systems" (KiS 126). Machines are thereby able (taking into consideration what Klaus has said about the abstraction process of cybernetics) to "learn." For example, they may learn to "propagate" themselves, and so forth.[294]

The category of causality and finality, also called the category of interaction, plays an extremely important role in cybernetics. "One can therefore, with some justification, designate cybernetics as the theory of interaction" (KiS 146). The feedback principle, worked out practically in various forms,[295] agrees with this category. Since feedback must also deal often with eliminating external influences, it is at the same time an application of the category of necessity and contingency. In a certain sense, the relation between the system and its environment is also a form of interaction. "It is clear ... that the category of feedback is closely related to the general philosophical category of interaction" (KiS 145).[296] A game is to be observed, as it were, going on between the intended conduct or result of a cybernetic system and the influences at work on that system. And that inter*play* is subject to analysis along the lines of mathematical game theory: "the dialectic in its original, Heraclitian form," it appears, is "a struggle, a battle" (KiS 161).

Within the cybernetic system itself, the category of necessity and contingency also discloses itself. "Information conveyance is itself a process which always involves the unit of necessity and contingency, of useful information and noise" (KiS 160).

An *ultrastable system* is a cybernetic system equipped to react to various disturbances, sometimes simultaneously and sometimes not. In it are to be discerned the category of necessity and contingency as well as the category of quantity and quality.[297]

A *multistable* system is one made up of ultrastable component systems which sometimes function independently for long periods. The dialectic of the whole and the part is present in a multistable system. And the whole is more than the parts here; that is to say, the total system is more than the sum of the behaviors of the separate, isolated parts.[298]

In this way Klaus attempts to make it clear that a philosophical justification of cybernetics is possible only on the basis of dialectical materialism. The mechanical materialists, who tried to explain everything on the basis of causes, are inadequate to the task. The "laplacian demon," which is a piece of speculation, according to Klaus, must then provide a way out. Dialectical materialism, by contrast, does justice to the spontaneous and the coincidental. For the explanation of such things – for example, the phenomenon of life – it does not need to appeal to supernatural forces or metaphysical final causes, as idealism does.

Thus it is apparent that Klaus regards these categories as controlling not only technological systems but also the whole of reality, which is to say that reality is to be viewed as a "whole" composed of cybernetic systems, each of which may in turn be duplicated in a technological system. In appraising such a duplicate system, the problem of abstraction as referred to in the preceding subsection must of course be taken into account.

3.4.6.2 The method of cybernetics

The categories above determine the content of the method of cybernetics and provide a criterion for evaluating cybernetic systems.

The method of cybernetics consists of the *mathematical description* of the behavior of a cybernetic system, the *principle of the "black box,"* and finally the *model principle* or *analogy principle*. I will say something briefly about each of these components in turn.

A *mathematical description* of the behavior of dynamic, self-regulating systems is given in the dialectical unity of quantity and quality. "The functions are descriptions of qualities" (KiS 179). The mathematical description is not formal logic, then, but touches the essence of reality, according to Klaus. However, the essence of reality is not fully grasped thereby. Therefore a quantitative approach must always be complemented by a qualitative approach. "The only important thing is that neither of the two is adequate for a determination of the case, but from the quantitative determinations one can draw conclusions as to the qualitative ones, and vice versa" (KiS 182). One must continue to keep in mind, however, that it is the *behavior* of the system which is in view. Only by maintaining due regard for this abstraction can a behavioristic or idealistic derailment in the appraisal be avoided. The matter and structure of a system are not reflected in a mathematical approach, although it is possible to incorporate in the mathematical description certain constants that point toward the matter and structure of a specific concrete system.

The *method of the "black box"* as introduced by Ross Ashby, together with all the related conceptualizations, constitutes the *core* of the cybernetic method. With this method an attempt is made to first handle complex systems theoretically, so that a model can then be made on the basis of the theory. The system is initially regarded as unknown and fed certain "inputs," with the results or "outputs" being registered as precisely as possible. The relation between input and output is translated into mathematical terms. This method of working is especially suited to explicating biotic and social systems, and to controlling them eventually as well.

In earlier times simple elements were used for explaining complicated systems, and when they proved inadequate, people turned to supernatural explanations to round out the picture. With the black-box method, which assumes the complex system, no supernatural explanation is needed. As a result, the possibilities for applying cybernetics with this method have been greatly expanded. The method, in short, is as follows: "In its elementary form it consists in this, that certain magnitudes working into the system, together with the results following from their working through the system, are brought into functional coherence. It is of course assumed that the inputs and results are observable, measurable, and so forth. It is true that the internal structure of the black box is not unequivocally specified by the transitional function thus brought about, but the behavior of the system concerned is established in relation to the magnitudes involved" (KuG 140).[299]

The questions this generates are about as immense as the black-box method itself is simple and synoptic. I will let the matter rest here. It is sufficient to note that Klaus believes that only dialectical materialism can provide correct answers. The various categories of dialectical materialism – and thereby of cybernetics – which disclose themselves in this method are Klaus's assurance of that. Taking these categories into account obviates all speculation.

On the basis of the black-box method and of the mathematical description of systems, which is part of that method, it is possible to make *models* of the investigated systems. While the model concept in cybernetics need not lead to a cybernetic machine, I will limit myself here to such a development, for as Klaus also says: "The black-box method is above all a general construction method for cybernetic machines" (KiS 233).

Elsewhere I have emphasized that Klaus warns against allowing Gestalt psychology, which is a justifiable abstraction in itself, to degenerate into a mechanistic or behavioristic philosophy. Moreover, he also warns against idealism, which denies that pronouncements about the qualitatively distinct fields of objective reality can be made on the basis of cybernetic models.

The model is the bridge between theory and practice, says Klaus. It is theory on the basis of the abstraction employed; it is praxis to the extent that it must function practically. "Thus the model method, which is very often applied by cybernetics, is not just one epistemological principle alongside others; it is the principle uniquely suited to establishing the connection between theory and praxis" (KiS 277).

Klaus elucidates what he means by using the example of a model of the human brain. In the background of his view there always remains, as we noted earlier, the fundamental relation between matter and consciousness. "For in the final analysis it is man who builds these machines and gives them their program" (KiS 267). A model of the brain would behave like the brain. Because there is a dialectical unity between behavior and structure, the model must also reveal something about the structure of the brain.[300] The structure of the digital computer is similar in certain respects to the structure of the brain. The similarity involves discontinuity in functioning, for example. However, the "all or nothing" (yes or no) principle of the digital computer is apparently not to be found in the brain. Influenced by chemical factors, the neurons function rather in the form of what we might call a "more or less."[301] Moreover, the functioning of the brain is not simply discontinuous; it is also characterized by the dialectical unit of continuity and discontinuity. In analog calculating machines, the continuous processes are duplicated. Even an interplay of digital and analog machines does not lead to identification with the brain. Apart from abiding differences in material composition, Klaus demonstrates, with the help of mathematical evidence, that any such interplay between digital and analog machines only enlarges the differences with the brain. "The construction of a complete brain model from component models produces an ever greater difference between the model and that which is being modeled" (KiS 254).[302]

Klaus reaffirms that he is not of the opinion – an impossibility for him, since he is a dialectical materialist – that the brain is incapable of duplication. The *homunculus* is indeed possible, he asserts, but not in the field of machinery. "Whoever admits that matter brings forth consciousness must also admit that the repetition of this process is possible in principle" (KiS 255).[303]

According to Klaus, dialectical materialism is the only philosophy to shed a correct light on the cybernetic model. All talk of machines that can think and that not only equal but surpass mankind in everything issues from philosophies that serve reactionary forces – i.e. capitalism. Capitalists encourage the fear of modern machines in order to hinder technological progress and keep the control of technology in their own hands. Marxism must therefore warn against the incorrect terminology that is now threatening to become ingrained. "The fact that this false reasoning is thoughtlessly repeated in hundreds of books and periodicals can, perhaps, if not resisted, lead to this false reasoning slowly becoming ingrained; it cannot, however, make true something that is fallacious in principle. One of the tasks of a scientific philosophy is to prevent such language from becoming ingrained" (KiS 286).

3.4.7 The development of cybernetic machines

Klaus believes that a correct view of technological development can be attained if the development of the technological panoply of tools and instruments is viewed in historical perspective. Moreover, the societal situation in which technological development occurs must be taken into consideration. On the one hand technological development may alter the societal state of affairs, but on the other hand the societal situation may amount to an impediment in principle to the progress of technology.

The tools of the old technology were an extension, an improvement, a reinforcement of human organs. Modern technology, however, is characterized especially by the machine. "Each *machine* is something of an automaton, to the extent that it replaces certain human activities" (KiS 382). Initially the machine performed work for people, but people had to guide the production process completely. This stage of technology is termed *mechanization*.

Automation commences when guidance, too, is taken over by machines. In this case human beings preset both the end result and the production process; the work itself is automated. In the present stage of modern technology, only the final product is still determined by people ahead of time; the machine itself chooses, depending upon circumstances and bound by the given possibilities, the way of production. "Importantly... the issue is one of liberating humanity from repetitious patterns of not only physical labor but also intellectual labor" (KiS 385).

The final stage of automation is approaching full development and will, among other things, "have an effect upon the future forming of our material and intellectual being" (KiS 386). Just how it will do so will depend upon the societal structure. "Where socialism has triumphed and will triumph, these new lines in the development signify essential factors in the transition of humanity to the communistic societal order" (KiS 386).[304]

With the application of the cybernetic principle of feedback, the final phase of automation has dawned. Previously, machines were constructed on the basis of linear causality – a special instance of circular or dialectical causality! (I should mention here that Klaus overlooks the fact that James Watt employed the principle of circular causality in constructing the famous regulator.) Klaus claims – correctly, as I see it – that the construction of machines on the basis of linear causality must become more and more precise as the allowable tolerances in the deviations of the product become fewer and fewer. Between cause and operation there must exist a precise, controllable, functional relation. "It satisfies the spirit of the demon of Laplace" (KiS 306).

The high demands and the precision that must be set for a construction based on linear causality impede technological progress. This is all the more the case because machines constructed on the basis of this principle function best when they are brand new. The tolerances are subsequently increased by wear and tear. Such machines, produced to rigid norms, are ultimately unstable. The production process itself is not flexible enough to allow adjustments to new technological developments and societal exigencies.

Cybernetic machines have quite different characteristics. They are not sharply specialized and so do not immediately fall victim to technological development, let alone impede it. "The transfer machine satisfies the metaphysical spirit of mechanical materialism, the spirit of linear causality; the automatons with more than one objective, on the other hand (an example is systematic application as adaptive automation), are in some measure dialectical" (KiS 396). The adaptive machines are not rigid in construction and are able to operate ever more precisely. This implies increasing economic benefit. The new machines are a production force of the first order. In contrast to the "transfer machines," moreover, they do not require mass production but are ready for individualized production. This is true specifically of the learning machines.

According to Klaus, the basis for this development is dialectical materialism: the stated categories of dialectical materialism find their application in the construction of cybernetic machines, which might therefore also be designated dialectical machines.[305] To the extent that people build these machines in the West, they make unconscious use of the intellectual property of dialectical materialism.

The manner of constructing these machines is as follows. A system that people in modern technology would like to duplicate is regarded as a black box. Using the black-box method, the system is broken down into components, each representing a function. Each function can be performed by a simple machine. Each elementary machine can be so constructed that it is able to optimize a particular function. The linking of such machines or components leaves us with an entirely new system, one that is extremely versatile.

The advantage of such construction is that no special apparatus has to be built. Moreover, the separate elements can be mass-produced. By adding new elements and by altering the linkage of such elements, a great number of different objectives can be achieved. In other words, with the help of standardized subprograms, an unstandardized total program can be built. "This means, however, that from now on the structural problems, the problems of linking the basic elements, assume increasing importance" (KuG 115). This is especially so when the basic elements, as in the computer, perform an irreducible function and the linkage elements increase in number. Linkage logic here provides a means of discovering meaningful linkages. A result of this development is the very scientific construction of these machines. Experience, intuition, and genius are found dangerous and are eliminated. Logic replaces experience and intuition and accordingly becomes a production force par excellence.

The development sketched here has been carried farthest in the construction and use of so-called strategic machines. These machines have a heuristic program. They do not have a complete algorithm as their foundation; rather they are based on a program that seeks the optimal solution to the problems posed. Subprograms that can learn and hence can improve themselves are joined with the aid of the trial-and-error method and tried out in the search for a solution to a specific problem.[306]

These strategy machines are utilized to simulate social games like chess and checkers, but their practical significance is to be found especially in the fields of economics and warfare. These machines will also have tremendous significance in space. All the planets have the same natural laws, even though each planet has unique and only partially known characteristics. "Rigidly programmed automatons would have to surrender, and only heuristic game machines would be in a position to successfully withstand the game against the natural conditions prevailing on these planets" (SiS 325).

These machines will eventually assume complete command of the state (so-called governing machines), of the army, and of economic life. The peril of these machines is to be equated with that of the hydrogen bomb. Although people have made these machines, the results they will produce cannot be predicted. It is clear that limitations will have to be imposed upon the machines if dangerous consequences are to be avoided. The question, however, is whether we can foresee all the possible dangerous consequences. "The positive and negative possibilities of such machines, which are technologically feasible, are tremendous in magnitude. They are as enormous as the positive and negative possibilities of the application of thermonuclear energy" (SiS 329).[307]

With this, an entirely new problem is brought into view. Klaus has persistently upheld man's superiority to the machine. But is humanity dependent upon learning machines, and are the machines employable for various strategies?

To resolve this problem, Klaus again returns dogmatically to his starting point in dialectical materialism. "Only when we have the courage, in scientific abstraction and relying on the principles of the materialistic dialectic, to take the lines of development now clear to us and extrapolate them into the future will we be in a position to meet the demands of the future" (KiS 392). In all of this, marxist philosophy has the task of being "not a chronicler of what has been but the herald of what is to come" (KiS 396). It must contribute to establishing the cybernetic way of thought, based on dialectical materialism, in the worldview of modern, socialist humanity. Moreover, "The matter is one of so co-ordinating the efforts of humanity as to bring about a societal situation in which hydrogen bombs and learning automatons remain subject to human control" (KuG 77, 78).

On the one hand, the capitalist world will seek to impede technological development as Klaus has sketched it, in order to keep the reins in its own hands and to exploit the workers. On the other hand, technological development cannot be blocked by the capitalists. Therefore, as this becomes apparent to them, they will send the workers home unemployed (exploitation in *optima forma*) and utilize the newest machines for waging war – which implies the self-destruction of humanity.

Klaus never takes this threat quite seriously, since technological development would appear to be overthrowing capitalist society with a degree of necessity that would have surprised even Marx. Technological progress dooms the capitalist system. The capitalists try to blame technology as a demon, but when they reproach it, they are only indicating how threatened they feel.

Capitalism and modern technology are incompatible, whereas communism and modern technology belong together.[308] Communism creates the right conditions for the use of computers, which, given their information content and information processing, can be a tremendous power. Capitalism is dependent upon this modern Golem because the conditions under which the computer works are the conditions of capitalism. Capitalism seeks to maintain existing labor relations, while modern technology itself refutes them. The computer cannot and may not serve some group interest. Only in a communist society is it meaningful to use computers. "The character and manner of production finally determine the nature of the societal order in all its facets" (KiS 390).[309] Engineers, mathematicians, philosophers, politicians, economists, and sociologists will collectively determine the conditions under which the computer ought to be employed. The basic principle here is "that everyone should so act that the principles of his action might at the same time be general principles of action" (SiS 330-1).

3.4.8 Humanity and labor: the relation between mechanical or schematic work and original or creative work

In view of the development of cybernetic machines, the nature of people's work must change. Because work is essential in marxism, because it constitutes the very essence of humanity, there is clearly a need for reflection on the changes in the world of work wrought by technological development.

One of the most urgent questions raised by automation is that of the relation between schematic work and creative work. Schematic work is increasingly preempted by machines. Does this mean that the worker is liberated to engage in creative work?

Marx has shown that the worker in a capitalistic society becomes alienated from his *work* and therefore from himself: "Capitalism, the capitalist way of producing, separates the worker from his product, that is to say, from the item in which he has externalized his own human being" (KiS 405). In capitalism, labor is forced labor. The worker has no part in technological development. "The essence of the alienation investigated by Marx and Engels consists precisely in this, among other things, that the productive work of the producers in capitalism, that is to say, of the proletariat, does not bear the stamp and the motives of this class, the working class, but must, to the contrary, serve alien objectives, alien motives" (KuG 51). That the capitalists urge their own motives on the workers as being equally good for them does not restore the situation but aggravates it instead.

On the one hand capitalism resists automation, and workers must toil on the production line since that is temporarily cheaper. On the other hand, the inexorable development of automation in technology affords only a limited number of jobs – jobs for those workers whose task it is to provide direction and who therefore come to constitute a labor aristocracy. While the present economic boom forestalls unemployment on a large scale, capitalism will not be able to avoid it in the long run. The subsistence of the unemployed, however, will not be imperiled: the capitalist system will see to it that the workers originally utilized like machines are kept alive like animals – the ultimate in alienation.[310]

The alienation already present in the capitalist system will be deepened by automation. At first the worker was still able to sell his labor, even though he had to surrender what he had made to others. Given automation, however, there will soon come a time when he will not be able to work at all.[311]

On this point, I believe, Marx was clearly mistaken. Marx claimed that the capitalistic society would become increasingly dependent on its workers, and that one of the fundamental – in fact, prerequisite – conditions would thereby be laid for the rise of a communist society.

Wiener, according to Klaus, has correctly understood that American monopoly capitalism wreaks havoc on the systematic working and creative capacities of large segments of the populace. But Wiener is mistaken in believing that communist society, by way of contrast, is just one gigantic system of coercion and unfreedom. If that were the case, argues Klaus, how could one explain the fact that Russia has achieved such tremendous technological development? Everyone is involved in that development, because even with progressive automation and exploitation of the possibilities of cybernetics, people must preserve their freedom to work as the first necessity of their lives. In a communist society it is not possible for human beings to lose their work through advancing automation. But their work does change in character: creative work increasingly replaces mechanical work. "The automation of physical and intellectual labor raises the problem of the relation between creative activity and schematic activity – with hitherto unimagined acuteness" (KiS 435). To the extent that there may have been some alienation from work in communist society, it will now be eliminated. The machine will do what machines have the capacity to do, and people will be liberated for creative work. "The work ... which is really the first necessity of life for people is creative work, creative work that can be performed naturally in exceedingly diverse fields. Only the pursuit of creative work fully eliminates humanity's alienation" (KiS 430).[312] Mathematicians, physicists, engineers, highly qualified workers, and so forth together constitute a collective in which each can develop. The differences among them will only be ones of degree.

On the one hand technological development creates these possibilities, but on the other hand the creative work of the communist society must foster even more technological development. Technology liberates people for creative work, and creative work remains the condition for continuing technological development.

To understand this, it is necessary to establish firmly what the relation is between creative work and schematic work. Creativity is a historical category. "When a problem appears for the first time and is solved, the solution is creative. Once solved, every subsequent instance – in principle, at least – is routine work" (KiS 407). Through the preparation of an algorithm, the work can be done by machine.[313] Because of this development, everyone must receive opportunities for scientific and technological training. Whatever the remaining differences between physical work and intellectual work, the real dichotomy between them is eliminated. Dynamic technological development produces dynamic persons, but it also *requires* them. Thereby humanity first comes into its own. The human being is an "unfinished person" whose learning process will increasingly have to be directed toward the learning of methods to develop new methods, methods that are subsequently transferable to machines. "Whoever ceases to learn can in the long run not remain creatively busy; whoever wishes to be creatively employed must learn" (KiS 398). A person is free when he is creatively employed; to be free implies being creatively busy. The contrast between work and leisure is thereby eliminated in principle. The situation in capitalism is entirely different: "In capitalism, work discipline and personal freedom are inherently contradictory" (KiS 444). If the West is ahead in practice, it is behind from the point of view of principle; in principle it is trailing the communist world.

Meanwhile, the distinction between creative work and schematic work, regarded as a line distinguishing the human being and the machine, does not do full justice to technological development. The machine performs its tasks under conditions set by people. When heuristic principles are incorporated into these conditions, is it still correct to contend that the machine lacks the capacity for creative work?

285

Answering this question properly requires making correct distinctions. It is true that machines can assume more and more human functions. When the activities of the machine are carried out on the terrain of object-language, human activity will be found to be on the terrain of meta-language. Should machines assume human functions on this latter terrain, then it will be necessary for the construction of those machines for people to utilize formulations from the meta-meta-language. The human being always remains the master of the machine: "...no machine can assume the task of the higher types" (KiS 414). Thus man's creative work, abetted by algorithms or heuristic principles, can be transferred to the machine. The original relation between creative and schematic work is thereby altered, or "eliminated," as it were, to reappear on a higher niveau. The basis for this development is that a machine can never break through its own niveau; the human being must always be there to introduce changes into the conditions under which the machine works. Gödel's theorem provides the evidence for this view.[314]

In principle the machine cannot do creative work, since it cannot break through the conditions laid upon it. But within these conditions it can perform more than algorithmic tasks if it is equipped with a contingency mechanism. "Once it is possible in principle to imitate such a human skill with a machine, then the machine, in the process of its further development, will also always surpass the human being, since it is able to bring greater celerity, precision, and quantitatively greater capacity to this qualitatively new task" (KuE 258). Where creative thinking is the expression of the dialectical unity of necessity and contingency ("Creative thinking is ... a contingency process" – KuE 272), there is ultimately no objection to saying that machines, too, which work according to the trial-and-error principle, do creative work. That Klaus arrives at this conclusion, even though he gives due consideration to the qualitative difference between man and the machine, is a result of his dialectical materialism. According to Klaus, creative work has no transcendental origin. "Creative thinking is no more irrational or transcendental than the trial-and-error method, to which creative thinking can in large measure be reduced" (KuE 259).[315]

Klaus's conclusion, in summary, is: "Creative thinking can always be formalized and mechanically imitated. The possibility of constructing trial-and-error machines now allows also for the mechanical imitation of creative thinking..." (KuE 281).

Cybernetic machines that can learn might one day constitute a danger to humanity. However, we must face up to this danger and prevent it. Only on the basis of dialectical materialism can these hazards be prevented and overcome.[316]

3.4.9 The man/machine symbiosis

In contrast to Wiener, Klaus does not introduce the symbiosis between man and the machine only at the moment when the machine places man in jeopardy.[317] According to Klaus – and he is correct here – the symbiosis is already present between man and the implement and between man and the mechanical device. The character of the symbiosis is sharply altered first by mechanization and subsequently by automation. "The symbiosis of man and the machine is a theme traceable throughout the whole of the history of the development of production forces" (KuG 118).

In the mechanization stage, the human being functions as the regulator and steerer. At the outset of automation, when machines work according to algorithms, the human being is still the regulator. His role as the regulator changes sharply when machines seek their own way toward objectives set by people. This occurs when a machine can optimize its own behavior, or when it functions according to a heuristic principle, or when it is equipped with a learning matrix. In the extreme case, the machine selects its own strategy and thus makes decisions.[318] "Everything can be reduced to normal logical relation plus strategy and tactics in the sense of game theory" (KuG 312).

The relation between people and tools has always been symbiotic. In the new situation, it is not only the human being that is a cybernetic system;[319] the machine is too. At the same time, the relation between man and the machine, as a cybernetic relation, is severed in any *practical* sense. This breach contains dangers for both the individual and humanity. "Not only does automation threaten the jobs of millions of artisans and lead in capitalism directly to the devaluation of qualified skilled labor, the tremendous additional danger exists that reactionary forces will utilize the attainments of automation, just as they have utilized atomic energy, for their aggressive military interests" (KuE 323).[320] "It is not just the fear that machines now exist that are capable of making very many, and very essential, things better than human beings can; it is the fear of robots that are capable of making independent decisions – decisions that might turn against people. It is especially the fear of such decisions in conflict situations that provides the primary motive for a negative attitude toward such automatons. It is at this point that many people first feel threatened in their innermost sphere" (KuE 371).[321]

Indeed, the latest possibilities of cybernetic machines have brought people into an entirely different relation to the machine. With the appearance of cybernetic machines, the symbiosis takes on another character: "The symbiosis between mechanical systems and man will acquire an entirely new form" (KuG 160). The dialectical relation between man and the machine as it exists in the old symbiosis is overcome. People no longer need to adapt themselves to the machine, nor does the machine need to adapt itself to people. "For man there remains, finally, the area of objectives and motives" (KuE 68).[322]

This entirely new situation was foreseen by Marx when he offered the suggestion that with the progressive development of the means of production, people might take up a position alongside the production process in order to be the creators and lords of that process. The human being is no longer forced to be a part, psychically and physically, of the technological apparatus.[323] On the one hand capitalism wants to resist this development, but on the other hand, as we saw, the worker loses his employment to it.[324] Since societal alienation persists, individual alienation increases in these circumstances. Capitalism offers only pseudo-solutions. It denies that work constitutes a fundamental human necessity. "Every attempt to situate the essence of humanity outside labor (and this is unavoidable if work is a physical and psychological vexation) is itself a sign of human alienation" (KuG 157).

In communism, this alienation can be abolished. "The basic form of alienation is societal alienation..." (SGF 195). Whenever this alienation is eliminated, technological alienation is eliminated as well. The human being in his freedom is himself again and can deploy the force of his being in creative activities. "Through conscious social guidance of the development of automation in the sense of a constant elevation of the culture of the working class, technological alienation will be eliminated in what, historically speaking, is a short time" (KuG 151). Within communism, every human being has the possibility, given progressive technological development, of participating in all areas of culture – science, dramatics, music, sports, and so forth. The disappearance of certain chores does not lead to idleness and unemployment in communism, as it does in capitalism. Alongside the possibility of expanded cultural activities, there is an increase in the number of vocations – for example, among repairmen ("who in a certain sense play the role of [medicine's] general practitioner in the world of the robot" – KuG 155), production leaders, production improvers, constructors, and scientists. In summary: "The elimination of technological alienation creates the conditions under which humanity's future work will be creative work, freed of any constraint in the direction of schematic work" (KuG 156).

The qualitative alteration of the symbiosis consists not only in man's setting the objectives and motives for the development of the machine, but also in his doing so as a social being – thus, in cooperation with others. Robots in a communist society abet humanity's becoming welded together in a collective in which every person occupies an equal place. In other words, the symbiosis is increasingly societal in character. The relation between societal and individual consciousness is therefore of the greatest importance in this new situation.[325]

3.4.10 The relation between individual consciousness and societal consciousness

The development of technology is of such a nature that only collective control can forestall the impending dangers. In principle robots can develop themselves further. "It is self-evident that here a dangerous moment confronts humanity" (KuE 256). To forestall these dangers, societal norms must be taken into account in a collective control of these machines. "Within a cybernetic total system (and human society is in a certain sense just such a total system), component systems cannot be permitted deployment and developmental liberties after their own fancy. A cybernetic component system that desires freedom at the expense of the total system is the capitalistic social class, and also the exploiting and oppressive classes of all times" (KuE 262).

From the development of the man/machine symbiosis, we learn that where complete cybernetic machines are involved, the human being is required to set only the objectives. Should he fail to do so, however, the new machines would create dangers to humanity, given their propensity to *autonomous* development. Resisting such development is the hierarchical structure: it runs from the machine via the individual to society. Just as the individual with his consciousness of self once stood above the machine, so societal (German: *gesellschaftliche*; Dutch: *maatschappelijk*) consciousness will now have to appear as regulator above the whole made up of machines and individuals. Societal consciousness is the regulator of social relations, thereby imposing norms on the individual consciousness, enabling the human being to unfold himself as a societal being.

Earlier we saw that Klaus does not mean to imply by "consciousness" anything irrational or transcendental. Consciousness is a property of matter in motion. Just as individual consciousness is secondary to matter in a genetic sense, so societal consciousness is secondary with respect to the consciousness of self. And just as the individual consciousness transcended machines, in an evaluative sense, so the individual consciousness is transcended by societal consciousness. In the dialectical process, this means that society and societal consciousness fundamentally influence and give direction to the individual and his consciousness. "Societal being and societal consciousness, in their totality, form a complex cybernetic control system ... " (KuG 43).[326] Societal consciousness, which is expressed objectively in language, is the result of the relation between production forces and production

289

relations. Whenever the dialectic involved here is abolished, societal consciousness comes on as regulator. "Societal consciousness develops itself in its character as regulator on the basis of a specific form of reciprocity between humanity and nature, system and surroundings" (KuG 325). Societal consciousness penetrates the individual consciousness whereby the human being, as a societal being, is enabled to impose norms on the newest machines. In this way the results of the new machines can be harnessed for the service of humanity. "The task of the peace forces of the whole world, and in particular of the socialist camp, is to see to it that these possibilities are not misused for a crime, which would be the greatest crime against humanity and the greatest crime in the history of humanity" (KuE 373).

In a communist society, the dialectic between the individual and society is eliminated. "In this way it is possible for individual experiences to become societal experiences; individual experiences can take on a societal character" (KuG 330). The individual consciousness is not denied in the eliminated dialectic. Societal consciousness enters into the individual consciousness and receives from the individual consciousness a particular stamp. Nevertheless, "Personal experience is ...imbedded in societal experience and is permeated and formed by it" (KuG 334).

3.4.11 Technology and the future

From the preceding subsection it is clear that technology and its development occupy an important place in dialectical materialism. Technological development is the condition and motive for every advance, although there is this restriction, that technological development must conform to the principles of dialectical materialism, which are also the principles of cybernetics: "One of the intellective conditions for the construction of tomorrow's world is the rapid development and multifarious application of a new science – cybernetics" (K ix).

Marxism flaunts its pretension of being able to predict the future in a scientific way. This pertains to capitalistic society as well as to communist society. The development of technology is the decisive factor.[327] "That the development of the production forces can be regarded as the highest criterion of societal progress follows logically from the basic perceptions of historical materialism. The foundation of all societal development is material life activity, the struggle of gregarious humanity with the forces of nature" (SGF 140).

Klaus's view concerning capitalistic society can be summarized briefly as follows. In capitalism, technological development is controlled by the profit motive of the capitalists. The workers are societally alienated, which implies an enlargement of their human alienation in technological development. On the one hand the capitalists would like to perpetuate this situation. Therefore they set themselves against continued technological development. On the other hand, the existing technological development is so unstable that the possibilities of cybernetics must be applied. Yet, this application holds great dangers for the workers – in fact, for humanity. Cybernetic machines can be extremely dangerous if they are not subjected to communal, societal control.

This ominous development necessitates a revolution to convert capitalistic society into socialistic society. "The historical inevitability of societal progress in the direction of socialism and communism consists presently in this, that the working masses must carry on the struggle to break open the limits set by the capitalistic production relations for the development of production forces, since the loss of their own physical existence looms before them" (SGF 159). The inevitability of the revolution does not, however, imply fatalism, according to Klaus. "From an analysis of the process of societal development, marxism drew the conclusion that historical necessity is more and more a necessity of the acting of the societal person" (SGF 183). Given technological development as a historical necessity, socialistic society is a task – a task that can be completed only through *conscious* human activity. Yet, is the necessity of historical development not shifted here to the necessity of human action? Does marxism not simply obscure its fatalism with this removal, instead of "abolishing" it?

In socialism or communism, technology is at the hub of thought.[328] The profit motive has yielded its place to the motive of increasing production and thus of increasing consumption.[329] Societal alienation is eliminated because the community owns the production process. This implies the end of technological alienation. In view of the possibilities of cybernetic machines, this means that the human being is liberated for creative work in which he can be completely himself.

It is alleged, Klaus says, that marxism acknowledges a final goal of history. Nothing could be further from the truth.[330] Within communism, historical progress knows no end. Only alienation is ended. The "kingdom of freedom" is not a transcendental or immanent final goal of history; freedom itself is subject to the law of historical development. "For there exists no freedom in an absolute sense ..." (SGF 126). Material relations also unfold further within communism. This implies an increasing freedom accompanied by a growing unity of production means and production relations.[331] "Decidedly compatible with – yes, logically inferable from – the fundamental principles of historical materialism is the proposition that the societal process itself is goal-oriented" (SGF 153). "The goal can better be seen as immanent to the total societal process, and in each phase of societal development it is realized in a form appropriate to that phase and is at the same time reproduced at a higher niveau" (SGF 163).

Although the dialectic between production forces and production relations is eliminated and the class struggle no longer exists, there are still contradictions in communist society. Thus the development of communist society is not without conflict; by means of struggle, the old will ever and again have to give way to the new. That brings *progress*: "...the societal progress of socialist-communist society is based on the conscious 'abolition' (*aufgehoben*) of that which has been created by earlier generations and has been valuable for the further development of humanity" (SGF 217).

Cybernetic principles, incidentally, do not only control nature: societal relations, politics – in short, society, too – behaves in conformity with cybernetic principles.[332] Society is a dynamic, self-regulating system – thus, a cybernetic system.[333] This implies that societal development can be brought under control just as effectively as the means of work can be. "In this sense the politics of the marxist-leninist parties of the respective socialist states can be regarded as a piece of applied cybernetics..." (KiS 316). Even history consists of systems that strive together toward an optimal, societal situation. "Historical science, to put it cybernetically, is the description of the possible and actual behavior of societal systems and subsystems, with the individual person as the smallest societal component system" (SGF 86). In planning, this knowledge is put to use. The objective is the constant optimizing of human possibilities on the basis of technological development. Cybernetic machines are at the service of planning. The societal system can be ever better and more fully simulated with the help of the computer. The societal system is a multistable system composed of ultrastable subsystems. The whole optimizes itself in such a way that the various component systems (among which are individuals) possess the maximum possible freedom with every state of the means of production. The component systems are dependent upon the great system, but in that dependence they are free. "In a multi-stable system, the partial freedom of the subsystems, of individuals, and so forth is not only possible but also inescapably necessary, for only when the ultra-stable component systems of the multi-stable system are able to make their contribution to the attainment of further development, that is to say, only when they can develop themselves as individuals or component systems respectively, is the optimal functioning of the total system possible" (KuE 141).

This cybernetic control machine that regulates the whole societal process represents democratic centralism, says Klaus. In democratic centralism, societal consciousness regulates the consciousness of individuals, who in their turn so foster technological development that the accumulating dangers are avoided. Humanity constantly progresses in freedom because technology advances. "The meaning of history consists in progress" (SGF 244) and "the progress of history is the ever-expanding range of controlled and regulated acts" (KiS 317).

3.4.12 Critical analysis

3.4.12.1 Introduction

Klaus's view of technology and its development, when compared to that of Wiener or Steinbuch, proves to be much more consistent. The reason for this is that he has taken dialectical materialism as the starting point for his outlook. And this philosophy is characterized by inner unity and closedness, even if the inner unity is based on a dialectical bifurcation.

From the descriptive section of this treatment of Klaus, it is already apparent that I find myself in agreement with his view of technology at a number of points. The points involve his outlook on the thinking of machines, his answer to the question what information is, the relation of schematic thinking to creative thinking, the social (*sociale*) disclosure of the symbiosis between humanity and the machine, and his rejection of the autonomy of technology. In all these areas, Klaus has come further than Wiener or Steinbuch.

In my critique I will focus especially on the background of Klaus's thought (dialectical materialism), but I will have to limit myself. Thus I will forego asking whether Klaus remains within the limits of this philosophy at all points, or whether, under the influence of his perspective on technology, he adds new elements. I will follow him critically when his notions about technology are led by dialectical materialism. For Klaus, dialectical materialism is in the first instance philosophy of technology; thereafter – but this line I will have to leave almost completely untouched – it is also philosophy of economics. What I will focus on especially is the importance of dialectical materialism, as philosophy of technology, for society and its regulation and control. The central question in this area is whether the benefit brought by Klaus's reflections on the relation between humanity and the machine are not lost at the societal level. In other words, can it not be said that the human being, who was found to surpass cybernetic machines when a comparison was made, becomes their victim when these machines are utilized as governing machines?

I will inquire first into the deepest motive of dialectical materialism, in an effort to determine how this philosophy provides a general context for Klaus's reflections on technology. Within that context, the significance of Wiener and Steinbuch for Klaus's thought must be dealt with, together with his critique of them. After more light is shed on Klaus's view of the thinking of machines, I will show how, according to Klaus, cybernetics offers the possibility of clarifying the significance of dialectical materialism for the special sciences. For Klaus, every science is a component of cybernetics, and therefore at the same time a production force.

As in the case of Steinbuch, I will inquire into the real significance of Klaus's philosophy for the future of society as I proceed with my critique. Klaus champions the freedom and creativity of the individual. But do freedom and creativity not go by the board when society grows into a cybernetic technocracy?

3.4.12.2 Dialectical materialism

Klaus has shown that dialectical materialism as a philosophy offers more possibilities for approaching theoretical questions than does mechanical materialism. The latter, with its linear conception of causality, soon collides with reality. But dialectical materialism, with its circular conception of causality and its dialectical laws and categories, offers a possibility of accounting for more elements of reality. The mechanical materialists' conception of matter is enriched and deepened in dialectical materialism.

Meanwhile, dialectical materialism *remains materialism*. The world is a material world that develops in a dynamic dialectical process. Beyond this process there exists nothing: every form of transcendence is rejected in this materialism. The laws controlling the dynamic dialectical process are material, and as such are objectively knowable for mankind.

Dialectical materialism is accepted by Klaus in a dogmatic manner. "We do not claim ... not to set out from a pre-established philosophical standpoint. We stand for Marxism-Leninism, and so for science" (JGM 13). At no point does Klaus bring his choice of a starting point up for discussion.

Within the context of this study, it is impossible to engage in an extensive critique of dialectical materialism. The following questions, however, resist suppression. If objective material reality reflects itself in human consciousness as knowledge of objective laws, how is it possible that there are other philosophies besides dialectical materialism? If matter reflects itself objectively in consciousness, why are we still faced with the task of coming to objective knowledge? Is there not in these questions an indication that Klaus's starting point in marxism is not a scientific starting point, as he claims, but a pretheoretical and supratheoretical starting point that pretends to be the only scientific starting point?

On the basis of a scientific explanation of the relation between matter and consciousness, there arises a scientific worldview and a general method. The latter is an instrument of expanding knowledge and practical change. At the same time, this materialism offers a possibility of foreseeing future developments. There is clearly a value judgment involved here. Dialectical materialism sees itself as the consummation of the history of philosophy, and since philosophy is the reflection of historical material reality, marxism sees itself as the terminus of dialectical history. Thus it avoids any fundamental relativizing of its own outlook, for, if matter is dynamically dialectical, would not the reflection of this material reality be equally historical and, accordingly, as knowledge, be subject to relativizing? Such historicizing is rejected in a dogmatic manner: dialectical materialism regards itself as the end of the development.

I will pass over the following questions entirely: Has the course of societal development, of history, both in the East and the West, not been different from what Marx had imagined? Has there only been progress? Has there not also been decline? How can a representative of one class know the other one?[334] Moreover, there is this question to be considered: Has the concept of "dialectic" not become a magic word which may appear to solve every problem but which can be said with equal justice either to obscure it or to charm it away?[335] I will return to this last question in connection with the relation between freedom and necessity.

These are not the only questions that might be raised. The most critical point for our study is that dialectical materialism regards history as being governed by the dialectic between production forces and production relations. All other factors, the most important of which are the intellective and religious, are seen as being dependent on the relation between production forces and production relations. Meanwhile, the actual relation between production forces and production relations has been *selected* from among many other facts and *interpreted* as an expression of the marxist *belief in inevitable progress*.

The dialectical relation between production forces and production relations unfolds in accordance with the three laws of historical materialism. These laws prescribe that history as movement is self-contained, and thus self-sufficient. This means that the meaning of history is *immanent*-historical in character. The highest form in this history is the societal person. The human being as societal person is the highest product of matter – and thereby also the ultimate *point of support* for dialectical-materialist philosophy.[336]

3.4.12.3 Dialectical materialism and technology

Dialectical materialism and technology belong together, according to marxism. In the absolutized relation between production forces and production relations, the production force is the foundation. The idea of the production force governs history, according to the marxists. The production force is composed of human beings and the means of production – the technological panoply of tools and instruments. Technological development has consequences for the production force and, as a result, for the production relations as well. "Plekhanov concludes ... that the development of productive forces must be regarded as the 'general cause' of the historical progress of mankind, since these forces determine the social relations of men."[337] According to Klaus, too, only dialectical materialism can do justice to technological development, since technology is the basis of this materialism. This must be given special emphasis, now that cybernetics is soaring so magnificently.

Technology has been of great importance to marxism from the very beginning. According to marxism, humanity originally lived in need, pressed by nature. The human being, as *homo faber*, as a producing being, can escape his subservience to the "necessity" of nature, to its vicissitudes, its caprices, and its threat to human existence. Armed with technology, humanity can save itself from scandalous servitude. In work, humanity produces itself as being *free from* nature and *free over* nature. Capitalist production relations are a hindrance to this liberation. Yet, because history is the history of self-becoming, these obstacles will vanish in the revolutionary abolition of the dialectic between production forces and production relations.[338]

The meaning of history is the increasing dominion of people and of humanity, first over nature and subsequently over society with its relations and its future. The meaning of history is progress and, as such, the progress of technology. In cybernetics as guidance and control capability, this progress achieves its consummation. The result will be a "kingdom of freedom" as a being free from nature and as societal freedom with increasing prosperity.[339] The unavoidable question is whether this "kingdom of freedom" will be ruled by humanity or by technology.

To answer this question, which will be more fully elaborated in the subsections to follow, we will first have a look at the interpretation of marxism presented by J. Hommes, which is an interpretation and implicit criticism of marxism with which I concur.

As a nonmarxist, Hommes regards marxism as the philosophy of technology par excellence. According to him, marxism regards itself as the basis of the "age of technology."[340] Hommes maintains that all

296

manner of societal phenomena, including economic relations, are construed by marxism as functions of the technological production force.[341] Everything is dependent upon technological development. "All existence (*Dasein*) may be nothing other than a logical unfolding of the production forces embodied in the tools."[342] The significance of this is that technology overcomes every obstacle that might hinder technological development. And this in turn implies that through technology, the human being can arrive at his authentic self, technology being the only valid attitude of humanity towards reality.[343]

The human being, as *homo faber*, is ruled by technological eros. His essence is work – "that is to say, the immediate relation of the human being to the whole of reality, which gives him purpose and mind in his technological creativity, in such a way that he decides to view the whole of reality as the body and stuff of his own creative force."[344]

Yet, humanity has lost itself in technology; the human being is alienated from himself. "Man, in his use of the world, has lost himself; the world has become an alien and hostile power against him; he must appropriate it anew."[345] In other words, in work the essence of the human being expresses itself not only positively but also negatively. This negation is not abolished as long as the product of work remains an item for sale and the worker does not dispose over the product of his own labor. "The item produced by work, the product, appears as something alien, as a power independent of its producer."[346] In the capitalist situation, the product of work does not belong to the worker but remains alien to him. Work reduces him to slavery; the worker is estranged from his own essential being, from nature, and from his fellow human beings.[347]

The elimination of this alienation is only possible in communist society. On the one hand humanity finds itself again, but on the other hand the elimination of self-alienation signifies the surrender of individual-being. "The re-appropriation of the objective world, the world of things, which had escaped humanity, signifies for Marx the surrender of individual-being ... ."[348] Only in relation to other people is the human being himself and does he become himself. The being-oneself of humanity is essentially a being-societal. "Through its dialectical abolition (*Aufgehobenheit*), existence is for Marx essentially societal."[349] In dialectical materialism, *anthropology is identical with sociology.*[350]

In communism, technological development is all-controlling since it overcomes all obstacles. Hommes says of Marx's philosophy: "The first trademark of his philosophy is in our view precisely the technocracy, to which he would only give a new, communistic form in place of the form based upon private property which it has had up until now."[351] In other words, in communism the human being, as a societal being, is a technocrat. "Being a technician means here: being a ruler by performing technologically, technocracy."[352]

Since every science becomes a production force within communism, the thesis that anthropology is the same as sociology can be completed as follows: "*Sociology is thus technology*"[353] and *technology is technocracy*.[354] In communism, humanity is exhausted – or completely accounted for – in technological activity. In short, in communism "existence is understood *exclusively in its artificiality*, thus as belonging utterly to humanity itself, or as being nothing more than human self-generation; it is admitted as being only the expression of the production force, i.e. of the procreation force . . . ."[355]

In summary, in marxism humanity as a "species" (*Gattungswesen*) *liberates* itself, in and through technology, from all oppression and bondage, whether natural or societal. "The unadulterated technological-active-historical society thereby becomes the authentic being of mankind."[356]

From the preceding it is apparent that for the marxist, technology becomes a *religion*. Such a person *believes* that technological development brings progress that will issue in a "kingdom of freedom."[357] The reverse side of this "kingdom of freedom" is the "elimination" of the individual. Therefore the freedom attained can never be anything more than societal freedom. We shall go on to find that this "freedom" is severely restricted by the possibilities of cybernetics.[358]

3.4.12.4 Klaus, Wiener, and Steinbuch

In my critique of Wiener and Steinbuch, I showed that the effect of their thought in practice is the loss of human freedom. Before proceeding with a critique of Klaus, it should prove instructive to examine his criticisms of Wiener and Steinbuch.

Although he has concluded that cybernetics can develop only on the basis of dialectical materialism, Klaus often appeals to Wiener and Steinbuch. They have made considerable use – without knowing it, in his opinion – of dialectical-materialist lines of thought. They have perceived that cybernetics has more to do with philosophy than any other science does. Klaus agrees with their conviction that cybernetics requires no "supraphysics" (*Ueberphysik*) to explain human and societal problems. To all three of them, the world is a closed world. There is no transcendence, and to the extent that it is still even pondered or discussed (Wiener), it is a "transcendence" dragged into the field of cybernetics to be cybernetically accounted for. Moreover, all are agreed that cybernetics will assume tremendous importance for society and its future. They are united in favoring more consideration for technology in education and training: technological thought must be put at the center.[359] We might expect to be able to conclude from such unanimity that since Wiener and Steinbuch leave no room for freedom, neither will Klaus. To Klaus's way of thinking, however, the great difference between himself and the other two concerns precisely this point.

Despite that great esteem he has for Wiener and Steinbuch, Klaus is driven to conclude that neither of them has any appreciation of the fact that people within the capitalistic system are in bondage. Within a capitalistic framework, humanity is the slave of technology, a condition which the development of cybernetics is more likely to aggravate than ameliorate. Because the worker within capitalism is not emancipated from all authority relations, the path down which he might become his own master, given advancing technological development, is cut off to him. According to Klaus, both Wiener and Steinbuch are oblivious to this problem. Moreover, they perceive other problems where none exist.

Klaus can agree with Wiener's pronouncement that information is neither matter nor energy.[360] However, when Wiener concludes from this that cybernetics has thereby triumphed over the old controversy between materialism and vitalism, Klaus cannot possibly join him in this conviction. Wiener contends that materialism never gets finished with the results of cybernetic machines, and that vitalism thereby has something on materialism. Yet Wiener also states: "... now it appears that organism functions which were hitherto viewed at the level of vitalist interpretation as being in principle immaterial are now able to be repeated at the machine level, and that vitalism has thus in some measure been experimentally refuted" (KiS 66).[361] Klaus, too, rejects vitalistic speculation. He believes, however, that Wiener's rejection of materialism embraces only mechanical materialism, and that Wiener has passed over dialectical materialism. This oversight avenges itself: Wiener's cyberneticism leads to new speculation. He draws irresponsible conclusions from the possibilities of cybernetic machines, and he has no notion of what information really is. Information, according to Klaus, is derived from consciousness and, as such, is consonant with language. Wiener makes the mistake of attributing "information" to cybernetic machines. However, "Pure physical functions which appear in a closed control system have no semantic" (KiS 82). Wiener puts information on a par with negative entropy, something that Klaus contends can only be done with the physical carriers of information, not with information itself. For these physical carriers, the signals, Shannon's laws obtain.[362] "Information does not attain another rank until an undifferentiated conception of information is extended to arbitrary physical processes, as Wiener, in our opinion, has inadmissibly done" (KiS 85).

I will not comment on the fact that Klaus calls the processes in cybernetic machines physical processes. They are technological processes. For the rest, I am in agreement with Klaus's critique of Wiener up to this point.[363]

The source of Wiener's error lies in his positivistic behaviorism, says Klaus. Wiener reduces everything to observable behaviors. He has discarded the fundamental relations between matter and consciousness.[364] It follows that Wiener will never be able to gain a correct perspective on the significance and possibilities of cybernetics for society. Wiener supposes that in capitalism, cybernetics will lead to unemployment, and that the creativity of many will be left unutilized. Klaus shares this outlook. He objects, however, when Wiener maintains that in communism, cybernetics leads to a totalitarian, coercive system. Nothing could be further from the truth. Within communism, the possibilities of cybernetics will liberate humanity from schematic work, and the creativity of many will bloom. To be able to see that, however, one must appreciate the fundamental relation between consciousness (of self) and societal consciousness. This relation is understood only by dialectical materialism – not by mechanical materialism. Wiener failed to see it, and therefore he ended up on the wrong track.[365]

It was chiefly from Steinbuch that Klaus learned the significance of cybernetics. Without hesitation he adopted Steinbuch's ideas about learning machines and learning matrices.[366] Nevertheless, Klaus differs with Steinbuch just as he does with Wiener; he differs with him on the significance of what cybernetic machines "can" do and also on the application of cybernetic principles to society.

Although Steinbuch distinguishes information from signals – which Klaus considers a step in the right direction – information remains for him not a derivative of consciousness but a derivative of signals. Steinbuch fails to see that consciousness, as the source of knowledge, converts signals into "information."[367]

Steinbuch's protest against "supraphysics" comes down to this, that he regards matter as *physical* matter. Consciousness, according to him, can be accounted for by (physical) information theory and then duplicated.[368] It is self-evident, Klaus observes, that Steinbuch is left without any defense against the evolution of learning machines that will proceed to surpass humanity.[369] Although Steinbuch is inconsistent in this view – according to Klaus, it would be impossible to be anything but inconsistent here[370] – he has nevertheless neglected the significance of feelings, of the subconscious, of society, and especially of societal consciousness, that is, their significance for thought.[371] When Steinbuch assumes that even human motives can be accounted for by physics, he is bogged down in the problematics of the Enlightenment, says Klaus. "Refusing to see that science is inextricably interwoven with the societal debate, he carries on arguing like some eighteenth-century philosopher for whom all difficulties of science and society stem from a lack of enlightenment" (KuG 295).

Steinbuch believes that humanity is impelled by the motive of the survival of its kind. Here he has made an abstraction from the class struggle and thus from the fact that man is a gregarious being. "It is precisely the fact that there are not only natural laws but also societal laws and norms . . . that creates the possibility of an accommodation of cybernetic systems to their surroundings, thereby creating possibilities of learning and knowing. In an environment not subject to norms, there is no learning and no accommodation" (KuE 139).

Steinbuch's view of cybernetic machines can be of great value – but only within dialectical materialism. Steinbuch will have to dispense with his notions of physicalist explanations and give due regard to societal consciousness, which governs consciousness of self. This does not imply a renunciation of materialism. On the contrary, society is materialistic society; its norms are intrinsically materialistic. "Even the development of ethical motives is a form of adaption of humanity as a species to the environment" (KuG 290). In rejecting a physicalist explanation of the economic, for example, Klaus does not appeal to the supernatural; rather, he requests consideration for the enriched and deepened conception of matter in marxism, which recognizes that matter displays leaps that give expression to the differences in quality of separate fields of matter.[372]

Meanwhile, it has already become problematic at this point whether dialectical materialism *as* materialism leaves any room for human responsibility.

Klaus believes that once cybernetic machines are given a dialectical materialistic basis, they acquire significance for society. In marxism the methods of cybernetics can be applied to society.[373] Steinbuch favors that, but he is hindered at every point by this physicalistic materialism. In describing society as one great cybernetic system, Steinbuch begins with the individual, seeking to build up that great system out of this littlest component. The presupposition here, Klaus claims, is typically capitalistic. It is the individual consciousness that determines societal being in Steinbuch's view, whereas in marxism the relation is precisely the reverse: societal being determines individual consciousness.[374]

Because Steinbuch has gotten this fundamental order reversed, his societal system as a cybernetic system has a perplexing number of regulatory and feedback systems. Even then it remains a hodgepodge. In the planned economy, by contrast, it is not the individual but the *state*, as the supreme regulator, that provides the starting point. The various existing dialectics can then be eliminated: " ... a consistent application of cybernetics to economics and to society in the broad sense must build upon the most important results of historical materialism" (KuG 35).

To summarize: neither Wiener's positivistic behaviorism nor Steinbuch's physicalistic materialism are adequate, according to Klaus, if we wish to arrive at a correct estimation of cybernetic machines or apply cybernetics to society properly. They unwittingly do capitalism a favor and foster unfreedom. Only with due regard for the matter/consciousness/society relation does cybernetics lead to liberation from all oppression and to unheard-of prosperity.

3.4.12.5 The thinking of machines

Considering the unique place technology has occupied within marxism as the means of deliverance from natural and societal oppression, it is understandable that Klaus should resist the notion that machines can think and that they might one day surpass humanity. Such machines, should they appear, would usher in a new slavery. Within a marxist society this is inadmissible, and hence unthinkable. Marxism has no room for "technology as crisis." Technology and technocracy are wholesome because they serve humanity. This is abundantly clear from how Günther quotes the Russian intellectual Novik: "A kingdom of machines, even self-reproducing, cannot become independent, self-contained, without depending on man as the prime mover of cybernetic machines.... The automation is no more than a link in a closed chain: man-nature. This link can become progressively longer and more complicated, but does not become the entire chain. The automation cannot occupy any other space in the universe except between man and nature. The space of automata can become progressively wide but it cannot cease to be only an intermediate place.... Always nature will be below the automation and man above it ...."[375]

We have already seen that on the basis of his dialectical conception of matter, Klaus raises objections to Turing's hypothesis that machines can think and also to his allegation that "The whole mind is mechanical" (CT 30). Disregarding the fact that in comparing the human being and the machine Turing considers only *behavior*, it is clear to Klaus that the physical matter of the machine must be distinguished from that of the human brain. Turing's comparison allows an analogy at best, and certainly not an identification of the human being and the machine. There can be no identity. Turing has failed to consider that the construction of cybernetic machines involves an abstraction from the basic problem of philosophy – that of the relation between matter and consciousness. It is unscientific to go on and ascribe consciousness to the machine. Klaus would speak of an identification of the human being and the machine only if, in addition to behavior, the result, the function, the structure, and the matter of the machine and the human being were the same. In that case the *homunculus* would be a fact.[376]

By this Klaus does not mean to imply that the human being cannot be *fully* accounted for by science and then mechanically duplicated. "Whoever acknowledges that matter brings forth consciousness must also admit that the repetition of this process is possible in principle" (KiS 255). This is consistent with his materialistic standpoint. At the same time, it sheds light on the question whether the human being is a cybernetic system or not. Klaus answers in the affirmative. Does the matter not come down to this, that the human being is simply more complex than the machine? He is totally subject to the objective, determined laws of dialectical materialism. And with that, does the much-vaunted freedom not go by the board?

It might also be pointed out that Klaus gives the false impression that everyone in the West believes machines can think. He devotes one-sided attention to the behaviorists and the physicalists and then suggests that only dialectical materialism denies that machines can think. However, Turing's view has also been vigorously *opposed* in the western world. Shannon, for example, notes that in considering the *behavior* of machines, Turing has already abstracted from the specifically human. In his opinion, Turing overlooks the fact that without the human being, machines are inexplicable: "...the machine does only that which it has been told to do. It works by trial and error, but the trials are trials that the program designer ordered the machine to make."[377] This is also true of whatever "decisions" machines may make. "In short, the machine does not, in any real sense, go beyond what was built into it."[378]

Von Neumann, too, has delivered a telling critique of Turing's claim that machines can think. First of all, he says, Turing, when comparing the human being and the machine, has ignored the fact that the processes of the human brain are a composite of both digital and analog processes.[379] Secondly, Turing is relying on a hypothesis when he states that neurons function according to the yes and no principle. Von Neumann does not exclude the possibility that this hypothesis will eventually be discarded on scientific grounds,[380] and in this regard he is one with Klaus. Thirdly, the brain restores itself following some kinds of damage, and impaired functions can often be taken over by other parts. The complexity of the brain should not be underestimated.[381] Finally, on the basis of mathematical demonstrations, Von Neumann shows that the operating precision of a digital computer diminishes when the number of bits is extremely large. A digital computer with the memory capacity of a person would therefore be of diminished reliability.[382]

Klaus maintains that only dialectical materialism gets the relation between humanity and the machine into correct perspective. Even if there is opposition in the West to the ideas of Turing and Steinbuch, it is always couched in idealistic philosophy, in Klaus's opinion, and that idealism always goes hand in hand with capitalism. For this reason he also rejects Ashby's view that human thought is primary and not subject to scientific explication. The primacy of thought, according to Ashby, is the presupposition of every science.[383]

Although Klaus accepts the human being as a "transcendental" condition for all mechanical activity, this does not prevent him from maintaining that the human being is a cybernetically explainable cybernetic system. The human being must remain superior to the machine on the one hand, and is able to be fully accounted for by science on the other. Dialectical materialism solves "the mystery of life" with the feedback principle and its own categories.[384] Humanity does not transcend scientific analysis, as Ashby thinks. Notwithstanding the sound distinction Klaus makes between creative and schematic thinking, he also says: "Creative thinking is . . . a contingency process" (KuE 272) and "We designate it a basic form of dialectical thought, and for this reason, that it best illuminates the dialectically dichotomous unity of necessity and contingency" (KuE 273). The adverb *best* is indicative of Klaus's unwillingness to reserve "originality" for humans alone. Machines equipped with randomizers can also, in a certain sense, do "creative" work. In that case Klaus speaks of the relative consciousness of the machine.[385]

Earlier I posed the question whether, in view of Klaus's dialectical-materialist starting point, human freedom is not lost. He now seems to answer this question in the affirmative, even while continuing to speak of human creativity and originality (now in the sense of contingency). This offers science the chance to renew its hold on "freedom." He goes on to say that with the aid of randomizers, "freedom" can be simulated in the newest machines, so that he can speak of the (relative) consciousness of these machines.

Meanwhile, one will note that the identification of "freedom" and contingency implies that Klaus does not really know what authentic freedom is. Authentic freedom is the ground and boundary of science. It is in freedom that humanity transcends science. Freedom is the crux of human creativity. The *new*, which is not scientifically predictable, arises in freedom.

Furthermore, although there are points at which Klaus introduces the argument for human freedom, we must not allow ourselves to be misled either by his advocacy of a change in the symbiosis between the human being and the machine as the machine gradually takes over more human tasks, or even by his declaration that this symbiosis must become a social or collective symbiosis to counter the dangers of the autonomous functioning of such machines.

However much we may appreciate Klaus's viewpoint, its basis cannot satisfy us. Freedom, for Klaus, is not freedom beyond the reach of science; it is not a boundary and presupposition of science. No, freedom as contingency is the dialectical counterpole of the category of necessity. The human being as "free" person is the highest product of the dialectic in matter undergoing development. To Klaus, this is the only basis for arguing that humanity occupies a special position in the order of things.[386] On the one hand humanity regarded in this light cannot lose its "freedom" to the machine, but on the other hand that freedom can indeed be lost to the *homunculus*. In fact, humanity has already lost its freedom if some people dominate others through technology. In either case, humanity is not itself but is alienated from itself. Marxism has understood these perils. In marxist philosophy, therefore, the human being is not an individual but a societal being ruled by societal consciousness. Societal consciousness forestalls the appearance of the *homunculus* – even though one could be realized – and overthrows any dominion of one person (aided by technology) over another. Thus Klaus comes out for *freedom as societal freedom*.

3.4.12.6 Dialectical materialism as cyberneticism

Klaus initially argued that cybernetics is not in conflict with the laws and categories of dialectical materialism. Later he went over to the attack and stated that dialectical materialism is the only philosophy "which emerges confirmed and enriched from the confrontation with cybernetics and information theory, to which all philosophies are presently exposed" (KuE 206).

We have been looking at the significance of cybernetics for technology and its development. Applications of cybernetics (cybernetics as production force) increasingly free humanity from schematic work and put people in a position to do creative work. Moreover, the development of the computer has transformed the man/machine relation into a societal relation. All of this fits in with communism, a system in which technology is venerated as the motor of progress and the human being as a societal being is primary.

Nevertheless, the significance of cybernetics extends beyond technology precisely because cybernetics is founded on dialectical materialism. Cybernetics is important *in the first place* to all the sciences. The exactness of cybernetics touches all science. The existing cleft between the natural sciences and the so-called human sciences can be bridged by cybernetics. Cybernetics has brought about a single-pronged methodology for all the special sciences.[387] This is all in harmony with Marx, who declared that "a science is really developed only when it has come far enough to be able to utilize mathematics" (KuG 352).[388] The category of quantity and quality assures, however, that even with the growing unity in their method, the differences between the various special sciences will continue to exist. The primary benefit of cybernetics for the sciences consists in its affording a better insight into areas hitherto approached only with the greatest difficulty. "Many a field is too difficult, too complex, to be understood apart from the application of mathematics" (KiS 354). The normative sciences can be made exact sciences through cybernetics. This is possible in marxism, because norms are in principle material and are incorporated by the area for which they obtain.

In the second place, cybernetics has become important to human society. This society behaves in harmony with the laws of marxism; thus the order of this society is a cybernetic order. Knowledge of this order makes the entire societal process more transparent, and the application of this knowledge helps the societal process to run more efficiently. Just as cybernetics became a production force in technology, so it is applicable to every other special science as well, including, of course, the social sciences. In cybernetics, marxism has gotten an instrument in hand with which to make good its claims. "It [cybernetics] is a synthesis of knowing and doing, and it has taken, as it were, Marx's eleventh thesis on Feuerbach as its motto: What matters is to change the world" (KuG 4).[389] Cybernetics does not integrate only the sciences; it transforms culture into a single whole of human deeds. "Cybernetics is an excellent means for enriching and establishing the unity of human culture" (KuG 4).

In capitalism, too, people utilize the possibilities of cybernetics for the social sciences. However, they start with subsystems and strive for integration. That approach, according to Klaus, will not work. Only when the starting point is society as one great system – and this is the case in communism – and when component systems and subsystems are given their correct place with respect to the whole does cybernetics foster society's being controlled from one center, the state.[390]

Society as one great cybernetic system bears a striking resemblance to the construction of a computer. The component systems and subsystems coincide, Klaus claims, with the industrial concepts of "standardizing" and "normalizing" construction elements and with the principles of the division of labor. "A series of organizational principles for scientific collective work, just as for any collective work whatever, is thus isomorphic with certain principles for programming calculating machines, especially the principle of linking together a total program from component programs" (KuG 69). That is also the reason why the computer can be used to simulate, guide, and control societal processes.

The conclusion could be drawn that dialectical materialism is enriched by cybernetics. Sometimes Klaus even creates the impression that cybernetics displaces dialectical materialism. Can society as a whole not ultimately be called a cybernetic system? Cybernetics has at that point been transformed, due to its absolutization and its imperialistic character, into *cyberneticism*. Just as materialism is governed by immanent materialistic laws, so society, as a single, great cybernetic system, is governed by immanent cybernetic principles.

In the relation between humanity and the machine, first the individual and then people in cooperation appeared as the regulators of the cybernetic system. Society as a cybernetic system is governed by societal consciousness itself. Societal consciousness – a concept whose content is never quite made clear – makes sure that everyone occupies his proper place in the great system and that no one rules anyone other than himself. All self-alienation is eliminated. Humanity itself is the master in a society that is the highest product of dialectical-materialist development.[391]

The question that arises, however, is whether a cybernetics based on marxism does not call forth a technocracy in which every individual lives in chains.

3.4.12.7 Technology and the future: the technocracy

In view of what was said above, it is necessary at this point to take a critical look at the question of the future that is to come with and through technology.

On the basis of the fundamental relation between matter and consciousness, Klaus rejects every notion of machines that might think and therefore surpass and overwhelm mankind. I leave aside his idea of a *homunculus*, which is inconsistent with this.

Consciousness of self stands above matter since it has issued from it. Yet that consciousness has brought forth societal consciousness and is in turn dependent upon it. That means man is only man as a *societal* being. It is from this second fundamental notion that Klaus gets the idea that cybernetic machines are ultimately regulated and controlled not by the individual consciousness but by societal consciousness. The leading idea of marxism, namely, that the human being is a societal being, thus receives strong support in the development of the latest machines. With the view that cybernetic machines require the cooperation of all the people involved I can happily agree, as I stated earlier. Nowadays technological development requires the social disclosure of work more than ever before. A community of labor is an essential condition for breaking through autonomous technological development. However, the question is whether a community of labor is a sufficient condition. Moreover, we should ask what *sort* of community of labor is required. Klaus has no doubts. To him, a *communist* society assures that technological development will bring progress and liberate humanity. According to Klaus, therefore, the community of labor must satisfy the norms of marxism. Those are materialistic norms, which means that communist society suffices. The meaning of everything is the self-sufficient, communist *society*; in other words, society is absolutized in marxism. Society becomes the common denominator of everything.

I can also agree, as I noted earlier, with Klaus's perception that the computer increasingly relieves people of schematic work, freeing them for creative work. Technological development really can work in a liberating way in this regard. To Klaus, however, this development is a matter of course. This is so in the first place because marxism says it is so, and also, I suspect, because Klaus has too little regard for the relatively independent functioning of cybernetic machines. Those machines can "learn"; they can fine-tune themselves to the greatest degree attainable, and in solving a problem they can select a tactic different from any tactic that might be predicted. When these problems are not given sufficient reflection, great dangers may threaten, as Wiener has demonstrated. Klaus's contention that these dangers can be avoided in a communist workers' collective is based on an *absolutizing* – and especially a *leveling* – of the social aspect of reality. Responsibility is nothing more or less than societal responsibility. Klaus contends anew, very dogmatically, that in a marxist society no drastic mistakes are made by human beings or, therefore, by the machine.

This absolutizing of the social political community has tremendous implications for Klaus's view of freedom. The benefit of the thought that humanity stands above the computer and is increasingly liberated for creative work is lost, I believe, when we see people subjected to hierarchical, central planning. This kind of planning is characterized both by cubernetics and by cybernetic machines.

The basis for it is given with marxism. In marxism the human being is a societal being.[392] Society determines what the individual does. De George says: "Individual acts are not determined, but the choices available to an individual, as well as the pattern of historical development as a whole, are determined."[393] In marxism the individual cannot and may not oppose the *given* societal conditions, let alone break through them. And is that not precisely the significance of "historical formers"? "The character of an individual is a 'factor' in social development only where, when, and to the extent that social relations permit it to be such."[394] Marxism sacrifices the individual to society. To Klaus this means the attainment of freedom and the abolition of self-alienation. He is therefore fully in the line of Marx, who said, following Rousseau: "... in the genuine community, the individuals attain their freedom at once in and through their association."[395]

Freedom, to Klaus, is societal freedom, and as societal freedom it is an accommodation to the *necessary* marxist arrangement of society. Freedom is insight into objective necessity.[396] This is the consequence of applying the dialectical category of necessity and contingency to society: " ... in the consciousness of the acting subject, necessity coincides with freedom and freedom with necessity...."[397]

We have already seen that Klaus regards society as one great cybernetic system that steers and regulates itself since there is nothing superior to societal consciousness. That great cybernetic system guarantees societal freedom, he says. Every person is but a small component in that great system. Whatever the number of individuals, that's how many equal subsystems there are.

With this knowledge it is possible, for the first time, to rule history and gain control of the future.[398] Mechanical materialism also attempted to get the future in tow. If it had succeeded, Hobbes's *Leviathan* would have been the result. However, "In a world in which there is contingency and in which there are dialectical contradictions, it cannot exist in the long run" (KuE 136). No, only the knowledge of dialectical-materialist cybernetics makes the central control of society possible. Freedom must be guaranteed within such a system as that. Freedom is even necessary; otherwise the cybernetic system would not be stable, but unstable. "Freedom in the cybernetic sense means that a cybernetic system relies on freedom, being in its behavior *less than totally* determined by the pre-established program" (SGF 141).

Central control of society is possible on the marxist basis because politics, the economy, and the state are regarded as fused into one. At the apex of the cybernetic hierarchy is the state. The state does not control persons, says Klaus – just affairs.[399] In ruling affairs, room is left for societal freedom. Planning aided by cybernetics is at humanity's service: "In the socialistic societal order, the goal function is fixed as the maximum satisfaction of the material and cultural needs of the working masses" (SGF 229).

Naturally, it could be questioned whether the government of affairs may or can be separated from the government of persons. Is the government of persons not clearly discernible in centralized societal planning? Is it not the case that human freedom is only possible here to the extent that some cybernetic system allows it and requires it? Klaus answers these questions himself, affirmatively, when he says that even ethical principles are incorporated into that *system*.[400] And does he not say as much again when he states that "the progress of history is a constant enlargement of the sweep of guided and controlled acts"? (KiS 317)

The preceding is made even clearer when Klaus draws a parallel between a centrally governed society and a multistable cybernetic system. "A multi-stable system reveals – and to this extent it is an image of a societal order structured in accordance with the principle of democratic centralism – that it is possible for the component systems to secure a certain measure of freedom without the planning and the unity of action of the total system being jeopardized thereby" (KuE 137).[401] From this quotation it is clear that Klaus's freedom is a restricted and determined freedom – "a certain measure of freedom." Human freedom, even when construed as societal freedom, can be described mathematically and simulated by the computer[402] – possibilities which are entailed in the categories of quantity and quality and of necessity and contingency.

Klaus's conception of freedom is a very limited one. In part he has *reduced* freedom, making it in effect space for societal movement, and in part he has *distorted* it, making it contingency or interference – the kind of thing that has a detrimental effect in or on the system and must be eliminated through the feedback principle. Freedom is rendered a "freedom" *determined* by the cybernetic system. Freedom as societal room to move in is the *necessary* condition preventing the destabilization of society as one great cybernetic system. The next question – whether humanity, given control of the future with the help of cybernetics, is not *forced* to be free – is hereby answered affirmatively.

Naturally, Klaus denies the coercion involved here. Societal consciousness guides and controls the development of society and is for Klaus *the* manifestation of democracy. With cybernetics, the state as a government ruling persons dies off. Only affairs are controlled.[403]

Klaus believes that every individual is controlled by societal consciousness and that he will therefore put all his activities at the service of communism. What the collective wants, he will want. He regards it as impossible that an arcane elite – indispensable even for Klaus in the background of central planning – should ever design plans of its own and so be able to control the people.

In my opinion, the "freedom" of the people is no more than the "freedom" of the cybernetic system, the goals of which must necessarily be determined by an elite. Klaus has no doubts about this elite. Misguided intentions, to his mind, are ruled out. That very technocratic evolutionism that brings progress is for Klaus the deepest foundation of hopeful expectations for the future.[404] His religion is the religion of technocracy. It blinds him to the dangers of technology and leads him, under the guise of "freedom," to reduce the human being to a "planned animal." Democracy degenerates with these possibilities of cybernetics and under the leadership of such an elite of planners. In the name of the people, democracy becomes a dictatorship over the people. The free communist society turns out to be a myth.

The origin of the reduction of freedom within the communist society is given with the dialectical-materialist starting point. Society is constructed after the analogy with technological control of nature. Freedom is not perceived as the *condition* for constructing communist society; it is posited only as the *prospect*. Freedom will be attained when society has been constructed fully in harmony with cybernetic principles. "Freedom" can accordingly never be anything but "planned freedom." It is not understood – and on the materialistic standpoint such opacity is inescapable – that the construction of a free society requires free, responsible people.

To Klaus, Karl Marx's original notion of a "kingdom of freedom" seemed at the start to be achievable through the possibilities of cybernetics. Klaus thought it possible, given a cybernetic system, to arrive at a synthesis of societal regulation and human "freedom." Ultimately, however, humanity for Klaus is *imprisoned* within a hierarchically structured cybernetic system, a "total system of planning."[405] The idea of a "kingdom of freedom" has brought forth total technocracy.

The deepest ground of this development is the secularized worldview of marxism and its secularized conception of freedom. Society is perceived as a total, closed society. The norms are cybernetic principles. With cybernetics, society shuts itself off definitively and becomes completely self-regulating. The promised freedom thereby issues, as I see it, in a new slavery, in self-alienation.

This self-alienation can be eliminated in principle when society is construed neither in an individualistic way – marxism rightly opposes this – nor in a sociologistic, technocratistic way. Marxism is guilty of the latter. The great diversity of human contacts and relationships is leveled in the one great cybernetic system of society, and the uniqueness of these human contacts and relationships is thereby disregarded. The richness of society's manifold human associations and their connections is undervalued. Only when supraarbitrary normative principles are taken into consideration as the transcendental conditions for all these associations and their connections do we have a condition *necessary* for the meaningful disclosure of society. Cybernetics could be of great importance here.

This condition is not sufficient by itself, however: the people who form these associations must understand them and live them out in a fellowship directed not at the things of this world (*Diesseits*) but at those of another (*Jenseits*). They ought not to seek the fulfilment of the meaning of things in the things of this *created* world. Rather, they must acknowledge before God the fulness of the meaning of all things in Jesus Christ. In this attitude all the aspects and things of this world retain their specific meaning and are not successively lost in one or another absolutization.

4

A Liberating Perspective for Technological Development

4.1 Introduction

In the last two chapters I have presented and critically examined the views held by a number of transcendentalists and positivists concerning modern technology. I dealt with them at length in order to make it as clear as possible that their thought constitutes a coherent whole. I deliberately let the emphasis fall on what these thinkers, given technological development, expect of the future.

In general we found that the *transcendentalists* are inclined to be pessimistic. They are certainly of the opinion that we could not live without modern technology: even to propagate their negative ideas about technology, they have to make use of technology. But a meaningful future based on modern technology is not to be expected, according to the transcendentalists. Whenever they engage in reflection about the future, they get bogged down in a resigned passivity, pleading for a return to nature, turning speculatively to Being or "Speech," reconciling themselves to development even when they can only view it as ominous, or engaging in an attempt to transcend the "autonomy" of technology and thereby curtail it. In short, they flee before modern technology, essentially leaving undisturbed what they see as nihilistic development.

The views of the *positivists* appear to be stimulated above all by the

313

development of cybernetics, especially the possibilities of the computer. Their optimism toward the future contrasts with the pessimism of the transcendentalists. They believe that with the help of the computer, people will be able to reinforce their power over the present and over the future. They expect cultural progress if technological-scientific methods – especially the method of cybernetics – are put to use on a wide scale.

I showed that under the influence of the positivist view, the future will be determined as a technological future. Wiener seemed the most aware of this. Steinbuch shut his eyes to it by regarding human freedom as a pretense; in this way he opened the door, in principle, to an all-embracing technocracy. Klaus, by contrast, allowed himself to be guided by the ideal of the liberty that is to be realized through technological development. In the meantime, he became so enmeshed in technological thought that the result is a technocracy managed by the state. It seemed to escape Klaus that the result would be the very opposite of what he intended – the absence of freedom. People become the servants of modern technology.

The views of the transcendentalists and the positivists alike imply an unpromising, somber future. The reason for this, as I see it, is that both categories of thinkers espouse an autonomous philosophy or, as is the case with Ellul and Meyer, an attempted synthesis with autonomous thought in science and technology. In the autonomy of thought lies the ground of the tensions that pervade their philosophies, as well as the reason for their inability to indicate a way of escape from the real problems that modern technology entails.

From my critique of certain thinkers, it is already apparent that I do not advocate autonomous philosophy but reject it instead. It is in autonomous philosophy and in the autonomy of technological-scientific thought that the origin of the problems of our "technological" culture is to be sought. Such thought cuts off the possibility of a meaningful, liberating perspective for technological development.

In this chapter I will begin with a recapitulation of the views of the positivists and the transcendentalists. After showing that there are still areas of agreement between these two groups of thinkers, despite all their differences, I will make use of christian philosophical thought to indicate the main lines along which a meaningful perspective for technological development is possible.

I realize that the problems with which modern technology confronts us are enormous. Furthermore, pat answers in every situation are not available. A primary reason for this, of course, is that although we do not want to adjust to the present dislocated situation as though it were unavoidable, we do not wish in the least to deny or ignore it either.

Rather, we must relate to it because it is *real*. The actual state of affairs involves a technological development that has grown up distorted and malformed, influenced by the motives that rule people in technology. These motives, too, must therefore come under our scrutiny.

After explaining the basis of christian philosophizing, I will discuss the import of the ground of christian philosophy for a number of general themes, such as the meaning-character of reality (being-as-meaning); meaning-*dynamis* as normativity; freedom in responsibility; dynamic and dialectic; and the future as progression. Then, in order to make it clear what technology is, I will introduce a number of necessary distinctions, such as science; the application of science; the scientific method used in the changing of reality; and research. Also, futurology and the problematics of quantification will engage our attention briefly.

With that, the prerequisites will have been met for philosophical reflection upon the demarcation of periods of technological development and upon such derivative problems as the motives that drive man in technology, the autonomy of technology, and computerocracy.

All these themes will be dealt with in a more or less polemical manner and will thus have to be viewed against the background of the discussions above.

This chapter will be brought to a close with a perspective on technology as it discloses meaning and a summary of the idea of technological development.

4.2 Recapitulation

In this section I will present a short summary and overview of the ideas of the transcendentalists and the positivists who were dealt with in the preceding chapters. In the process I will focus attention particularly on the actual perspective they offer for the future. While the transcendentalists agree among themselves in more than one respect, I will also point out their differences. With the positivists I will do the same.

The points of agreement and disagreement between the transcendentalists as a group on the one hand and the positivists as a group on the other will then be taken up and explained in the section following this one.

According to *Friedrich Georg Jünger*, the development of modern technology is fed from two sources – the technological mentality and the will to power. Increasingly, technology has become detached from humankind and has assumed dominion over man. It has become an unrivaled, demonic, universal power that exploits nature and robs people of their freedom. Technology leaves a trail of destruction behind on the earth's face as it marches through world history.

At first Jünger believed he could do no more about the future than alert people to the impending disaster. Later he advocated a "return to nature." Still later he sought inner consolation in a highly speculative view of *Sprache* (speech, language) as the origin of technology. Taken together, these "avenues of escape" mean accommodation to and flight before technology. Thus Jünger certainly has no meaningful perspective to offer on the development of technology, which he sees as a nihilistic development.

It is noteworthy that as early as 1939 – much earlier than the other transcendentalists – Jünger had already put his finger on a number of the problems brought on by modern technology, including the growth of technocracy and the threat of environmental problems. In this respect Jünger was ahead of his time, for it is only in the last decade or two that these two questions have become matters of general public concern.

In *Being and Time* (*Sein und Zeit*), his book of 1927, *Martin Heidegger* was still subjectivistic and anthropocentric in his view of technology. Death was the result of every technological activity and undertaking – the preeminent absurdity. After his "reversal," Heidegger rejected this subjectivism dominating technology. Man (humankind), he now said, is an isolated subject in which all being is concentrated and from which all being, once "represented," is ruled. And man is alienated from himself, according to Heidegger, in technological development. The problem, as Heidegger saw it, is not only that technological development as an autonomous power subordinates man to itself but also that man surrenders to technology. The reason for this is that man has "forgotten" to inquire after the "ground" of both himself and technology. Man has become estranged from "Being." Therefore the technology he has brought into being is his greatest danger: he can neither keep it within his purview nor control it.

Heidegger calls for "devotion" to the Being to which the being of man (in freedom) belongs, and from which technology, too, has come forth. On the one hand man has "progressed" away from Being – the history of modern philosophy is evidence of this, says Heidegger – and on the other hand technology as a power over man is settled upon man as his fate from Being.

A reversion to Being is required to escape technology as the greatest danger. The difficulty here, however, is that this conversion to Being is itself a contingent "destiny of Being." And the question whether Being will disclose itself afresh following conversion is left still unresolved. For man there remains nothing more than a *hope of deliverance* from technological development alienated from Being.

In the meantime, the meaningless development of technology continues. Nevertheless, "prophesies" Heidegger, the greater the looming danger, the nearer salvation's dawn. Being contains latent possibilities of deliverance. Even so, the problems do not disappear, for every "revelation of Being" as deliverance brings with it a "concealment of Being." Thus it turns out that Heidegger, with his metaphysics of *Being* and his speculative *hope* of deliverance, finally winds up in a vicious circle. The liberty of the "revelation of Being" *necessarily* entails an absence of freedom, a powerlessness, a "forgottenness of Being."

More than others, Heidegger has seen that where subjectivism is the dominant motive in technology, technology becomes a power opposing humankind. However, in rejecting the rationalistic form of subjectivism that dominates technological development, Heidegger, thinking he has thus overcome subjectivism, ends up in another subjectivism – not an "outward" subjectivism but an "inward" subjectivism. The philosophy of Being he points to as the precondition for deliverance from technology, from the "forgottenness of Being," is his own, and it always remains a subjective, autonomous view. Thus the remedy he proposes is surely no better than the disease he seeks to cure. Heidegger fails to indicate a viable, authentic way of escaping the destruction of culture by technology and the technological mind.

Jacques Ellul raises the problems that pertain to technology and society. For him, our culture has become a technological culture in which freedom must compete with an already determined technology and at last be destroyed in the competition. He traces the cause of this state of affairs to the alliance of science and technology. By making this relation the center of his attention, Ellul has succeeded more than anyone else in noting the characteristics of modern technology that are of great importance for understanding modern technology's meaning for culture.

Along with freedom, the meaning of human life is also expelled. Technology is not concerned with the "whys" and "wherefores"; rather, it is ruled by know-how and efficiency. The future which technology seeks to usher in is therefore a universal concentration camp in which human existence is fully encapsulated.

Ellul does not wish to return to nature, for he regards such a prospect as impossible and unrealistic. Initially he resigned himself to the advent of a worldwide technological collectivist society. Later he exhorted people to stop venerating technology, proposing to restrict technology to so-called material technology. This might open a perspective for transcending technology. By limiting technological power, Ellul would safeguard freedom as freedom that transcends technology.

According to Ellul, technology (which he defines in very general terms as a scientific method of dealing with people and things) is driven forward by an inner necessity, thereby making everything subject to an integral and radical technocracy. On the one hand humanity becomes the victim of this technology, and on the other hand humanity subjects itself to technology and worships it as a god. To the extent that humanity does not accommodate itself to technology, the "human techniques" (biotechnique, psychotechnique, and sociotechnique or social engineering) will see to it that humanity is suitably transformed or made "technological." Although humanity thinks it can attain freedom in and through technology, it is drugged by technology and becomes addicted to it. Human "freedom" is only an *apparent* freedom.

As a Christian, Ellul warns against accommodating the christian view of life to technology, which he feels theologians and philosophers are too prone to do. Yet, he is not interested in integrating christian belief and technology; rather, he stands for their separation. As a result, he is unable to offer a meaningful alternative to the autonomy of technology. The fact that the idea of the autonomy of technology is called forth by the idea of the autonomy of man eludes him at bottom. His "separation from the world" implies that Ellul denies that the Christian should assume responsibility *for* technology and so experience his freedom *within* technology. Ellul indicates no normative direction for technological development. Harmony between technology and culture is impossible, as he sees it. Even to strive for such harmony would be meaningless.

The significance of *Hermann Meyer* consists above all in his having shown that the subjectivism of Descartes and the natural-scientific method of Galileo effectuated the mechanization of our world-picture. He is aware that it is human autonomy that has brought about the autonomy of technological power. He is of the opinion that the *absolute* origin of modern technology is given with the rupture between faith and science. For this reason he is also convinced that modern technology is inevitably accompanied by a purely secular, this-worldly culture. Via the technologization of nature and human relations, humanity itself becomes a mere technological object: people are reduced, leveled, and prepared, so that they can function as components in the great machine of culture.

Meyer is undoubtedly right in claiming that an autonomous humanity brought forth the meta-human technological power and that that fact is the basic reason for man's loss of meaning in our time. But to him this all seems inevitable, given modern technology; technology and secularization, he says, are joined indissolubly. He fails to understand that despite humanity's *pretensions*, human autonomy can never be realized. People in their autonomy can never escape the reality that is *given*, for reality continues to make its demands. Despite all their pretensions to autonomy, people remain *creatures*. Human autonomy, then, is a fiction. This fact becomes apparent too. Humanity directs its pretended autonomy toward something in this creation – in this case, technology. Man projects his "autonomy" toward technology, which then turns against him.

In contrast to other transcendentalists, Meyer still speaks of a *normative* development of technology as he looks to the future. The basis of his norms is the *free personality*, which must be preserved as technology continues to develop. Since this is not happening in the present situation, Meyer, like Ellul, would like to limit technology to material technology. But because this material technology is still independent, it, too, must be restricted, he argues. To this end he would like to introduce norms for the *use* of technology. He regards technology itself as *neutral*, as it were; the *use* of technology is presumably another matter.

In short, Meyer would like to clear and preserve space for the free person. Because he accepts the autonomy of technology, however, this space remains under fundamental attack. Moreover, his normativity does not apply to technology itself but to its use. Thus his personalism is not fruitful in resolving the questions related to the themes of humankind and technology, and technology and the future.[1]

The main tendency in the views of the positivists runs counter to the main thrust we find in the views of the transcendentalists.

Norbert Wiener, the father of cybernetics, has designed self-regulating and self-guiding machines. Such machines will assume more and more human functions. The development of the computer, especially, brought certain questions to the fore: Can the human being be replaced entirely by the machine? And can the machine surpass the human being? Wiener initiated this discussion.

Cybernetics seemed so successful in the field of technology that the suggestion was soon made that its principles and methods might be made fruitful for many sciences. Wiener even launched the notion of integrating the various sciences into one great whole with the help of cybernetics as a philosophy.

Wiener's philosophizing about the possibilities of the computer and cybernetics exhibits a dualistic character. Whenever he gives free rein to his thought, he expects much from the computer and cybernetics, but whenever he turns to the practice that follows from theory, he fails to get beyond a pessimism reminiscent of that of the transcendentalists.

For example, with the aid of the (future) possibilities of learning, self-replicating machines, Wiener would like to throw light on religious questions about God's activity in creation, God's omnipotence, and God's providence. With the possibilities of cybernetic machines, he hopes to explain what has thus far been inexplicable in religious experience. To this end he reduces everything to "information." God and humanity are data-processing systems that can lose in a contest with cybernetic machines. In other words, the future belongs to the machines, which will be very powerful. Along these lines Wiener clearly absolutizes cybernetic thought.

Yet, what he does in practice is the reverse of what he does in theory. Wiener fears that in the future the mighty machine will no longer serve humanity, but that humanity will instead be the victim of technology. People will lose out to the computer, widespread unemployment will be unavoidable, the dictator supported by technological power will make himself strong, and massification will burden humanity. Wiener is of the opinion that something can be done about this unhealthy trend in the short run. He calls upon people to keep technology in hand and not to surrender to it. But in the long run catastrophe cannot be averted, as he sees it.

We must understand this pessimism in the light of Wiener's absolutization of the concept of information. He conceives of information in a physical sense, calling it the "negative of entropy." Data-processing machines contribute to the disappearance of differences of structure, since the quantity of information levels off as information is being processed. Thus the quantity of information decreases as time goes by, which means that entropy increases. And this will inevitably lead to chaos.

Wiener's conception of science is dualistic. On the one hand he is rationalistic. His expectations of the possibilities of modern technology are great, but he must ultimately conclude, since he understands "information" in a physicalistic sense, that the distant future will mean the end of humankind. On the other hand Wiener's view of science is affected by his confrontation with practice, which makes demands of its own. As a result, Wiener is less optimistic at the end than at the outset.

The ground for this dualism, briefly stated, is not only Wiener's failure to incorporate man into his view as free and as responsible for technological development, but also his construing "information" in a physicalistic sense and attaching to this construction some far-reaching consequences. At bottom he has ignored the distinction between culture and nature. This is why it can also be said that by virtue of his informationism, his cyberneticism is physicalism, which also explains why his scientific idealism with respect to the possibilities of cybernetics changes into its opposite, cultural pessimism, a pessimism heightened by his recognition of the possibility that the computer and cybernetics will be misused.

Karl Steinbuch is an unadulterated optimist with respect to the future of technology. He, too, absolutizes the concept of information. Yet, although he regards "information" as belonging to physics, he does not tie it in with entropy.

Cybernetics, according to Steinbuch, will be the science of the future. It delivers what it promises. Steinbuch strongly opposes Christianity and any philosophy of a "background world," for he believes that such outlooks do not direct our attention toward the future. Instead they resist technological progress.

Steinbuch advocates a futurology based on the natural sciences and technology. The advance of technology is Steinbuch's source of inspiration. This advance will be an advance of humanity and society if the method of modern technology and the principles of cybernetics are generally applied.

Steinbuch's reverence for the computer is especially striking. From the computer he expects many surprises for the future. With its unlimited possibilities it will overshadow humanity. In principle the computer and people are cybernetic systems, but in practice people will be no match for the machine. Steinbuch even suggests that in the future, machines will be able to think and to say of themselves that they have consciousness. He begins by humanizing the machine, but he ends up "thingifying" people. He compares man to the machine, which puts him on the wrong track. The proper comparison to make is man *without* the machine versus man *equipped with* the machine.

Steinbuch states clearly that from a scientific viewpoint, freedom and responsibility are pseudoconcepts. We use such terminology outside science to indicate that people are all programmed differently. On the basis of this view, he would like to use cybernetics to bridge the so-called gap between the natural sciences on the one hand and the human sciences on the other.

Finally, Steinbuch advocates a cybernetic world state. Influenced by the ideas of Herbert Marcuse and Jürgen Habermas, he connects his idea of a cybernetic world state to the idea of direct democracy. It is essentially information technology that offers opportunities in this direction. But if such a future is to be realized, the present generation of political and economic power-holders – who do not understand cybernetics and its possibilities, who resist it and make their decisions on an irrational basis – will have to be replaced by engineers. What stands before us as the goal to be realized is not a rigid technological state but a cybernetic state in which everyone will have his "planned freedom."

Despite Steinbuch's appeal to Marcuse and Habermas and his idea of direct democracy, there appears to be a fundamental difference between Steinbuch and these revolutionary thinkers. Steinbuch is driven by the thought that the future can be completely controlled by cybernetics. We saw that even his notion of direct democracy must serve this end, since the pronouncements of this democracy are predictable on the basis of statistical theory and thus can be taken into account as the future is being planned. Marcuse and Habermas, though they have their differences, seek a technological development that is "free of controls." They agree that priority must be given to the realization of practical goals rather than the solution of technical problems. In short, although Steinbuch has borrowed a few ideas from Marcuse and Habermas, his ideology of technology stands opposed to their ideology of freedom.

Georg Klaus shows that in dialectical materialism, technology as the means of production forms *the* basis for social development. He is of the opinion that what Marx championed in his philosophy, namely, the realization of a "kingdom of freedom," can and will be brought about with the possibilities of cybernetics.

Modern tools and instruments liberate humanity from heavy physical and mental labor. In a physical and psychical sense, people are increasingly free. They set the goals of technology and are the masters and creators of the production process. The continuing development of technology brings about a situation in which people are more free to undertake creative work.

In contrast to Wiener and Steinbuch, Klaus ranks the cybernetic machine below the person in principle – that is to say, below the *communist* person. Modern technology and communism go hand in hand. It is only in a communist society that the dangers of the development of modern technology can be escaped, since this society promises the possibility of societal control. In the capitalist world, the various interest groups turn technology into a menace. Societal alienation is still the order of the day in capitalist society, and so technological alienation is reinforced instead of being abolished. At the same time, this means that the capitalist world hinders the advance of technology, while communism assures it. Everyone will share in the fruits of greater prosperity – more consumption and greater freedom.

According to Klaus, cybernetics has made full control in a communist society possible for the first time. This society is a cybernetic system in which the state as the central organ embraces all other societal functions. It is possible, says Klaus, to simulate a communist society with so-called *machines à gouverner*. That the fruits of technological development are equally distributed can be determined and assured from one control center.

In communism the meaning of history is immanent: humanity extends its dominion first over nature and then over society, including its structures and its future. The meaning of history is progress as the progress of technology, which will bring freedom. Meanwhile, it turns out that the freedom to be had in the cybernetic, communist society is a planned, appointed, determined freedom. It is the freedom of a certain space to move in, which clearly represents a reduced notion of freedom. Insofar as freedom is more than societal space to move in, that "more" is construed by the great cybernetic system as interference or as accident, as something to be intercepted and eliminated by regulative intervention from the top. Those at the top – the elite – represent the people (they call it a central democracy!) and can set up a technocracy over the people in the name of the people. The hopeful expectation of a "kingdom of freedom" by way of technological development turns out to be an all-embracing technocracy in which force rules in the place of freedom. In other words, in marxism technology forms not only the basis for society but its apex as well. As a result, the relation between humanity and the machine, though viewed correctly at the outset, is destroyed again in the end, as communist society becomes technocratic society.[2]

The origin of the reduction of freedom within communist society is to be found in the fact that this society is built up as an analogue of technology's mastery over nature, and also in the fact that freedom is not the *condition* for building the communist society but rather a *prospect* held out for the future. Freedom is to be attained only when society has been erected in a manner entirely congruent with cybernetic principles. In this case, then, freedom can be nothing but "planned freedom." It is not recognized and acknowledged that building a free society requires people who are responsible and free. If this norm were respected, it would indeed be possible to have societal development based on technological development in which human freedom was opened up.

Because marxism derives its norms for the building of a communist society from the possibilities of technology, its promise of a "kingdom of freedom" cannot be made to come true. Rather, the opposite will result. The free communist society will prove to be a myth.

4.3 Similarities and Differences between the Transcendentalists and the Positivists

The transcendentalists and the positivists are agreed that modern culture bears the stamp of technology. The transcendentalists find that a single world culture arises along with modern technology, and the positivists strive to attain this world culture by way of modern technology. Whereas the transcendentalists *complain* that technology has already subordinated economics, sociology, politics, ethics, humanity, and religion to itself, the positivists regard this as ideal. The difference, then, is that the transcendentalists believe that technology has already assumed the dominant position, while according to the positivists technological power is just now poised on the threshold of its true development. Heidegger, for example, says that modern man does too much (technologically) and thinks too little, while Steinbuch and Klaus argue to the contrary, following Marx's eleventh thesis on Feuerbach, that there is still too much thought and too little technological accomplishment. In short, the transcendentalists believe that the ideal of the positivists has already been largely achieved. Moreover, they are convinced that this state of affairs is disastrous for humanity and for human culture.

Heidegger finds that humanity in modern technology is alienated from itself. This alienation is fundamental and cannot be abolished by abolishing private property, as Klaus supposes, following Marx. For Heidegger, modern technology and self-alienation belong together, quite apart from any consideration of societal structure.[3] Therefore Heidegger regards technological development as a nihilistic development – just as the other transcendentalists do, although from another vantage point.

Also part of the picture is the fact that the transcendentalists have an eye for the past, since they would like to discover the origin of the havoc-wreaking technological development. The positivists, however, pay scarcely any attention to the past, preferring to direct themselves toward the future, which they believe will be a technological future.

In summary, we could say that the transcendentalists describe the present culture as a technological culture in which there is no place for human freedom, while a technological culture is just what the positivists are striving to achieve. What the former are opposed to is the technocracy of the latter. Therefore they take no satisfaction from the resolution of such great, conspicuous technological threats as those deriving from atomic energy, nuclear weapons, and environmental pollution. All the campaign against these dangers signifies to the transcendentalists is that technology itself is no longer seen as *the* great danger. Technology, for them, is a structural problem, and therefore they inquire into its basis.

However many differences of nuance there may be among the transcendentalists we have dealt with, they all agree that modern technology is not to be understood apart from modern science.

In the background of modern technology are modern science and, more importantly, subjectivistic humankind – people who wish to use science to subject reality to themselves. The transcendentalists regard the tendency toward continuity in modern science as being extended and carried forward in the autonomous power of modern technology.

It turns out, however, that at this point the transcendentalists fail to do justice to the facts of technological development. They fail to see that the continuity of technological development certainly can be broken, as it is by inventions, and that development can acquire a new perspective.

Although the transcendentalists reject the imperialism of technology because its superior power banishes human freedom, they do not see through the autonomy of technology and perceive that it is an illusion. Because they fail to undertake a critical examination of this autonomy, they are never in a position to indicate a perspective for the future in which technology can assume a meaningful place of its own. Technology and freedom exclude one another, according to them. What they propose to do is to save freedom and ignore technology as such. Thus some of them appeal for the future to "Nature" or "Being," while others manufacture a synthesis of some kind or other between technology and human freedom. There is not one of them who perceives that freedom *in* technology is possible, provided that the development of technology is guided by supraarbitrary norms.

At bottom it eludes the transcendentalists that the *absolutization* of technological-scientific thought is the source and origin of the nihilistic development of technology. Because this development excludes freedom, they rally to the cause of freedom at the expense of technology. They absolutize freedom and judge technology in terms of it.

The positivists maintain the contrary position. They derive their norms for technological development from the absolutization of the technological-scientific method, especially the method of cybernetics. Thereby they exclude freedom in the proper sense of the word. For them, technology as control of nature serves as the model for control of society, and even of humanity itself. This technicism in the theory becomes an all-embracing technocracy in practice. The transcendentalists take appropriate note of this situation, but without perceiving its source, as we have noted.

There are great differences, then, between the views to which I have called special attention. The two sets of views are even contradictory. Yet the controversy is never complete, for as we saw, the transcendentalists who reject modern technology on the one hand are necessarily forced to accommodate themselves to it on the other. That proved to be the inner tension in their thought. For the positivists, the problem is just the reverse. Although they absolutize technological-scientific thought and so deny or exclude freedom, freedom is what turns out to be the great problem in their views.

The absolutization of technological-scientific thought can no more be carried through to completeness than can the absolutization of freedom. Both these views are therefore characterized by an inner dichotomy and outward struggle: these views are at loggerheads not only with each other but with themselves. Perhaps the matter can be stated as follows. Each of the two views is intrinsically dialectical in character: the absolutization of technological power summons up the problem of freedom, while the absolutization of freedom cannot be severed from the problem of "autonomous" technology. Also, there is the dialectic between the two views: while they are mutually exclusive, they also evoke one another.

Neither the positivists nor the transcendentalists can give an integral, harmonious view of the relation between humanity and technology. As a result, they are without a meaningful perspective for the future. This brings us to the matter of the fundamental agreement between the transcendentalists and positivists. Their agreement is grounded in the *pretension to autonomy*. It is only in the respective directions in which this pretended autonomy is worked out that they are opposed. The self-worship of the positivists is directed outward, while the self-worship of the transcendentalists is directed inward, in the flight before technology. Both set up their own laws, and neither acknowledge any suprasubjective laws or normative principles.

Secularized, technological culture exhibits an inner dichotomy. Its dominant motives are those of the positivists and transcendentalists. The tension between the thought of the transcendentalists and the thought of the positivists has insinuated itself, as it were, into our culture. As technological development led by technological-scientific thought unfolds, thereby reinforcing the captivity of people in technology, this tension increases. Moreover, such technological development acquires the attribute of automatism, and the reactions against it invariably increase. These reactions take on many forms, the most far-reaching of which is barbarism, the destruction of technology.

In view of the dichotomous character of technological culture, the question that confronts us at this point is whether a meaningful, liberating perspective for technological development is still possible.

326

Can technological power be joined harmoniously with freedom? If so, on what basis?

4.4 The Basis for a Liberating Perspective

It seems to me that the positivists and the transcendentalists, because of their closed humanistic worldview, are not really able to offer a perspective for the future in which science and technology have a legitimate place and in which human freedom is neither excluded nor made absolute.

The source of this failure is the human pretension to autonomy: humanity legislates for itself and therefore is self-willed. The fundamental choice upon which this pretension rests is a radical one and thus is religious in character. Although it precedes all autonomous philosophizing, this choice is seldom accounted for in such philosophizing – even though theoretical thought will always be wholly controlled by the choice for autonomy. The contradictions in the thought and the (often obscured) dearth of outlook for the future are attributable to this religion, in which – regardless of the variety – mortal humanity is itself the *alpha* and *omega*.

I believe that autonomy must be rejected in principle, whatever its form. As I see it, philosophy can be serviceable in indicating a meaningful perspective for technological development only when it is anchored in religion, a religion in which it is confessed that reality is a creation of God, that God is the Origin of all things, that He binds the creation to His laws, and that the history of created reality, in which the mutual relations and coherence of all things are fixed, is led, controlled, and brought to its consummation by Him. The christian religion acknowledges that God accepts humanity as a partner in all this in that He makes people in their freedom responsible for the progress of history. Humanity, having fallen into sin, receives salvation in and through *faith* in Christ. In technology, too, people may work again (although their work may be accompanied by crises or "judgments") at the disclosure of the creation, working toward the building and coming of the Kingdom of God, in which the creation will be fully opened up and redeemed from all the consequences of the fall into sin.

This *confession of faith* that reality, including humanity, is radically and integrally *dependent* upon and *involved* with God should be the hallmark of philosophical thought. It should also be the basic human motivation in technology. A philosophy fed by the springs of christian faith, which is a liberating, saving faith (saved from the delusion of

327

self-sufficiency, and therefore free to serve God), will be serviceable in indicating a meaningful, liberating, normative perspective for technological development. Cultural forming is not to be condemned – perish the thought that flight from the world should be propagated! Being busy with technology should mean being busy *serving* God. This requires the rejection of every form of autonomy and the acceptance of the status of the bond-servant; it means the repudiation of self-worship.

Given such a view, it becomes possible to assign technology its appropriate place. At the same time, it turns out that technology and freedom, or power and freedom, need not exclude one another and chronically engage each other in dialectical conflict. In faith humanity is joined to Christ, in whom power and freedom are harmoniously united. In Him, freedom and technological power are joined to God as the Origin of all things.

This means that technological development may neither be judged according to the idea of an absolutized freedom, as was the case with the transcendentalists, nor dominated and guided by absolutized technological-scientific thought, as with the positivists. Thus the norms for technological development may not be derived from the freedom or power of humankind. The normativity for the direction in which technology should be developed is *given* in the form of normative principles that are supraarbitrary and suprasubjective, and thus independent of humankind. It is man's duty to bring technology further along, and to do so *in freedom*, in believing obedience, that is, in subjection to the normative principles of the law of God for the disclosure of culture.

It is this awareness that opens a meaningful perspective for culture, especially in our time, when the significance and consequences of technology are increasingly impressive and dislocation is increasingly severe because of the influence of the secularized motives that now rule people.

When humanity gives up its pretension to autonomy and submits in faith to the revealed truth that is in Jesus Christ, there will once again be perspective for culture.

4.5 Philosophical Explication of Several Main Themes

I will now examine the basis of the liberating perspective sketched in the preceding section and work it out briefly in a number of main themes. These themes will serve to clarify the depths of the transcendentalist-positivist problematic. They will also serve to make possible a personal, thetical elaboration of a christian philosophical view of technological development.

I will deal respectively with the meaning-character of reality; meaning-*dynamis* as normativity; freedom in responsibility; dynamic and dialectic; and the future as progression.

4.5.1 The meaning-character of reality

In the reformational philosophy, the idea of the meaning-character of reality or of "being-as-meaning" occupies a central place.[4] What is meant by this is that everything is created, that everything is dependent for its existence upon God the Creator as the Origin, and that everything finds its destination in the Origin. All things are from, through, and to God. An enormous dislocation occurred in this relation as a result of humanity's fall into sin; the entire creation was affected. Nevertheless, in Jesus Christ as God-man and Mediator between God and humanity, the original relation is restored. In Christ the whole creation, including humanity, is joined once again to God. There is nothing within the creation that possesses independence, nothing that can exist in a self-sufficient way. The destiny of the entire creation is at bottom the honor and service of God in and through Jesus Christ.

Two directions are to be distinguished in connection with the meaning-character of reality. First there is the "vertical" direction – the involvement of everything with, and the dependence of everything upon, the Origin. Subsequently we also speak of a "horizontal" direction – the history of the creation from beginning to end. This distinction may not be permitted to lead to a dissociation, however, for "verticalism" fails to do justice to history and its meaning, while "horizontalism," by contrast, leads to a secularized worldview in which the relation to the Origin is denied and in which the meaning of history is made immanent. Rather, the "vertical" and "horizontal" directions of being-as-meaning form a unity: everything in the history of creation is taken up in directedness to God.[5]

In being-as-meaning there is a *diversity of meaning* so immense that it defies description. This is to be observed in the great number of *entities* (such as things, plants, animals, people, facts, and events) and in the diverse modes of being, the *aspects* of the entities (namely, the numerical, the spatial, the kinematic, the physical, the biotic, the psychic, the analytical, the historical or cultural, the linguistic, the social, the economic, the aesthetic, the juridical, and the pistical aspects).

Every variety of meaning points beyond itself and refers to all the remaining being-as-meaning. All being-as-meaning is embraced by history as the dynamic *coherence of meaning*. This coherence is then concentrated in the *fulness of meaning*, the Radix or source of power of all meaning.[6] Here the Bible directs us to Jesus Christ as the incarnate Word of God, who called the creation into existence and propels it on its way.

Humanity is the culmination and apex of creation. All being-as-meaning is related to humanity. It is in Jesus Christ as God-man that every meaning finds its fulness. Jesus Christ brings this fulness to its rest and destiny, which is in God.[7] All things are directed to Christ as the Lord of history, and in this way all things are related to God as the *Origin of all meaning*.

The meaning-character of our reality, with its two stated directions and further distinctions, is *given* – given, that is, prior to any involvement or interference of humankind in this reality. Every involvement of humankind with reality is to be qualified according to one or another modality of meaning and has its lawful place in the coherence of meaning that is directed at the fulness of meaning on account of the Origin of all meaning.

Seen in this light, christian philosophy is, first of all, *listening*. In this listening, wisdom grows and enables one to respond to the problems of complex, structured reality.[8] Assent to the meaning-character of reality, to being-as-meaning, is not a matter of course. People can set themselves against it or submit to it. When opposing it, they direct all meaning toward some creational given and often toward themselves; they divert all meaning away from its true destination toward something that is absolutized. Meaning is then meaning for and through humankind.

Submitting to the *given* meaning of everything signifies that humanity subjects itself to and lets itself be led by the meaning-*dynamis* as the law, as the norm for all being-as-meaning.

4.5.2   Meaning-dynamis as normativity

History is the all-embracing, dynamic coherence of meaning, a coherence in which humanity too is taken up. In history humankind must make a radical choice of direction. This choice is religious in nature. This is to say that humanity, as full, total humanity, must direct itself to God, an act which has become possible again through Jesus Christ, and that in this directedness to God humanity lets itself be led in history by the meaning-*dynamis*.

Humanity's choice of direction in history is a choice made in faith. In *faith* humanity must give itself to Jesus Christ, the Radix of all meaning, subjecting itself to, and letting itself be led by, the meaning-*dynamis* as normativity. Humanity must agree that God establishes

330

all meaning and must assent to the meaning-character of reality by responding in this way. If humanity does not submit in faith, if it does not conform to the meaning-*dynamis* but instead attempts to establish the law in autonomy and thus to be the giver of law on its own, it sets itself against the meaning-*dynamis*. Humanity then falsifies it and brings about dislocations. Yet, this choice cannot in fact be isolated from the meaning-*dynamis*, as we shall see, any more than the consequences of this choice can be isolated. The consequences soon manifest themselves in disturbances and dislocations.

In its diversity, its coherence, and its fulness, the meaning-character of reality is not disorderly but lawful. The meaning of all things is established by God's law. God, as the meaning-giver, is at the same time the law-giver. It is His law that brings all being into existence as meaning. God upholds and rules the whole creation in its history in Christ as the fulfiller of the law. It is to Him that humanity ought to submit, and it is to Him that humanity owes obedience.

This obedience is centered in humanity's acceptance of being-as-meaning with all its heart. (The heart is the religious center of persons.) The heart is directed to Jesus Christ as the fulfiller of meaning.[9]

On the one hand humanity is taken up in the all-embracing scope of history, but on the other hand it *must choose* direction within history. Seen in this light, human responsibility consists first of all in submitting wholeheartedly to the Radix, and then also in conforming to the superior force of the meaning-*dynamis*. Thus humanity must abjure haughtiness and self-will.[10]

It is out of the heart, as the religious center for all human beings, that every issue of life comes forth. This means that the fundamental choice of faith is decisive for every subsequent act,[11] which accordingly will bear the stamp either of submission and response to the call directed to humanity to open up the creation in its meaning, or of resistance to it.

What the foregoing means with respect to technological-scientific thought is that this thought is itself contained within the coherence of meaning, and that there it has its own limited but meaningful place. The direction of technological-scientific thought and of the consequent development of technology is guided, at bottom, by the religious choice and conviction of humanity. When this choice and conviction are nourished by attentiveness to the meaning-*dynamis*, every technological activity can help to disclose meaning, and technological-scientific thought can render its meaningful service. However, when technological thought and activity are set against the meaning-*dynamis* from the beginning, humanity may maintain a pretense of opening up a meaningful future, but its resistance means (as is brought to light, for example, in the absolutization of technological thought) the actual sealing off and locking up of meaning. And this, in turn, is the same as humanity's pulling back from any meaningful perspective for technological development.[12]

4.5.3 Freedom in responsibility

The central, religious choice of humankind is the beginning of every human occupation. Viewed from humanity's side, this choice is the beginning of the process of disclosing the meaning of reality – a task for which humanity is responsible. Every act of commission and omission ought to be a response to the meaning-*dynamis* as normativity.

The results of human activity originate on the basis of the so-called historical or cultural modality of acting.[13] In other words, these results have a historical or cultural *foundational* function. Their *qualifying* (guiding, destinational) function can also be the historical or cultural function – which is the case with technological objects[14] – or a function above the historical, such as the linguistic (a book), the social (an association), the economic (a business), the aesthetic (a painting), the juridical (a traffic regulation), the ethical (a marriage), or the pistical (a confession of faith).

Obedience to the meaning-*dynamis*, which implies acknowledgment of the responsibility of humanity for the disclosure of the meaning of the creation, at the same time signifies *freedom*.

The question what freedom really is, is very difficult and complicated. Freedom is a *limiting concept*, and the problem of freedom lies at the frontiers of human thought. We can say, in any case, that freedom is freedom in responsibility, that humanity is called to freedom, and that freedom is a *being free from* every kind of autonomy and *being free for* the service to which God calls people in Jesus Christ. Freedom, in other words, is freedom from all decisive bondage to any constituent element of the being-as-meaning of reality – and that through faith.

Although we recognize that in speaking about freedom we are dealing with a limiting concept, we can still make certain distinctions. Freedom is in the first place religious freedom. This freedom is central and total. Thereafter freedom is also modal. Humanity's central, religious freedom comes to visible expression first in the historical modality. Thereafter it differentiates itself in various differently qualified activities, including social, economic, aesthetic, ethical, and other activities.

Modal freedom is not the choice between right and wrong; it is *responsible* choice from among a multitude of possibilities in a certain situation, all of which satisfy the modal meaning by which the situation is qualified. Thus not just central religious freedom but also modal freedom is freedom in responsibility. The difference is that the first freedom can only exist as surrender to the Radix. That this choice is not always the one adopted does not mean that modal freedom is thereby abolished: people can continue to choose from among various alternatives. Yet, because people set themselves against the meaning-*dynamis* as the law of liberty, they always end up, along one path or another, in unfreedom. I will say more about this in the following subsections.

4.5.4 Dynamic and dialectic

Humanity, as we saw, is taken up in history, which is the dynamic meaning-coherence of all being-as-meaning. Taken up in history and driven by the meaning-*dynamis*, humanity is brought into confrontation with the choice either to submit to the meaning-*dynamis* or to resist it. Submission signifies that people understand and accept their responsibility in their task as cultural formers. This is the most fundamental prerequisite for a harmonious, dynamic disclosure of the meaning of reality. Radical resistance implies that some creational given is taken, elevated, and made the basic denominator of all meaning. In an *abstraction* from the dynamic reality of meaning, some element of reality that is set apart and made *independent* is made to fulfil the function of the Origin and Fulness of meaning.

In western culture it is philosophical-scientific thought, especially, that has set its course toward some abstraction or another, elevating such an abstraction to independence and entertaining the pretension of establishing all meaning upon such a basis.[15] Resistance of this sort to the meaning-*dynamis* which has been *given* can only result in disclosure of meaning being pushed aside by disturbance of meaning.

Whenever the meaning-character of reality is not acknowledged, and whenever part of this reality is absolutized and set up on its own, so that all remaining reality is interpreted in terms of it and related to it, no harmonious disclosure of meaning is possible. Humanity, sad to say, lets itself be led by this absolutization as it engages in its cultural work, thus fixing its religious faith upon it. Disharmony then arises. Tensions appear in such a development of culture because the development originates in resistance to the true meaning-*dynamis*. That resistance is torn and divided internally from the very start. Although it is utterly impossible for humanity to separate itself from the given meaning-*dynamis*, it is indeed possible for people to resist the *given* direction of the meaning-*dynamis* and thereby try to bend the course of history.

That which is absolutized is only *imagined* to be absolute, and cannot in fact be made to be so. Because it is itself meaning, it cannot be the basic denominator of all meaning. Moreover, that which is established through the absolutization is only established in *appearance*, for it is actually taken up in its meaning-character in the diversity and coherence of meaning and directed to the fulness of meaning. Thus the absolutization is not successful – the original meaning prevents this. This being the case, people go on to fall into some other absolutization. Whenever humanity fails to submit in faith to the meaning-*dynamis*, one absolutization gives rise to another. But each absolutization is no more successful than the one before. This is the origin of the dialectic.

An example taken from the views of modern technology presented above will make this clear. We saw that the absolutization of technological-scientific thought leaves no place for freedom, which is accordingly driven out. Yet the exorcism proved always to be less than completely successful (for the simple reason that it is impossible). The problem of freedom reappeared time and again. Nevertheless, when freedom was absolutized in turn, no one seemed to know what to do with technological power. While viewing it on the one hand as hostile power, people necessarily accommodated themselves to it on the other, for humanity would no longer be able to live if bereft of technology.

When humanity fails to submit radically and integrally to the Radix of all meaning and to the meaning-*dynamis* of the creation, that is, when it fails to let itself be guided by the norms for the development of culture, when in pretended autonomy it sets its own laws and goes its own way in conformity with these laws, then the original dynamic *of* the creation turns into an original dialectic *within* the creation – with consequences which, despite the fictions being perpetrated, can only prove disastrous.[16] The dialectic resists the disclosure of meaning; it dislocates and stifles development.

It should not be forgotten that the dialectic always makes use of material for which humanity pretends to be the giver of meaning, whereas in reality the meaning is not established by people at all. The antecedently given diversity and coherence of meaning assert themselves here – and they *prevail*. The resistance of people has its effect, but in their resistance they are nonetheless propelled by the *dynamis* of the creation. In fact, it is precisely the meaning-*dynamis* that makes human resistance possible as resistance. The person who fancies himself autonomous is a parasite on the meaning-*dynamis*. At the same time, he perverts history. The effect of his doing so is not enduring, but it can bring about far-reaching and appalling consequences.[17] And these consequences, which in one way or another entail the penalty of the disturbance of meaning, inevitably compel one to backtrack. The pretension of autonomy must once again yield to the superior power of the meaning-*dynamis*.

While this "reversal" can lend a relative stability to historical development, it need not be interpreted as signifying that humanity thereby allows its activities to be established in submission to the normativity of the meaning-*dynamis*.[18] Rather, the turning can be a sheer *reaction* to the dislocated development.

It is difficult to foresee when a reaction to a dislocated development will occur. The limits to the disruptions of meaning are unknown to us. Sometimes it seems as though these limits are very sharply fixed, and then again they sometimes seem to hardly exist at all, for evil can assume immense proportions. Here, at bottom, we stand before a mystery – the divine mystery in history,[19] a mystery discernible in the permitting of evil, in the "turning" from it, and again in the disclosure of meaning. It is God who leads history in Jesus Christ. Dominant in history is the disclosure of meaning. Although it may sometimes seem as though the dislocation of meaning is in the ascendant, God does not permit the boundless dislocation of meaning.

Unless it involves tuning in to the given meaning-*dynamis*, a reaction against a dislocated development brings no real gain in the long run because the reaction, too, leads to dislocation. More often than not, this is not noticeable at the outset. Thus it is not the mere fact of humanity's turning against a wrongly directed development that is decisive; more important is the conviction from which human resistance arises. What is important is not that people set one absolutization against another – think of the dialectical relation between the transcendentalists and the positivists – but that humanity submit to the meaning-*dynamis* of the creation and cast autonomy aside. Only if this is done can we hope for a meaningful future.

4.5.5 The future as progression

In the actual course of events, we have two things to keep separate in our minds. *In the first place* there is the disclosure of meaning. This exists whenever humanity submits to the meaning-*dynamis* of the creation and believingly confesses its own powerlessness, thereby freely accepting its responsibility for the disclosure of meaning. Humanity then directs itself *in the present* toward the future of the Kingdom of God as the completion and fulfilment of all disclosure of meaning. *In the second place* there is resistance to the direction of the disclosure of meaning. The direction of this resistance is at cross-purposes, as it were, with the disclosure of meaning: it is the direction of the locking up and sealing off of meaning.

We would be resting content with the mere *appearance* of things if we concluded from actual developments that some last or most recent direction had finally gained the supremacy. Such developments could never materialize, as we have seen, if it were not for the Kingdom of God, which forges its way ahead in this world through all disturbances and dislocations. The coming of the Kingdom is the true progression of history.

It is good to be reminded once again that we stand here before the divine mystery in history. Our thoughts about it must respect fundamental bounds. While a full view of the relation between meaning-disclosure and meaning-disturbance is granted neither to ourselves nor to anyone else, the possibility – and, indeed, the necessity – of having a *perspective* from which to appraise responsible meaning-disclosure correctly becomes a matter of real importance. This is true with respect to the warping of meaning-disclosure by such positivistic cultural optimists as Steinbuch and Klaus, but it is also true with respect to the waiting game of the cultural pessimists, including the transcendentalists we examined.

Whenever humanity subjects itself in the present to the meaning-*dynamis*, the future opens itself as progression.[20] What this means concretely is that humanity submits to the given normative principles which obtain for the cultural and supracultural modalities, and that people positivize these principles, or work them out for the concrete situations in which they find themselves. Submission to the normative principles is the beginning of human responsibility. These principles motivate people and make them wise to the perspective of the fulness of the meaning-disclosure toward which all the structures of creation point.

Science and technology can also be taken up in the disclosure of meaning. The person engaged in science and technology can render important *service* to the true progress[21] of history. This future is equally real, of course, for someone who cannot penetrate the complexities of science and technology but whose activity nevertheless conforms believingly to the *dynamis* of creation even while he is unable to anticipate and command all the consequences of that activity – something which, truly, no one can fully do.[22]

When the person of science or technology seeks to make the future safe on the ground of pretended autonomy, that intended future will most assuredly be frustrated. With the help of his analytical and creative methods, the self-willed person seeks, insofar as this is possible, to dissect whatever exists or can be foreseen and to build the future from the dismembered elements.[23] The "future" becomes a calculable or technological object and is thereby fundamentally mutilated.

When humanity acts on the basis of its own putative authority, matters are simply reversed in principle. Contrary to their intentions, people deprive themselves of the liberating perspective. This also goes for those who, in the name of progress, wish to replace the existing "powers" with other ones. Today's establishment gives place to tomorrow's. Moreover, even the goal that anarchism strives after, i.e. complete freedom, boomerangs. The result is its counterpart – the absence of freedom.

The one requirement necessary for the future is that humanity must not set its own wrongheaded goals for the future. Rather, people must look to the sure future of the Kingdom of God that has come in Jesus Christ and is coming. With respect to science and technology, this signifies that for the future the center of the struggle must lie in the rejection of any pretension that science and technology can somehow be helpful in effecting our salvation and our future.[24]

There are two things we may not forget. The first is that humanity, *regardless* of its self-chosen objectives, is subject to the future that has been set by Jesus Christ – the Kingdom of God. We saw that the dialectic borrows its very possibilities from the meaning-*dynamis*, if only to resist it, but that it is accordingly brought to judgment.[25] The second is that although the achievement of the goal is not exclusively dependent on humanity, people nevertheless do bear responsibility for the disclosure of the meaning of the creation. To look on as spectators is not their lot. Too much is at stake in the dislocation of meaning and in the suffering it causes. People may not pass by on the other side and ignore it, and they may not acquiesce in it. They have a calling to resist the disturbance of meaning and to indicate a meaningful perspective. In this task, however, they do not get much beyond patchwork. Although they might like to break out of the dialectic, they remain far too deeply stamped by the powerlessness and presumption that result from blocking out the light of truth.[26]

4.6 Some Basic Distinctions

Before the philosophical themes just referred to can be explored with respect to their significance for technological development, we must make a number of distinctions.

We saw that the transcendentalists subsume the great diversity of the sciences and of scientific method under one designation – technology. Consequently, they are of the opinion that technology is an instrument for the control of nature, society, and humanity and its future. The transcendentalists experience technology as an autonomous power against which they offer resistance but before which they must eventually yield. An important cause of their predicament is their failure to distinguish adequately between science, technology, applied science, scientific method in technology, and so forth. These distinctions are also wanting among the positivists. As to the latter, the leveling of distinctions is inherent in their ideal, which is to command the future through technology. Everything that serves this purpose, including science, is viewed as an instrument of technological control.

A part of our task, then, is to clear a path through the tangle of diverse terminologies so that it will be possible to take a better look at the place and significance of technology for both the present and the future. Sound distinctions are necessary if the limits and possibilities of technology are to be recognized and, above all, if the responsibility of humankind in technology is to be placed in the proper light.

I will now proceed to devote a series of subsections respectively to science; the application of science; the intervention in, or alteration of, reality; scientific method as it can be employed in such intervention; and research. After that I will devote separate subsections to futurology as a multidisciplinary joint enterprise and the complex problems of quantification. Both subjects have been broached before. It is desirable to distinguish them from modern technology as such. At the same time, the nature of their connection with technology must be made clear.

4.6.1 Science

Science is concerned in a general sense with *knowing* reality theoretically – especially with knowing reality in its regularity or uniformity. To this end the scientist formulates laws which express the orderliness of reality in one way or another.

There are many sciences. What they all have in common is that they have one or more *aspects* of reality as their field of inquiry. The differences are due to the nature of the aspects.[27] A rough distinction in the sciences is that between the natural sciences and the human sciences (Dilthey) or cultural sciences (Rickert). I shall speak of the latter sciences as normative sciences.

In the natural sciences, the field of inquiry is ruled by determinism, while in the normative sciences people and their freedom occupy a central place in the field of inquiry. The laws formulated in the latter sciences have the character of norms, which in turn ought to be related to the normative principles that obtain for a particular aspect. An example may help make this clear: when an economist formulates the economic laws for an enterprise, the principle of saving should occupy a central place in those laws.[28]

The distinction between the natural sciences and the normative

338

sciences also involves the significance of knowledge for the future. Because the field of the natural sciences is ruled by determinism, the knowledge those sciences offer has validity for the future. And it is natural that great steps forward in scientific inquiry should result in the broadening and filling in of this knowledge. Yet this knowledge remains connected to that which is determined in the field of inquiry. For the normative sciences, by contrast, things are somewhat more complicated. In those sciences, humanity in its freedom appears in the field of inquiry – in this sense, that in connection with normative principles people positivize norms for the field concerned. Furthermore, for these sciences the problems with respect to knowledge of the future are already difficult in a static culture; in a dynamic culture such as ours, it is really extraordinarily difficult to say anything about the future in a scientific sense. The boundaries and limits that confront the normative sciences are related to God's control of the future,[29] His involvement with the future in the revealing and concealing of meaning, and human freedom (either in conformity with the meaning-*dynamis* or in resistance to it).

Within the framework of the limits mentioned above, which the normative sciences are forced to take into account, the following questions may quite properly be raised: What is going to happen? What ought to happen? Science extrapolates into the future, as it were, on the basis of the order of things that is already visible and tries to weigh its extrapolations in all their consequences against normative principles and the norms derived from them. This weighing and testing can lead science to formulate guidelines for the positivization of new norms. That responsibility in freedom which pertains to science at once entails both independence from practice and limitation of task. Science itself may not engage in positivization or bind people engaged in practice in the coils of its results. Science can only offer advice to those who bear the responsibility for practice. Scientific knowledge, which is very limited where the future is concerned, is of service to the free, responsible people of practice whose mandate it is to positivize norms.[30]

Overabundantly redundant though the observation may be, let it be noted again that scientific inquiry – the testing of given situations against norms, and not least of all the drawing of guidelines for the positivization of norms in the confrontation with practice – is fed from the deepest, religious convictions of the person of science. In other words, scientific work either will be guided by the meaning-*dynamis* or will oppose it, assuming a dialectical character as a result. In the latter case, scientific knowledge mutilates itself. Insofar as it then gains practical influence, it will contribute to the disturbance of the meaning of cultural development.

4.6.2   The application of science

Scientific knowledge is general, universal knowledge about uniform reality. Given scientific knowledge as a basis, the application of science aims at *knowing* the individual and the concrete. The concrete may consist of anything – even a free person or a particular situation in which people appear. The application of scientific knowledge involves particularized knowledge. Applied scientific knowledge is conditioned by both the nature of concrete reality and the freedom of the person who applies scientific knowledge. Applied scientific knowledge, accordingly, is not purely deductive knowledge.

Examples of the application of scientific knowledge are the location of a technical imperfection in some constructed object or the diagnosis of a particular malady from its symptoms. It is generally true that the application of scientific knowledge enriches human knowledge of the concrete and deepens human insight into concrete situations. This knowledge and insight, once gained, can then serve to help people anticipate the practical measures they should take. More on this below.

4.6.3   Intervening in reality, or altering it

The application of scientific knowledge can generate the insight that something ought to be done and can also provide knowledge as to the appropriate steps to take. Measures employed to intervene in reality, or to alter it, in accordance with the acquired knowledge and insight, may have either a singular character or a universal character. The former is the case in the correction of a technical error or in the results of therapy; in modern technology the latter is to be observed in the designing of production processes aimed at mass production or at production with universal characteristics.[31]

In both instances the connection between the application of scientific knowledge and intervention in reality is formed by the creative activity of people. The result of creative human measures turns out to be something new. (Depending on the circumstances, of course, the new may appear as new in varying degrees.) There is a difference between singular measures and measures that have a universal character: in the case of the latter, a distinct and separate preparatory phase is introduced between the application of science and the carrying out of the desired universal measures. In this preparatory stage, the so-called scientific method is used.

340

4.6.4 The scientific method in the alteration of reality

Whenever the intended alteration of reality is universal in character, the scientific method is used in the preparation of that change. As we have already observed, this is the case not only in modern technology but also in modern organization, in administration, in the conduct of business, and so forth.

The limits of this method are set by the nature of the area of its intended use. If people do not appear in that area to complicate matters, the limits of the method are set by the determined nature of the area. An instance in modern technology is the designing of a fully automated production process.[32]

The success of the technological-scientific method is so great that it inspires many to wish it could be applied everywhere. The clearest example of this that we encountered was Steinbuch. But limits of principle are transgressed in any such absolutization of the technological-scientific method. The scientific method, after all, is itself subject to norms. Whenever people appear in the area of its intended use, the limit of this method is thus the freedom of those people. They have to be respected in their freedom. We saw that this applies to modern technology too whenever people are involved in the execution: the freedom of people in the production process is a limit in principle for the technological-scientific method.

In general it may be posited that the scientific method is qualified and thus normed by the qualification of the area of its intended use – for example, the technological, the social, or the economic area. In these instances we should accordingly be able to speak of the social-scientific and economic-scientific methods in distinction from the technological-scientific method.

4.6.5 Research

In research, the human cultural activities just dealt with come together: science, the application of science, and the alteration of reality with or without the use of the scientific method converge, complementing and fertilizing one another in the process. The precise mutual relations and the character of the cooperation ought to be established by the *meaning* of the research. The inquiry might focus upon the enhancement of scientific knowledge, which would then need to be differentiated according to the various sorts of specialized scientific knowledge, or the aim of the research might be the application of scientific knowledge, or it might be the alteration of reality. Thus research can advance "technology," regarded as the giving of form to the nature side of reality, but it can also be serviceable in organizing an industry, for example.

4.6.6 Futurology as multidisciplinary cooperation

In view of the distinctions made above, it should be apparent that I reject any futurology in which technological mastery of the nature side of reality serves as a model for integral mastery of the future in the vein of Steinbuch or Klaus. Through technology, humanity is then said to be able to control not only material reality but the whole of reality. Whatever the differences between Steinbuch and Klaus, their philosophy, as futurology, is therefore to be characterized as technicism.

Indeed, futurology, as the science of the future, is an impossibility, first because the future cannot be a fixed "object" for this science, and also because the future exhibits all the aspects that reality has. The exploration of these many aspects is served by the various special sciences.[33] We have already seen[34] that every science has to do with the future in one way or another, according to its own qualification. This is also true of the application of science and, above all, of scientific method as it is used to alter reality. Tinkering with the future, an activity which eventuates in many different kinds of projects, can entail the use of the scientific method. In view of the abstractive character of science – it focuses on one or more *aspects* – a number of problems will be involved whenever the matter comes down to resolving concrete, practical questions. I should like to say a bit more about this.

The possibility of a special science's developing lies in its being able to abstract from concrete, particular reality and being able to focus on a single aspect of reality. Physics, chemistry, psychology, the social sciences, and economics come to mind, among others, as examples of such special sciences. In proportion to their development, these sciences are fruitful for the praxis from which their abstractions were initially made. That the fruits of science can be impressive is more than apparent from the connection between natural science and technology.[35]

The question that arises is whether the connection between practice and any one science is fully adequate to the resolution of practical questions. Can one science with its knowledge of one aspect do justice to practice, where all the aspects are discernible?

These questions arose when, with the development of technology through the connection with natural science, certain higher complex problems were engendered, problems for which no answer was to be anticipated by way of some single "aspectual science" as a special science. The problems pertaining to nuclear energy, space flight, milieu hygiene, and so on come to mind. No one special science suffices to solve such deep-seated, far-reaching problems. In fact, it would be dangerous to identify the knowledge of one aspect of reality with the knowledge of reality as a whole, or to allow it to become the controlling viewpoint. That is precisely what happens, however, when technological-scientific thought is not only absolutized but is invested with authority over everything else as well, so that it displays an imperialistic character. Viewed structurally, therefore, the problems just mentioned here cannot be resolved except through multidisciplinary cooperation.[36]

Until recently, little consideration has been given to multidisciplinary cooperation. In the dynamic development of contemporary culture, which is based upon the explosive development of modern technology, we shall be increasingly confronted with the fact of multidisciplinary cooperation. Such cooperation is being thrust upon the people engaged in practice by the complex problems which they face. Yet it would be preferable if they would foresee this situation. To my way of thinking, the need for multidisciplinary cooperation comes upon us structurally, since it is given in the connection that has arisen (in modern technology first, but elsewhere as well) between science and practice. One special science cannot suffice to guide practice scientifically.

Multidisciplinary cooperation will be most easily realized among specialists in the various exact sciences. But many problems will arise even in that theater. Multidisciplinary cooperation requires insight into the limitations of one's own knowledge, a capacity to listen and to cooperate, and responsible attitudes toward the achievement of compromises acceptable to all.

Multidisciplinary cooperation will nevertheless need to be broadened to include other scientists, such as psychologists, sociologists, economists, and ethicists. And even the philosophers, with their all-embracing, unifying approach to the various disciplines, will have to be included on the multidisciplinary teams.

The great benefits of multidisciplinary cooperation are the potential elimination of one-sided views of things and the emphatic confrontation of scientists from a variety of fields with the relativity and limitedness of special scientific knowledge – although the "multi-" is in itself not an adequate guarantee of the ultimate advantage of all this.

The most pressing question pertaining to multidisciplinary cooperation is whether or not there is agreement among the various scientists about the nature of their responsibility and the meaning of their work. Essential to such agreement, above all else, is the perception of a common norm – a high requirement, indeed, but one that is prerequisite to the success of multidisciplinary cooperation.[37]

4.6.7 Quantification

The distinctions I have made here do not appear in the thought of the transcendentalists or the positivists. In a very general way, they subsume science, the application of science, intervention in reality or the alteration of it, every scientific method employed in altering reality, and even mastery of the future (the subject of futurology) under the designation "technology." In general they subordinate everything to technology – a practice that engenders confusion in the formulation of ideas and leads in practice to a technocracy.

In the midst of all this, we encounter the problems pertaining to quantification. Many of the thinkers we examined identified quantification with technologization. Jünger, for example, senses in quantification the penetration of clock time as "dead time" into that which is first quantified and then, on that basis, manipulated. Ellul and Meyer are of the opinion that things – as well as plants, animals and people, if the method of quantification is absolutized – are reduced to what can be counted, measured, and weighed. The very essence of what is quantified is then identified with the quantitative. Nature is "conceived to be rational and then construed in mathematical categories" (TdW 47).

What the transcendentalists have observed – and reject – in connection with quantification is striven for by the positivists. This is clearest in Klaus's case. Klaus is of the opinion that, given the dialectical category of quality and quantity, every quality can be extrapolated in terms of quantities, processed mathematically, and then controlled.

Therefore it is not particularly remarkable that the notion is afoot that the method of quantification has become a worldview of sorts, and that it is identified with the worldview of technocracy. This view is palpable above all in the increasing use of the computer in the many sciences and in the various sectors of culture. L. Heieck accordingly concludes: "The reduction of reality to the weighable and measurable has recently been elevated to a worldview."[38]

There is a great danger that the rejection of the absolutization of quantification will be attended by the rejection of the method as well. That would indeed be wrong. Some serious reflection on the proper use of this method is called for.

First, it is wrong to identify quantification with technology. We saw that in technological designing, arithmetical treatment is resorted to because it is fruitful to abstract from the free person whenever the design objective is a fully automated production process. Yet, even in the designing stage, arithmetical operations sometimes have to be complemented by geometrical and experimental operations. Even *argumentation* is used when other means appear to be inadequate.[39] These operations involve factors other than quantification.

Thus technology and quantification are certainly not to be identified. The method of quantification has value in areas beyond technology. It can be applied whenever abstractions from humans may be made without objection and wherever the arithmetical treatment is determined from beginning to end. There, needless to say, the computer can also be brought into play. However, whenever this method is applied to a cultural area in which people in their freedom play the focal role, as they do in economics, for example, one ought to be cognizant of the limitations of quantification. Thus in econometrics, abstractions are made from the free person, as attention is paid only to such features as are commonly displayed by most people, to the end that the rules of statistics may be employed. This manner of going about things does have its limitations; it will certainly jump the rails if people fail to remain alert to the semblance of an economic determinism.

Perhaps what follows will shed some useful light on the question of normative thought about quantification as a method.

An arithmetical aspect may be distinguished in all things. The fact that the arithmetical is an *aspect*, however, signifies at the same time that the thing itself is not exhausted in that aspect. The other aspects of the thing can also be scrutinized from the angle of quantification. In the latter case, great care is warranted, since human subjectivity plays a great role in such quantification and since the terms upon which the quantification is made must be taken into account in the interpretation of the results. It is certain, for example, that a cow has four limbs and one tail, and that the economic value of the cow may perhaps be estimated at one thousand dollars. The latter quantification, however, is dependent on the judgment of the estimator, on economic circumstances, on the age of the cow, and on other factors. The economic value is therefore to be *estimated* rather than firmly fixed. Neither the settled fact that the cow has four limbs nor the estimate of its economic value may be permitted to seduce us into identifying the cow with these quantifications. Every quantification presupposes an abstraction, and because the abstraction can change, many quantifications are possible. Along with the result of the quantification, the "selected" abstraction must also be taken into account.

Quantification of human phenomena, which must be assayed against norms, occasions particular difficulties. Perception of the normed character of the phenomena must be reckoned among the prerequisites of such quantification, for without this perception it is scarcely conceivable that quantification could be engaged in in a responsible manner. In the comparison of similar phenomena, a scale of values can of course be employed. Yet, with human phenomena it is especially necessary to take into account the particular focus and the subjective character of any such scales of values and of the quantifications based upon them. In connection with the specifications of the lie detector, for example, investigators make a quantitative connection between physiological phenomena and truthfulness or un-truthfulness. This correlation is not susceptible to mathematical treat-ment itself but is assumed; truthfulness and untruthfulness are not identical with their quantification. Before proceeding to quantifica-tion, investigators must therefore know what falsehood and truth are and how both demeanors may be reflected in human physiological phenomena.

Once quantification occurs, many projections may be implemented within the mathematical model without difficulty or objection. Even the computer can be brought into play. But then we must remember that after quantification, and especially after the computer is engaged, the problems of quantification are camouflaged. Whenever we wish to draw conclusions from particular results – perhaps, for example, so that on the basis of these results we may undertake something with respect to the phenomena examined – we must therefore remember to again take into account the pertinent subjectivity, the determining assumptions of quantification, the normativity, and the limited focus. Quantification is *conditioned, relative*, and *limited*. It is conditioned by the assumptions of quantification; it is relative because it has significance only for the aspect being quantified; and it is limited by the extent to which quantification is possible.

These matters may be illuminated with a simple, arbitrarily chosen example. When the correlation between lung cancer and cigarette smoking is being established, it should be asked, before far-reaching recommendations are made, what a particular inquiry had in view and which (possibly important) factors were considered to be beyond its scope. Suppose that the quantification in question involves the norm for life – preservation of life. If the inquiry has not established the correlation between lung cancer and smoking for the various age groups, then the avoidance of smoking cannot properly be so much as advised, let alone forcibly imposed. While the correlation may point pretty well in one direction for higher age groups, the case may be otherwise for lower ones. For the latter, other causes of death may be present. Also, the question must be asked in the case of the higher age groups, where the correlation is clear, whether such people have not attained their greater age precisely through much smoking, and whether, for this same age group, abstaining from smoking might not have meant a much earlier end of life, through obesity, for example, rather than through lung cancer. And so we could go on.

In short, quantification is always based upon a *theory about* phenomena, a theory that abstracts from the nonquantifiable parts of aspects, even while it is based on hypotheses that fix the limits of the quantification. Yet, the real uniqueness of phenomena can never be apprehended through enumeration. This is true for physical phenomena, and it is all the more true for living, psychical, and all other, human phenomena. Quantification is a *one*-sided, and thus inherently limited, approach to a particular aspect of the phenomena. The approach is limited even further by the question whether or not a normative aspect is involved in the quantification. The assumptions upon which the quantification is carried out and the phenomena from which the abstractions are made must accordingly be taken into account, especially when arithmetical processing of the quantification is to be utilized in drawing specific conclusions intended as a basis for policy decisions.

As we have seen, quantification is most effective in relation to technology, for with the technological implementation of full automation, people can be eliminated. Quantification elevates the basis for technological design. From the fact that quantification (and thus the computer) has possibilities in other sciences and in various cultural activities as well, it should nevertheless not be concluded that technology is to rule these sciences and these cultural activities. By way of quantification and with the help of the computer, technology simply offers its special, limited services in these areas.

4.7 The Periodization of Technological Development

In this section and those that follow, I will return specifically and for the main part to technology itself.

We began our philosophical analysis of modern technology in Chapter 1 by comparing classical technology and modern technology. The most striking characteristics of modern technology (the modern panoply of tools and instruments, technological forming, and technological design) were dealt with in turn, and the discussion issued in each instance in some remarks about the possibilities for the future. We found that by far the most promising of all technological objects is the computer. Technological execution will become increasingly automated, and laborers will eventually either have to change their work or find themselves unemployed. Nevertheless, when new types of work based on the modern panoply of tools and instruments and the newest materials can be developed, it will become possible to transform unemployment into a revival of the free professions – a transformation that will doubtless be accompanied by institutional change. Alongside the great industrial enterprises, there will once again be place for scaled-down initiatives of a more private and independent character. In the meantime, automation itself is in need of an increasing number of workers trained in science and technological science.

An analysis of technological preparation – the matter of planning or design – led us to the conclusion that the computer will be used increasingly in this area too, since many design activities are determined and are thus subject to mechanization. Nevertheless, this means that many highly qualified specialists will be needed, for the final development is hardly in the works as yet and many a field still lies fallow, waiting to be broken by technology's plough. Moreover, because technology frees more and more people to take part in technological development, it will also furnish more opportunities for reflection upon itself, and upon its power and meaning. This is all the more urgent, now that technology is generating problems great enough to threaten to overwhelm mankind.[40]

In order to shed the needed light on the perspectives and problems of technological development, it will be necessary to deal specifically with technological development itself. This I deliberately did not do in my comparison between early technology and modern technology.[41] Having taken my point of departure in technological development's past, however, I will now be able to take a good look at its possibilities for the future. The question of humanity's deepest motivation in technology will still be reserved, as much as possible, for treatment in later sections.

A philosophical analysis of technology and its history has usually been considered adequate if it suggested or supplied some division of the history of technology. Yet the essence of an analysis such as the one I provided in Chapter 1 entails as a necessary consequence both arbitrariness and one-sidedness in the division proposed and the neglect of multitudinous nuances, including some that are perhaps of real importance.

More often than not, technological development is divided according to the kinds of materials in use, such as stone, bronze, iron, and synthetics. The kind of energy employed has also served as a convenient basis of periodization. After human and animal energy, the natural force of water and the wind was brought into play; still later energy was derived from coal, oil and natural gas; and now the use of atomic and solar energy would appear to complete this development.[42]

Yet another orientation is gained from the tools in use. Material

technology is followed by energy technology, and this gives way in turn
to information technology. Every later technology presupposes the
earlier ones and gathers them up, as it were, into itself. Information
technology is inconceivable without the technologies of matter and
energy.[43]

All these lines of division do indeed reflect characteristic traits of
technological development. Yet it is most remarkable that the periodization of the history of technology which they suggest is based upon
criteria borrowed from the development of technological objects, both
active and nonactive ones.

Although the divisions of technological development just mentioned
are oversimplifications, their existence is in some sense justified. It is
really desirable, however, that the relation of humankind to the
technological objects in these schemes be made explicit and fully
elucidated. In any periodization of technological development, I would
very much prefer to let the emphasis fall upon the place of humankind
in technology – particularly on changes in the relationship between
people and the technological panoply of tools and instruments. Far
from excluding the other schemes of division, such a concept embraces
them. For the rest, giving central attention to the relation between
people and their technological devices does not signify that the problems inherent in technological development are finally placed in the
proper light, for this approach is still too anthropocentric. It embraces
the question of the meaning of technology, but it is only when the
motives of humanity in technology are brought into the discussion that
we can hope to gain clarity about this question. Technology is certainly
related to people, but its meaning is not exhausted with them.

Hermann Schmidt[44] views technological development as being in a
certain sense a progressive objectification of human functions and
achievements. He arrives at a tripartite division of the history of
technology. In the first stage, with tools such as the hammer, people
supply the skill and the required physical energy and intellective
effort. In the second stage, the skill and physical force are already
objectified in the machine. In the final stage, the intellective effort, too,
is transferred to the machine. "With each of these three stages, the
objectification of the realization of goals by technological means is
advanced – even to the extent that the goal aimed at becomes attainable without recourse to physical or mental exertion, by means of
automatons. In automation, technology reaches methodical completeness. This concluding of the objectification of work in technological
objects (in automatons), which began very early, is a decisive characteristic of our time."[45]

Schmidt's periodization partially satisfies the requirements of a
periodization of technological development. People occupy the central

place, but not as *responsible* beings. It is for this reason that Arnold Gehlen later adopts Schmidt's scheme without objections and appends to it the conclusion that technological development indicates that the human being is an "incomplete being" who makes up his deficiencies through technology but at the same time, since technology is given power to shape history, falls victim to meta-human technological development. Fully automated technology as completed technology breaks loose from humanity.[46]

In this view, humanity's responsibility for technology has disappeared. This occurs because Gehlen, when he looks at people in their development along with the technological panoply, does so exclusively from a biological standpoint. And he is not alone in doing so. Portmann regards the human being as an untimely birth, as an uncompleted person who, having come too early into the world, attempts to round out with technology first his body and then his mind. Thus technological development becomes prerequisite to the attainment of human freedom – a reversal of the proposition that human freedom, and thus responsibility, is a prerequisite for technological development.[47] To this Hannah Arendt wittily retorts that the human being in such a case is no longer a mammal but a crustacean.[48]

Human responsibility is also excluded when, in the vein of Ernst Kapp, one construes the technological panoply to be an unconscious projection of human organs. Here people are thought to be able to understand themselves better through the cognizance that they have fashioned their tools, if unwittingly, after the image of their own organs. Technological development is accounted for in psychological terms.[49]

I would assent to Schmidt's notions about technological development, but with one fundamental correction, namely, that people fashion their tools and instruments in responsibility. Beyond that, moreover, Schmidt's periodization is entirely too general. He makes no distinction between implements that one holds in his hand, such as the hammer, and rigs like the pulley, with which one is able to accomplish greater things. Also, Schmidt wrongly regards technology as being completed in the stage of automation, since in that stage the intellective force of humanity is projected into the automatons. The phrase "perfection of technology" is found in others, too. Could it be that there is hidden within this thought the desire to anticipate the development of technology and thereby to control the future?

In a certain sense, information technology is indeed to be understood as the final stage of technological development, but then only as a kind of qualitative completion.[50] The development toward full automation is only beginning. This development will require of people an intellective effort unprecedented in the history of technology.

The development of the technological panoply is the heart of the history of technology. In that panoply people attempt, among other things, to project and objectify their functions. The objectified function is not only superseded in the panoply but also strengthened.[51]

The projection of human functions into the panoply is possible through the use of new materials, and again through the tapping of new energy sources, and quite often as well through the interplay of material and energy.

However, chemical technology (note the manufacture of synthetics) and electrotechnology (witness radar equipment) show us that technology is not finally to be accounted for as an objectification of human functions, since in these sectors modern technology is not to be thought of as the enlargement or improvement of human functions. Nevertheless, technology as disclosure of the nature side of the creation with the help of tools and instruments certainly involves a progressive objectification of human functions. A periodization of technological development oriented along these lines is therefore of definite value. Let us consider the matter.

At the outset people furnish the energy for wielding tools, and their skill determines the results that will be achieved. In the stage of rigs and tackle, human skill is already being channeled along certain lines, and human energy, given the construction of the apparatus, exhibits more potential than ever before. Only with the introduction of energy from nonhuman sources – first horse power, then the force of wind and water, and finally manufactured energy on an atomic scale – is the human energy function superseded. In the meantime, it is accelerated and strengthened through the process of transference. People continue to steer and regulate, but their skill comes to be projected into the machine. By means of the cybernetic principle of backcoupling, first the regulation and then the guidance of the automaton are transferred. Machines can be constructed that can learn to guide themselves better than people can guide them. In the computer, moreover, given the basis of analysis, even human "choosing" can be transferred.[52] And by means of sensors, the human function of observation can be duplicated in the machine and enhanced and surpassed in certain respects. Robots are characterized by their sensory functions.

It is to be anticipated that still more human functions will be transferred to the machine. In principle it is possible to transfer all precultural or pretechnical substrata of human activity to the machine. Freedom and responsibility, however, are reserved for the dominion of humanity. Through the tremendous possibilities of machines, the cultural activities of people can certainly be enhanced. This is apparent from the mechanical processing of signs and symbols, from data and word processing, from mechanical translation, from programmed instruction for education, from the typesetting of newspapers and books, from airline booking systems, from the support given human decision-making in the areas of society, economics, socioeconomic politics, and much more. In short, technology is the basis for many human cultural activities.

There is a decisive moment in technological development that has not yet been made explicit. The magnificent flight of modern technology is made possible by the scientific approach and basis of technology. There are, of course, many connections between early and modern technology. Modern technology is nevertheless not to be construed from early technology. Indeed, such a construction would obstruct insight into the very close connection between the historical development of modern science and the new character of modern technology – and the latter does differ enormously from the earlier technology. Various lines of development are intermeshed through the scientific founding of modern technology, and the transition from the one line to the other is gradual.[53] Taken as a whole, nevertheless, the impression is one of a revolution. In the middle of the last century, many developments came to fruition at about the same time. And technology has developed since then almost exclusively within the framework of the modern industrial corporations. It is for this reason that people speak of an industrial revolution. Given their beautiful interaction, the significance of science for technology, and the institutional setting in which the development occurred, this designation is acceptable. If there is to be talk of a second industrial revolution today as automation nears its completion, however, it should nonetheless be added by way of reservation that the present development, even though it is going forward more quickly than many had anticipated or find desirable, is only a consequence of the beginning made in the last century. The present technological development is characterized by very rapid change, but no less by continuity of development.[54]

If, alongside the industrial development of technology, we should one day have a parallel development through private initiatives and new crafts, it would be possible to speak of the existence of a postindustrial technology alongside the present industrial one.

As I suggested, a periodization of the history of technology related to objectified human functions affords a clear insight into the past and opens perspectives for the future. We have seen, however, that technology is not exhausted in the objectification of human functions. Broadly speaking, technology is the disclosure of the nature side of creation with the help of tools and instruments. But while the panoply of technological tools and instruments embraces the objectifications in question, the character of technology is not to be explained entirely in terms of these objectifications. The reason for this, of course, is that the meaning of technology is ultimately not anthropocentric. Rather, the meaning of technology transcends humanity as the meaning of a specific, formative modality belonging to the nature side of reality, which is created by God. This being the case, the meaning of technology is service to God – and therein service to one's neighbor and to the improvement of culture.

I should like to conclude this section with a number of remarks. The first is that we find the essential character of technology in the panoply of tools and instruments. When technology is taken so broadly as to mean disclosure of the nature side of creation and actualization of its technological side, however, there must also be room for reference to the form-giving that pertains to the physical substratum of plant, animal and human life and to the psychical (sensitive) substratum of animal and human life. In order to distinguish these kinds of form-giving from the technology we have thus far been dealing with, I would prefer to speak of them in terms of biotechnology and psychotechnology respectively.[55]

Unmistakably, the problems of biotechnology and psychotechnology will be a tremendous challenge to people in the future, especially insofar as they touch humanity itself.[56] Very generally speaking, it may be taken as a rule that these technologies will serve to render life less vulnerable, and that the span of life will be lengthened as a result. For the rest, this development demands penetrating reflection because these technologies may spring greater surprises upon us than we have imagined thus far.[57]

The second remark concerns the intellectual background of modern technology. Meyer, especially, but also Ellul, is of the opinion that with the rise of modern technology, a caesura has occurred in the history of technology. They both explain this on the basis of the fact that the development of modern technology goes hand in hand with secularization. They believe that modern technology implies secularization, and that this is so because of the influence of humanistic science on technology. They argue that a "this-worldly" culture has arisen, a culture that presents problems beyond humankind's capacity to cope.

It is undeniable – and in this I agree with Meyer and Ellul – that the pretended autonomy of reason has had a powerful influence on both technological development and its dislocation.[58] Nevertheless, there remains the question whether they are correct in thinking that this influence is inevitable.

In many studies of the development of modern technology, it is noted that this technology originated in western culture, where the influence of Christianity has been enormous. Having looked at history, Olaf Pederson concluded that modern science and technology are the fruits of a christian culture.[59] The influence of the practical calvinist ethic on the development of modern technology is likewise unmistakable, says Walter van Benthem, a Roman Catholic.[60]

It escapes Meyer that what one finds in the Reformation is an effort to return to the original command given humanity to dress and keep the creation. The Reformation views nature as *created* nature and has once and for all rejected any deified nature that might be reckoned untouchable. From a christian standpoint, then, all technology, including modern technology, pertains to the service and glory of God. In modern humanism, however, this view has been secularized so that technology exists for and around humankind. Is it not precisely this that has proven so disruptive, that as autonomous humanity progressively set its stamp upon technological development, technology became an autonomous power set against humanity?

In the history of western culture, modern humanism has gradually gained the upper hand. Yet, Meyer overlooks the fact that humanism made its way through Christianity, as it were, and that it therefore remains dependent on its christian origin and inexplicable apart from reference to this origin.

I will have more to say about this matter in the following section, where I will go on to discuss the meaning-disruptive effects that secularized motives have had upon technological development.

4.8 The Disruption of the Meaning of Technological Development

In dealing with the positivists, I was concerned especially to make explicit what the content of the future of technological development will be if positivist ideas about technology are applied in practice. The result would be a technocracy in which people are imprisoned, with their liberty exchanged for coercion and oppression. Such a technological development would be anything but rich in meaning; it would be a development the meaning of which has been dislocated and distorted.

Generally speaking, the transcendentalists are quite aware of the meaningless, nihilistic development of modern technology in instances where modern technology is being led by the motives of the positivists. This is not to say, however, that they are in a position to offer a meaningful perspective for the future of technology themselves. On the contrary, it is quite clear that their own views of the future remain equally knotted in irresolvable contradictions. Technological development would be derailed quite as effectively by several of their scenarios, among which are a "return to nature," acquiescence, and flight into Being or into existential freedom.

At this point it would be worthwhile to take note of a number of the motives that are presently at work in modern technology, disrupting its meaning, and to assess their tendency. Among the motives we shall consider are: the notion that technology is applied science; the idea that technology is the handmaiden of economic life; the conviction that technology is a neutral instrument; the view that technology manifests the will to power; and the impulse towards technology for technology's sake. It should be obvious that these motives will not be found in isolation from each other, and that what is said of one motive may thus be true to some extent of the others as well.

Following a very brief discussion of these motives, I will devote a separate subsection to the source of the various disruptions of technological meaning.

4.8.1 Technology as applied science

Those who conceive of technology as applied science are of course convinced that the so-called technological-scientific method should hold sway by itself. They make technological development a mirror of natural-scientific knowledge. This leads to the theoretification of technology, that is, to its being reduced to an exercise in theory, in the planning stage. The result is that human creativity as manifest in invention is precluded, and human freedom in technological forming is destroyed.[61] The trait of continuity in natural-scientific knowledge as knowledge of the determined aspects is projected, as it were, into technology, rendering technological development a determined development.

Technological development is stifled by the controlling influence of this *rationalism* in technology; the possibility of further disclosure is discouraged. As the designer's impulse toward theoretical control gains ground, labor declines in importance and status, meaningful initiatives are thwarted, the achievement of breakthroughs and new discoveries is rendered difficult or impossible, and the disclosure of meaning in technology is impeded. In other words, *the absolutization of technological-scientific thought resists the disclosure of meaning through technology.*

4.8.2   Technology as the handmaiden of economic life

Technology is often found in the grip of economic powers. In such instances, the economic value of labor comes to be implemented as the sole norm for the production process, since this criterion is the one that affords the greatest profit. Saving, as the motive of economy, receives exclusive emphasis, and the specific meaning of technology as the free giving of form to given materials is totally ignored. The enterpreneur and the enterprise control technological development with their great power. The enterpreneur so orders the power of his economic organ that it appears that every compromise will be made to his own advantage. More often than not, the enterpreneur takes recourse to complete rationalization because it seems to afford him the greatest profit. He also attempts to eliminate factors that might increase the costs of production. In this connection one thinks, for example, of the problem of milieu hygiene.

Whenever *economism* dominates technology and profits assume central significance, the production of goods ceases to be governed by the present needs of consumers. Instead needs are created by way of advertising, for commercial reasons. The products of technology are imposed upon people without their ever having sought them.[62] In addition to a superfluity of goods, this means one-sided mass production at the expense of individual production, with all the consequent drawbacks.[63]

Let it be noted in passing that these objections are not being leveled at mass production as such. Mass production deserves a hearty welcome for many reasons, the most important of which is that it makes the fruits of modern technology available to many on a grand scale. It does this by achieving relatively high quality at reasonable cost. The great objection, nevertheless, is that people confine themselves to mass production at the behest of economic considerations and reserve no place for dearer and less profitable individual production. The latter would help prevent leveling and monotony in technological forming, and also in the use of goods. Relief from the uniformity of modern technology is especially important where people occupy a central place in relation to the products of technology. Modern domestic architecture is an example.

Meanwhile, it is good to keep in mind that both rationalism in technology (see the preceding section) and economism in technology can actually be implemented for only a short time. People seek the largest possible profit through a fully determined production process, but in the long run their endeavor encounters obstacles and ends ultimately in the accrual of losses. A determined production process is not flexible enough to be adapted to the latest developments; it becomes obsolete quickly, and the competitive position weakens. In other words, a setback occurs sooner or later when, as though by inner necessity, the absolutization which is the cause of the meaning-disruptive development of modern technology in the first place boomerangs.[64] It remains an open question whether this setback is only temporary, and thus a short-term reaction, and whether technological development will be redirected toward the disclosure of meaning.

4.8.3   Technology as a neutral instrument

In another perspective, the central place of importance is accorded to humanity outside and apart from technology. Technology is then regarded as a neutral means of reaching ends appointed by people. Here the real meaning of technology is missed. In fact, the meaning of technology is reduced to its possible significance beyond technology; technology itself is not subjected to technological norms. Thus the view that technology is neutral precludes acknowledgment of human responsibility within technology itself.

It goes almost without saying that if people should use technology as a neutral means and permit themselves to be led by their own arbitrary and subjective norms, the consequences could be so sweeping as to prove fatal to humankind. This is all the more likely to become the case as people in technology shrug off their responsibility and persist in regarding technological development as neutral and automatic. Should disastrous consequences result from this situation – should a power-crazed dictator unleash the atom bomb, for example – people in technology will undoubtedly attribute these consequences to impersonal factors and forces, or otherwise disclaim responsibility, perhaps by blaming the mishap on people outside technology who will be said to have misused technology, the neutral instrument.

The merit of Donald Brinkmann's contribution is precisely his underscoring of the dangers of the idea of the neutrality of technology: "While this view in effect neutralizes the meaning properly belonging to technology, it concomitantly proclaims a fateful irresponsibility in technological work and allows all the terrible consequences of the technological development of modern times to be blamed on other agents." "The thesis that technology is a neutral means affords points of contact nicely suited to *nihilistic tendencies*, for in the absolutization of technology as a value-free system of means, every goal and value is placed beyond consideration."[65] In short, an "iron sense of duty" and "belief" in the neutrality of technology can result in the subjection of technological development to the influence of improper powers and in the advancement of meaninglessness and nihilism.

4.8.4   Technology as the will to power

We have seen how sensitive the transcendentalists are to the fact that when the will to power becomes the controlling motive, technological development jumps the rails. As people venerate technological power, they become its victims.

Inspired by Friedrich Nietzsche, Oswald Spengler claimed that technology as the will to power would deliver into the hands of humanity a world created by itself and obedient to itself: "A will to power, which mocks all bounds of time and space and makes the infinite and eternal its goal, subjects whole parts of the world to itself, finally embraces the whole globe with its communication and information technology, and *transforms* it through the power of its practical energy and the awesomeness of its technological methods."[66] But this technology as an expression of the lust for power leads in a perverse manner to dictatorship and the tyranny of technology over humanity. Many are doomed to slavish subservience. People inspired by the will to power use technology to support their evil intentions. Technological power then becomes *destroying power.*[67]

4.8.5   Technology for the sake of technology

When technology is carried on for the sake of technology – a situation which can be brought about by elitist thinking among engineers or by the expectation or naive belief that technological progress is simultaneously and self-evidently the progress of humanity and society – the norm presupposed for technology is that whatever *can* be made *ought* to be made. No questions are entertained concerning the meaning of such activity. The consequence of this *technicism* or this *technological one-dimensionality* is that the relation between people and technology is held in contempt. On the one hand people in technology become burdened, as a result, with an autonomous technology which they can no longer guide and control because they no longer apply the norms intrinsic to technology;[68] on the other hand, they produce an abundance of superfluous products or generate serious effects which can put pressure on people outside technology, which has the effect of endangering both culture and nature. Involved in all this is the unrestrained dynamic of technological development, which gets people in its grip and deprives them of their freedom. This dynamic is also responsible for the damage caused by noise, pollution and poisoning.[69]

4.8.6 The source of the disruption of technological meaning

The *deepest ground* of the disturbance of technological meaning is *religious* in character. Meaning-disturbance, which arises under the leading of various motives, is occasioned in the first instance by humanity's refusal to bow before the true meaning-*dynamis* of the creation, that is, by humanity's refusal to offer itself in freedom and responsibility. In pretended autonomy, humanity goes its own bull-headed way. People imagine either that technology will somehow or other save them from their limitations or that it will provide them with the means and opportunity to establish their dominion.[70]

It is often said (the formulations vary) that technology has become for many a substitute for the christian religion. "Technologically minded people turn the *christian doctrine of salvation* into a desire for a kind of 'do-it-yourself' salvation."[71] In the veneration of technology – whatever the form it may take – religion is made "this-worldly," and christian eschatology is exchanged for an *expectation of salvation through technology*.[72] Spengler corroborates this view when he says that the ruling motive in technology is "a small, self-made world which, like the larger one, moves by its *own* strength, and obeys only the hand of man. *To be God* – that was the inventors' Faustian dream, from which all designing of machines sprang forth."[73] But he also says: "Technology is eternal and unchangeable like God the Father, it saves humanity like the Son, it illuminates us like the Holy Ghost."[74]

Whenever people in technology are ruled by a secularized passion to be almighty and ubiquitous through technology and to be saved from the imperfection of the human condition,[75] the result contradicts their desire. We observed this in the case of marxism. Instead of accepting people in their freedom and responsibility as the necessary precondition for technological development, marxism proclaims liberation as an expectation to be realized through progressive technological development. Reality would have it otherwise, however, for the result of materialism as technological religion (Ellul) is a totalitarian technocracy in which freedom suffocates.[76] Thus I agree with the general assessment of Ortega y Gasset when he says that the person who must live believing in technology and nothing else misses his meaning. "Therefore these years in which we live, although they are the most intensely technological that human history has known, are at the same time the most empty."[77]

Belief in the autonomy of humanity is thus the source of the dislocations of the meaning of technological development; it is the source of the meaninglessness that humankind with technology draws down upon itself as a judgment. Because perspective on the "otherworldly" has disappeared, perspective on the "this-worldly" must in the long run disappear as well. "In the long run" – for secularized motives can certainly have "success" in the short run. It does not always become immediately apparent that meaning-disturbance has occurred.

I have used the terms *meaning-disturbance* and *meaning-corruption* to indicate that the tensions evoked by autonomous humanity in technology always remain subordinate to the meaning-*dynamis*.[78] People in their *pretended* autonomy cannot put this *dynamis* out of commission. People can bend the direction of meaning-disclosure and they can increase or slow down its tempo, but they cannot escape being subject to the meaning-*dynamis*.[79]

The secularized belief in salvation which people attach to technology presupposes the reality of technology and is therefore unable to detach itself from the meaning of technology. Human autonomy in technology perverts the meaning of technology. Yet this perverting borrows its very possibilities from, and parasitizes upon, the meaning-*dynamis* that is given. "Technology seems to develop into the very opposite of what is intended; it seems to end up in a meaningless, negative situation."[80] Yet the other side of the coin (pointed out by the same observer) is as follows: "A profoundly negative and destructive way-of-being of technology does not cancel technology's positive basic meaning, but presupposes a limit to the negative. Thus it is not possible to speak of an absolute, but only of a relative, intensity of the negative, and of the more or less serious consequences of the negative. Whenever the historical tendency of technology, which is positive in principle, comes to be realized and concretized in a relatively (surely enough) but nonetheless far-reachingly 'poor' or perverse form, the task becomes one of bringing the negative situation back into line with the positive meaning; the negative way-of-being must be converted into the positive way-of-being, so that meaning may progressively realize and fulfil itself."[81]

The fact that meaning-disturbance remains subordinate to meaning-disclosure should hardly serve to make us optimistic about the actual development of technology. As technological development persists in being controlled by secularized motives, it acquires an increasingly catastrophic character. It is as if some inescapable fate were bearing down upon the earth. However, what we are dealing with here in fact is the meaning-*dynamis* of the creation as a law of life; as life itself and the things of this technological world are exalted in human thought and convictions are given an independent and absolute status, the meaning-*dynamis* acquires a judgmental effect.[82]

On the one hand, then, the seriousness of the meaning-disturbance of technology in a secularized world is not to be underestimated. Yet, on the other hand no situation may be allowed to have a deleterious effect upon our discharging our responsibility to point out a liberating perspective for technological development.

4.9 Technological Meaning-Disclosure

The one essential condition for humanity's going on its way in freedom in technological development is that people submit in *belief* to the meaning-*dynamis* as normativity for the meaning-disclosure of creation. With respect to technology, this means first of all that people must acknowledge that the specific meaning of technology is contained within the coherence of meaning and directed toward the fulness of meaning and hence toward the Origin of all meaning. The meaning of technology may not be turned aside toward one absolutization or another. No point within created reality may be made the absolute target of technological development. That would occasion meaning-disturbance.

A liberating perspective for technological development opens up when it is understood that the specific meaning of technology ought to be led by the normativity of the various postcultural aspects of our reality, namely, the lingual, social, economic, aesthetic, juridical, ethical, and pistical aspects. Belief is placed last in this sequence not because the disclosure of meaning is rounded off by belief – on the contrary, belief is precisely the most necessary condition for disclosure – but because it is in belief that the utter insufficiency of technological development is made clear. It is given to belief to point toward the fulness of meaning in a transcendent manner. In other words, technological meaning-disclosure ought to be directed toward the fulness of meaning in the Mediator between God and humanity, Christ Jesus, and so to the final rest and destination of the fulness of meaning in God, who is the Origin of all meaning.

Technological meaning-disclosure thus implies *deepening* of meaning or *enrichment* of meaning whenever the postcultural aspects lead technological meaning. Since these aspects also bear the character of being-as-meaning, it follows that technological meaning-disclosure as deepening of meaning is not rigid in character. There are always possibilities for technology to unfold in such a way that the problems which confront it can be resolved. Thus the future of technology is in fact not determined, but open.

There are various points I should like to clarify concerning the state of affairs that now obtains in modern technology. We saw that the key to modern technological activity is to be found in designing – more particularly, in the engineer who does the designing. The engineer designs – or rather, projects – laws that are related to the fashioning of technological things and facts. This designing is also called the positivizing of laws or norms. The person who submits to the meaning-*dynamis* as normativity possesses the freedom to help, in responsibility, in the enlargement or extension of this normativity. The more finely attuned he is to the meaning-*dynamis* and the better he thus understands the normative principles which are given him to be worked out, the more adequately he unfolds this normativity.

What this means for the engineer is that he must allow himself to be led sequentially by the normative principles of reality's various modes of being; that is, in the human positivizing of technological laws for things and facts, the supraarbitrary normative principles of the respective modalities must be worked out in relation to what is technological. Should the engineer fail to attain a correct insight, either because he surrenders, self-sufficiently, to absolutizations and thus shuts the whole process off, or because of shortcomings or shortsightedness, then meaning-disruption will arise and, as we have observed, prevail.

Technological meaning that is disclosed is in the first instance meaning that is expressed symbolically. A design is indicated in formulas and drawings that signify the laws for technological things and facts and serve as guidelines for technological forming.[83]

When an engineer absolutizes technological-scientific thought and accordingly fully determines his design, he at the same time closes off and precludes all further technological meaning-disclosure. Yet, technological meaning ought to be disclosed and deepened in social meaning. To this end, designing ought to involve the exchanging of ideas and the sharing of concerns among all persons and agencies involved in technological designing and forming.[84] Social disclosure of the technological may be observed in the *teamwork* that goes on during preparation and designing; it is also apparent in the *pooling* and *attuning* of the interests of the designers, producers, and consumers. And the important place of the worker in the forming process should not be ignored here either.

Within the marxist sphere, one finds a considerable awareness of the social disclosure of the technological. Here, however, social disclosure is itself at once absolutized; a leveling and stiffening of technological development occurs when particular potentialities and technological achievements are passed over in favor of the social aspect.

In the economic disclosure of the technological, waste is prevented and people strive (among other things) for the optimal utilization of technological objects. Moreover, the individuality of designers and workers must be upheld and protected from all immoderate manipulation by extraneous persons and authorities. While preserving the individuality of all persons and authorities involved in technology, cooperation in modern technology should serve to satisfy the norm for the increasing integration that is now occurring in tandem with increasing differentiation.[85]

We saw that when the economic aspect of technological development is absolutized, a dislocation of meaning occurs: rapid production and mass production must afford the maximum profit, and this is gained at the expense of the individual person and at the cost of overlooking the dangers to the environment posed by waste products. For the rest, however, this is not to say that "profit" has no legitimate place in technological development. Making profits should be incorporated into the enterprise and disclosed in connection with service to God and service to one's neighbor.

What needs investigating is the extent to which the lopsided and crooked condition into which things have fallen can be straightened out. It is possible that we will have to pick up – in a new way – some of the old and forgotten traditions in the history of technology. In this connection, the so-called alternative technologies (the small technologies) should be given more than passing consideration. Modern technology should not be exclusively massive and gigantic; there should be room for a high degree of differentiation and diversity as well. Worthy of consideration in connection with this objective are technological products of greater durability that might be made with little capital investment and a minimum of energy – with natural energy and natural materials to boot. Personal, individual production yields satisfaction in labor, products reflective of settled cultural patterns, and a minimum of pollution. These forms of technology, despite their small scope, are decidedly difficult to realize. They require a great deal of technological fantasy and imagination. Cultural stability depends on these forms of technology, however, both in developed and in developing countries.

Technological meaning-disclosure, then, may not be permitted to founder on the shoals of economy, but must be worked along, in a transcendental direction, to still further disclosure. Aesthetic disclosure should appear in a harmonious development among all persons and authorities having some connection with planning and technological forming. It should also find expression in the harmonious selection of available materials, and in the harmony that should exist between nature and technology.[86]

Technological meaning-disclosure should also be led by the juridical aspect. First, legal institutions must see to it that technology is promoted. Secondly, they have the task of protecting people from the superior force of any misdirected technological development. On the one hand technology must thus be given its lawful chance, but on the other hand, justice should protect people from being crushed by either technology or its adverse side effects. Nor may technological development be permitted to make a wasteland of nature.

Again, justice is reflected in the reaching of compromises in technology. There are many factors that must be taken into consideration here, and many individuals influence any compromise. The purchaser, the contractor, the financier, the plant management, the designing department, the people who work with their hands – all these are participants in a compromise and all have to make concessions of greater or lesser importance. One example may illustrate this. As we saw, there is no possibility of complete theoretical control at a distance of technological forming because, for one thing, the complete elimination of the hampering resistance of the nature aspects is not possible.[87] And even if cybernetics should at some point afford the possibility of actually going over to a fully automated production process, it could prove meaningful to forgo making the switch while the chance remains that new inventions will quickly render the new production process obsolete. In the present situation, therefore, compromises should properly be reached between the research, designing, and manufacturing divisions. Compromises should also be made with a view to the client, especially with regard to the desirability of customized production, mass production, or some combination of the two. All these compromises should be adjusted, furthermore, as new inventions are forthcoming. And such adjustments are to be understood as fostering and stimulating technological meaning-disclosure.[88]

Herman Dooyeweerd has pointed out that meaning-disturbance manifests itself in the juridical modality too. When the earlier anticipations are not disclosed and when technological development is accordingly not tuned in to the direction of the meaning-*dynamis* of the creation, then meaning-disruption perverts history. And this disruption avenges itself in the development of technology. A stiffened or absolutized technological development (various forms of which were presented in the preceding section) cannot fail to bring judgment down upon itself, for whether there is submission to the meaning-*dynamis* as normativity or not, this normativity continues to hold sway. Nevertheless, limitations have been placed upon meaning-dislocation, which means that "historical judgment" will appear. People can attempt to absorb this disruption in some form of reaction, but they can also draw a lesson from it and submit to it – or tune in to the direction of the meaning-*dynamis* and thereby disclose technology in the proper direction.[89] A fine example of the latter would be provided by juridical-social action aimed at preventing the destruction of the environment.

After juridical disclosure, technological meaning-deepening ought to be carried forward into ethical or moral disclosure. The meaning-disclosure of technology is set in a good direction when technologists do their work in love. Love will express itself in a good choice and a proper use of available materials and objects and in a good, artisanlike finishing of technological products. But the care and responsibility of the technician ought to be directed above all toward the neighbor, who can occupy a place within as well as outside technology, and toward nature, of which the technological side is the one realized.[90]

Finally, technological meaning-disclosure ought to be led by the belief that humanity is called to the task of technology and that people are obliged to accept this mission as a responsibility before God. Belief is always the boundary-function and horizon of meaning-disclosure. The fastening of belief to something absolutized will inevitably result in meaning being shut off rather than disclosed.[91]

As the boundary-function of meaning-deepening, belief ought to anticipate the fulness of meaning and thereby be directed to the Origin of meaning. In its analysis of meaning-disclosure at the point where it involves belief, philosophical-scientific thought approaches a fundamental boundary. That philosophical-scientific thought is not self-sufficient is thereby made apparent. Theoretical thought can serve only to point to this *transcending in faith*; it can only approach its substance by *describing an idea*, as it were – *an idea* that ought to lead people in their work in technology.[92]

Before approaching the technological development idea for a closer look, I will devote the next two sections to themes that are closely related and require further elucidation within the framework of technological meaning-disclosure. These themes are the autonomy of technology and computerocracy.

4.10 The Autonomy of Technology

From various thinkers holding forth in a babble of keys, we have heard that technology has now become an autonomous power – that people contribute to its growth, to be sure, but that they are nonetheless fully subject to its dominion.

Now, this problem should be investigated. The various motives for regarding technology as autonomous should be marshaled and reviewed so that attention can be paid in the final analysis to the deepest ground of the origin of the idea of autonomy.

At the outset it is necessary to have within our purview the unique structure of modern technology, for that structure, in a remarkable way, gives rise to the notion that technology is autonomous.

The gap in modern technology between preparation and execution contributes to the working person's no longer feeling fully involved in technological development. That feeling is reinforced because only an abbreviated function in the production process is meted out to him. He is thereby deprived of an overview of the process as a whole and of insight into the purpose of the product.

Not infrequently, however, the engineer himself, as the designer of the production process, experiences that very same feeling of standing in the path of onrushing technology. Many design activities have been taken over by the computer, and the designer is accordingly less independent. As a result, designing is increasingly being done by teams. Design activities have become so multifariously demanding and complex that the individual designer is relegated to the background. The great advantage of the teamwork context is that much more can be accomplished. However, although individual potentialities can come to appropriate expression in the specialist, the teamwork context does imply that the individual will be increasingly less involved in design activity as a whole. The cooperation required necessitates the division of responsibility and leads to a situation in which every person, because of his limitations, must place confidence in others As a consequence, however, people can easily be deprived of the possibility of gaining a clear overall picture of their own areas of responsibility, and they can thereby come to feel at last like insignificant links in technological development.

This extremely difficult situation is aggravated by the powerful dynamics and planetary character of technological development. Those who work within modern technology can just as easily experience it as an overwhelming avalanche as those who work outside it. While people are increasingly occupied by technology, they have less and less say over it.

Every technological innovation requires new norm positivizations so that technological development will not get out of hand. Norms can only be positivized, however, after preliminary communal reflection has taken place, and it is precisely this prerequisite that appears to be canceled and precluded by the dynamics of technological development. Technological development seems to overrun people as though they were constantly off guard; people measure its latest advances against the yardstick of outdated norms that cannot suffice in new situations. Examples of this are certain problems generated by the computer, and the surprise so often occasioned by the environmental pollution caused by modern traffic and modern industries.

The trend toward autonomy is also reinforced by the structure of the modern panoply of tools and instruments. A case in point, again, is the computer. It works faster and more accurately than people, has nearly inexhaustible endurance, and works by itself. In this sense it is, indeed, *relatively* autonomous. Furthermore, it gives the appearance of becoming entirely independent of people. The fact is that while people always set the conditions under which the computer must function, the functioning machine runs quite independently of its makers. People simply cannot continually oversee the functioning of the computer. Therefore the results, within certain limits,[93] do have the character of a surprise. Moreover, a computer's user is not necessarily its programmer – a fact which again increases the gap between people and the computer. The gap is widened even further as one user replaces another. At last the user has not a glimmer of the criteria by which the contraption may be functioning. He *must* yield to it, trustingly.

This problem will grow in scope and seriousness in the future as machines that learn and machines that reproduce themselves are brought into use.

The notion that people no longer control the direction of technological development but are instead controlled by this development gains added currency from the fact that technology is being developed under the influence of the huge industrial-military complex, with the state appearing increasingly in the role of the great regulator. Technological power is linked in the popular mind to organizational and bureaucratic power, which comes to be regarded as the anonymous power of technology. Certainly the transcendentalists see it this way. The autonomy of technological power is expressed very well, they think, in the growing collectivization of modern technology, with its unavoidable tendency toward a one-world culture and a one-world state. The positivists seem to authenticate this perception, for they advocate both modern technology and a worldwide technocracy.

With the statement of these views, we have now arrived at an extremely decisive point. The feeling that modern technology is an autonomous power may find its *occasion* in the structure of modern technology, but its *cause* is to be found in humanity itself. After all, *people* are the ones who place their confidence in technology, who yield to it, who revere it and fear it as if it were a god. Humanity, deluded with notions of its own autonomy, is left with no other choice in the confrontation with technology. The option exercised becomes an inevitable one, apparently, since the communal sense essential to the responsible guidance of technological development is found to be missing. In short, the question of the complete autonomy of technology is a *religious* problem.

Secularized humanity has plugged up the fountainhead of meaning. In its pretended autonomy, humanity thought it could save itself, and it still intends to do so. Modern technology, through its success and progress, has proved especially inspiring to people. The other side of the story, however, is that humanity has come to follow in technology's train in a religious sense: in the points of reference afforded by the structure of modern technology in seeming support of the idea of its own complete autonomy, people believe they find grounds for allowing and aggrandizing this autonomy. People *permit themselves* to be *ruled* by technology. They forfeit the ability to see through the *appearance* of a technological and organizational automatism. And that carries with it the consequence that human beings go out and act as if they were technological things of some sort.

On the one hand, there are people who shove their responsibility off onto the impersonal, anonymous power of modern technology and allow themselves to be transfixed by the myth of material prosperity and the expanding possibilities of consumption. On the other hand, there are also people who continue to resist this situation. They refuse to be reduced to material to be manipulated in a technological sense. Occasionally these two mutually exclusive propensities are to be observed in one and the same person. More often, however, they are found in groups of people who oppose one another.

We can make the following rough distinctions. The *technocratic elite* champion the collectivization of technological power and invest this power with an exalted autonomy, following the perspective of a techno-logically streamlined society. The ordinary mass of people acquiesce in this vision either because they are powerless or because they are "happy" with the promised fruits. They never notice their own slavery, which is obscured by the anesthetizing influence of technology's possi-bilities. Nevertheless, small groups of people do set themselves against the all-dominating position of modern technology. They refuse to accept the abolition of human freedom.

In the last category we can place the *transcendentalists* we looked at, together with those who, not content to drop the matter at "flight," nostalgia for the past, or a speculative hope of deliverance, are commit-ted to taking action to recapture the freedom that has been lost, namely, the *modern revolutionaries*. Marcuse and Habermas are among them.[94]

The transcendentalists achieve no success because, as we saw, while they want to save human freedom, they continue to honor the preten-sion of their opposition, namely, the autonomy of technology. They stand up for human freedom and have their eyes open to the freedom-

368

devouring development of technology. But they never tie this freedom to human responsibility for technology.

The revolutionaries can less easily be characterized as homogeneous than the transcendentalists. Among them are some who would resort to destruction and sabotage in order to build on the rubble a world fashioned after their own vision. And the most moderate may advocate the renovation of democracy, perhaps by replacing indirect with direct democracy. Such moderates seek to subdue technological power through communal political influence and to bring it into dependence once again upon humankind. They wish to abolish the evil in technological development by exercising democratic control.

Quite aside from the deepest intentions of these reactions, it is an important question whether some common factor deserving consideration and respect may perhaps lie hidden behind them.

Technological development, as we have seen, is a very dynamic development. This fact contributes significantly to the spread of the idea of the autonomy of technology. It is in the nature of dynamic technological development to be surprising and overpowering. Its force can nevertheless be restrained at present, in some measure at least, by the influence emanating from growing democratization. Unimpeded technological development can certainly not be taken for granted, for political pressure and even apolitical activism are capable of occasioning delays. And deceleration of the dynamo has positive value when it serves to stimulate reflection about the direction of technological development.

Of course this is not to say that the problem of the autonomy of technology is to be resolved through democracy. Democracy provides no assurance whatsoever that what is wrong in technological development will in fact be resisted – especially when the people in a democracy refuse to abandon the idea of their own autonomy or when they allow themselves to be ruled by economism or materialism.

In addition to the value of the restraining influence of modern democratic tendencies, we should note the value of increased leisure time (free time as time free from wage labor) – especially when this time is used for reflection in general, or, by many, for reflection on the direction of technological development in particular. Leisure time so used is time spent neither in consuming the fruits of technology nor in being consumed by technology.

Another factor worthy of mention in this regard is the advocacy of a new form of asceticism, which seems to be gaining adherents nowadays.[95] One of the principles of this asceticism is that humans should not make *every* technological thing they can, but only those that are essential for either their neighbors or themselves. For those outside technology, this asceticism means people making conscious, discriminating choices from among the superfluous superabundance of technological products; it means resisting being swept away in the stream of the consumer society.

Democratizing tendencies, increased leisure time, and new forms of asceticism could bring about the conditions in which a technology that vaunts its autonomy could be brought back into line. Then it would be possible to take distance from technological development again. The unbridled dynamo could be checked, and people could again make choices, through deliberation of a more conscious character. These suggestions are important as conditions; their effect, generally speaking, can be applauded. Their effect will be minimal, however, if no communal criteria are employed – that is, if the spirit of community, which has been so nearly destroyed in our time, is not restored. If there is to be a good disclosure, the strongly integrating tendency of technology urgently requires that there arise among people a *new sense of religious community*. If there is no fundamental restoration in a religious sense, and if the erroneous view of humanity is not corrected, the effect of any breakthrough with respect to autonomy will be only temporary. Surely humanity only jumps from the frying pan into the fire when it exchanges the idea of an absolutized technological power for the idea of an absolutized freedom or an absolutized political democracy.

In short, it is necessary to the termination of the idea of the autonomy of technology that humanity give up the idea of its own autonomy and, consequently, its highhanded pursuit of freedom and salvation. Humanity must instead submit, in freedom and responsibility, to the meaning-*dynamis* of the creation.[96] "Until this calamitous quest for 'do-it-yourself' salvation is recognized as something which stands in the place of true belief, everything will remain as it has been."[97] Should this fundamental condition be fulfilled, however, there will be perspective again. Even so, the resolving of many practical problems could still require prodigious efforts and cause endless headaches.

4.11 Computerocracy

The integrating tendency of modern technology spills over into a collectivizing tendency and is strengthened by the introduction of the computer. The possibilities of an all-embracing technocracy are greatly enhanced by the computer. G. Zoutendijk, a Dutch computer specialist and parliamentarian, has warned that this development could be extremely dangerous if a dictator should come to power. In the past, dictators became all the more dependent upon people as their power increased. This meant the intimidation and often even the demise of the dictator. Given the capabilities of the computer, however, the power of a dictator might increase as his dependence upon others diminishes.[98]

For the rest, even without a dictator the dangers of the enormous possibilities of the computer are not eliminated. Computer experts can exercise enormous power with computers. With their knowledge of the computer, they preside over all the stored information, and they can alter a program to suit their own insight. If this situation were to be abused, the consequences could be far-reaching and very serious indeed.

The great difficulty in this situation is that the work of these specialists is carried on beyond the influence of the range of judgment of the hosts of uninitiated and unequipped lay people.[99] With the givens and results of the computer, the computer specialist (if brought in, for example, in connection with political policy decisions) can lay down the law to the politicians, since the latter do not command the desired information. Democracy becomes a chimera in such a *computerocracy*.

A number of practical, technological and legal measures can be taken in the face of this dangerous situation.[100] The most efficient and practical measure possible against one-dimensional policy-making by an American administration, to use the United States as an example, would be to provide the Congress with an adequate control apparatus of its own, independent of those employed by the executive branch – computer specialists especially being kept in mind. (The Zoutendijk proposal would equip the Dutch parliament with some such capacity.) In other words, legislatures must have at their disposal data banks and data-processing systems of their own.

Precautions also have to be taken at the juridical level to protect information stored about private persons. Everyone, for example, should be guaranteed access to all information stored about himself. National security requirements may at times conflict with this provision; for reasons of national security, certain information about certain persons might have to be made inaccessible. It will never be a simple matter to weigh competing interests fairly and to prevent the rise of a dangerous computerocracy on the one hand and unimpeded assaults on a nation's safety on the other. More dangers come to mind as one considers future uses of the computer. The confusion of the data-banked information of one establishment with that of another could

371

lead to an undesired violation of private interests, and could be a threat.

The many dangers which accompany modern computer development must be investigated at every turn, and all suggestions for averting these dangers must be given serious consideration. The dangers which arise from human weakness and shortsightedness are perhaps the most quickly and effectively solved or prevented. But what about the dangers that arise from wrongful inclinations and erring propensities in people themselves? For example, can a controlling agency offer any guarantee against a computerocracy when the decisive condition for every public policy is that it must not hamper the rise of material prosperity in any way whatsoever? Such thinkers as Jünger and Ellul have shown us that materialism indulged in as a way of life must give rise to a technologically streamlined society in which the individual is fully imprisoned, but to which, on the other hand, he submits as prisoner.

All the measures that will ever be proposed for the prevention of computerocracy will be of no real effect unless people predominantly allow the interest of their spiritual prosperity to take precedence over that of their material prosperity. The wishes and longings of autonomous humanity must not be allowed to lead people; rather, people must bind themselves, insofar as technological development is concerned, to suprasubjective normative principles. These principles will not let technological development stiffen into a computerocracy but will instead afford possibilities, even in a society using computers, for the individual to enjoy the freedom and responsibility essential to a healthy democracy. The development of every individual and of every social relation may very well occur within the context of increasing integration, but this integration need not lead to collectivization. "There is no question of a coercive development which leads inevitably to collectivism. The choice of a direction lies with us."[101] In fact, it is electronic technology, paired with information theory, that affords the possibility of substituting another direction for the tendency toward collectivization, a direction in which the coherence of the many industries and of government can find expression even while the individualization of the various branches of industry is being assured in the very presence of that coherence. In fact, the issue should be one of striking a balance between integration and differentiation.

If computer development can lead on the one hand to collectivism and computerocracy, it can lead on the other to anarchism through individual utilization, which is a danger as well, a danger that should not be underestimated. At the moment, the latter danger is being virtually ignored because it is less obvious than that of computero-cracy. Yet, it remains a live possibility, given the ongoing development of computers for personal use. With the power and possibilities of these computers, people may be able to disrupt the trend towards centraliza-tion and even destroy it. Marshall McLuhan, who initially followed Ellul's views of technocracy, later came to the conclusion that modern electronics destroys centralization: "electronics destroys the centrali-zation characteristic of the industrial era in all fields."[102] The following statement by Victor C. Ferkiss says something of the same sort about political development: "Instead of a hierarchical, mechanical power structure, politics is now becoming a total process in which the foci of power are everywhere and nowhere."[103]

In summary, we can say that computerocracy is the more obvious danger, and that the present trend in its direction must be countered through elaboration of the norm for integration and differentiation on both the political terrain and the industrial terrain. In the future, when the computer is in the hands of still more private parties, this norm will be a potential safeguard against both an anarchistic ten-dency in politics and an individualistic tendency in industry. The availability of the computer must be permitted to lead to neither the concentration of power nor the misuse of power. Even less may it be permitted to lead to chaos. In the background of the one line of development, absolutization of technological power is at work; in the background of the other line, absolutization of individual freedom is at work. The first possibility threatens to become reality. It in turn calls forth the second possibility, which could likewise become a reality whenever the possibilities of the computer should come to be employed more fully for private interests absolutized as such.

A harmonious development of the relation between humanity and technology lies open whenever technological development is led along lines conformable to the indicated technological meaning-disclosure. Then the freedom and responsibility of the individual and the unique character of the various human associations and relationships is respected, rather than leveled by some normless technological-scientific sway. Balance, as between the universal characteristics of modern technology and the unique person, would not be excluded. What is indeed demanded here is a high sense of calling and of norm. And this demand must be made because the balance in question, given the strong development of technology, should be a dynamic balance. While requiring greater responsibility and riper creativity of people, such a balance calls above all for a more intensive spirit of community. This demand applies equally to authorities, engineers, designers, contractors, workers, and consumers. All of them should be led in everything they do not by some absolutization or other, but by the fulness of the meaning of the creation that is in Jesus Christ.

We have now come back to the matter of the technological development idea which ought to inspire people in technology. I ended section 4.9, the section about technological meaning-disclosure, with a mere reference to such an idea. In the following section I will elaborate on it briefly.

4.12 The Technological Development Idea

We have seen[104] that in an analysis of the meaning-disclosure of technological development, philosophical-scientific thought ultimately arrives at *belief* as the boundary-function of the process of meaning-disclosure. Humanity should believingly transcend technological development in the direction of the fulness of meaning and thus of the Origin of all meaning. Humanity does not dispose over the fulness of meaning as the ground of the disclosure of the meaning of technology. This ground should rather be presupposed in all technological activity and in all reflection on it. Its meaning can only be approached – in the idea which ought to lead and inspire people in their work in technology, the idea through which people are filled with hope.

In approaching the technological development idea, we have come back again to the question of the ground for a liberating perspective for technological development.[105] This ground appears to be religious in character. I reject as unsuitable for such a ground both an absolutized idea of control or power and an absolutized idea of freedom, since such ideas presuppose a closed view of the world and of humankind. These ideas, furthermore, as I have repeatedly indicated, do not offer a meaningful perspective for the dynamic development of modern technology. On the contrary, they are the very source of the tremendous problems called forth by this technology and of the real dislocations that are presently making themselves felt in our culture.[106]

To get at the substance of the technological development idea, we shall have to *listen*, believingly, to the fulness of meaning revealed in Jesus Christ. The Bible speaks of it. When people listen, they understand more and more about God's purpose with the creation. As for the present subject, it is important to perceive to what extent God's Word reveals the meaning of technology to us.

As God's image-bearers, people received at their creation the command to be stewards of God's completed work of creation and to disclose that work. Contained in this calling is the task of technology as the disclosure of the nature side of creation and as the realization of its technological side. The final purpose of all this activity is the service and honor of God: this is the path along which humanity must unfold and fulfil its life.

The fall into sin broke humanity's power, and nature was cursed.[107] Now people no longer live in harmony with God's law, which obtains for the whole creation. Thus they have to live in a creation that is broken and dislocated by sin. Humanity is no longer in fit condition to administer and disclose nature. On the contrary, people are constantly threatened by nature and must defend themselves against it.

Restoration is given in Jesus Christ. He heals the brokenness of the entire creation and turns it again, in its fulness, toward God, the Origin. Jesus Christ came into a world broken by sin to undergo the chastisement of death for sin. He also fulfils humanity's task of having custody over creation and opening it up. Jesus Christ *saves* and *fulfils* creation.

It is especially difficult – and perhaps impossible – to discern to what extent the pristine creation is present in nature as it now exists. Certainly the original elements are dominant. Sin did not and does not have the power to lay creation waste in an integral way, to destroy it as a whole. In any case, it is clear in biblical perspective that Jesus Christ saves creation from the curse and turns it again toward its original destination. This is all the more astonishing when we consider that the creation still abundantly bears the effects of sin. In history, Jesus Christ has laid the foundation for the salvation and fulfilment of the creation. In Christ the meaning-disturbance resulting from the fall into sin is itself destroyed, and the meaning of all that is created is disclosed.

Through faith, humanity participates in the work of Jesus Christ. People are to acknowledge His leading in history and are to work with Him. It is given to humankind to know that in the groaning of the creation, a new perspective has opened up: the world is being propelled toward complete salvation and fulfilment, toward the consummation of the Kingdom of God. That Kingdom is forging a path right through the disturbances and dislocations of meaning occasioned by the technological development led by secularized motives and fraught, today, with far-reaching consequences.

When people live rooted in this conviction, they are able to accept their task in technology, freely and responsibly. A liberated technology will then be able to ease the difficult circumstances in which people live "by nature." It will afford an enlargement of life's opportunities, relieve the aches and pains and difficulties of work, resist natural catastrophes, conquer disease, improve social security, expand communication, multiply information, augment responsibility, vastly increase material prosperity in harmony with spiritual well-being, and abolish alienation from self, nature and culture.[108] Technology frees man's time and fosters the development of new possibilities. Given these possibilities, culture will advance to new disclosures. Technology also will create room for multifaceted work – for careful, creative, love-filled work. In all of this, humanity finds its share and portion of the *meaning of technology* in the disclosure of the meaning of the creation as a whole – a disclosure that must attain its final destination in the Kingdom of God, the re-created universe, and so come to rest in that Kingdom.

Notes

Chapter One

1. Klaus Tuchel, *Die Philosophie der Technik bei Friedrich Dessauer: Ihre Entwicklung, Motive und Grenzen*, p. 15.
2. I refer here to a school of philosophy developed at the Free University in Amsterdam in the 1930s by Professors D. H. T. Vollenhoven and Herman Dooyeweerd.
3. See, among others: H. Dooyeweerd, *A New Critique of Theoretical Thought*, 4 volumes; L. Kalsbeek, *Contours of a Christian Philosophy: An Introduction to Herman Dooyeweerd's Thought*; K. J. Popma, *Inleiding in de wijsbegeerte*; H. Van Riessen, *Wijsbegeerte*.
4. See H. Van Riessen, "The Structure of Technology," pp. 300f. For some views of the etymology of the word *techné*, see 2.3 below (Heidegger) and also the following: W. van Benthem, *Das Ethos der technischen Arbeit und der Technik*, pp. 56f; F. Dessauer, *Streit um die Technik*, pp. 130f; W. G. Haverbeck, *Das Ziel der Technik*, pp. 61f; J. E. Heyde, "Technischer Fortschritt," pp. 3f; W. H. Raby, *Über Sinn und Grenzen der Technik*, pp. 13-26; W. Schadewalt, *Natur, Technik, Kunst*, pp. 45-53.
5. Van Melsen calls this "technique in the narrow sense"; "technique in the broad sense" is that which is present as method in all human activity (see *Science and Technology*, p. 229; see also pp. 218, 247-8). [In the present study and elsewhere in the neocalvinist writings of Egbert Schuurman and Hendrik Van Riessen on the philosophy of technology, a distinction is made between a practical activity, *techniek*, which I always translate as *technology*, and a theoretical activity, *technische wetenschap* (Van Riessen uses the term *technologie* in this sense), which I call *technological science*, the *science of technology*, or *technicology*. See, for an example, note 34 below. See also note 51. – Trans.]
6. W. H. Raby, *Über Sinn und Grenzen der Technik*, pp. 90-1.
7. *Ibid.*, p. 76.
8. For comparisons between traditional and modern technology, see also: H. Van Riessen, *Filosofie en techniek*, pp. 559-64; A. G. M. van Melsen,

377

Science and Technology, pp. 218f, 232, 241f, 272f, 278; W. H. Raby, *Über Sinn und Grenzen der Technik*, pp. 66f.

9. See 4.7.

10. Unless otherwise indicated, the references in parentheses in the text of Chapter 1 are to H. Van Riessen, *Filosofie en techniek*; see also W. H. Raby, *Über Sinn und Grenzen der Technik*, p. 76.

H. Skolimowski understands by a technological object anything that is formed in technology, which is to say, anything that has a technological foundational function. He overlooks the destinational functions, which may be other than technological in character, thereby forfeiting insight into technology and into the coherence of technological objects. "By the 'technological object' I mean every artifact produced by man to serve a function; it may be a supersonic airplane as well as a can-opener" ("The Structure of Thinking," p. 375). This article appears in a special issue of *Technology and Culture* devoted to the topic "Toward a philosophy of technology."

11. Dooyeweerd speaks of an actualizing relation (or "actualization") in the same sense in which I use this term, but he distinguishes in addition what he calls the opening or unfolding relation and the intentional representational relation (*De wijsbegeerte der wetsidee*, Vol. 3, p. 121, and *A New Critique of Theoretical Thought*, Vol. 3, p. 148). These conceptions deserve comment for the sake of clarity. The unfolding relation, which is a subject/object relation, is defined by Dooyeweerd as the "*subjective unfolding* or *opening* in human experience of the closed objective thing-structure." The intentional representational relation, which is also a subject/object relation, he defines as an *objectifying* relation, or as the "objectification in the structure of a real object [thing] of the intentional object, as the latter is conceived in the design of the thing" (italics in original). I do not believe these definitions make it clear precisely what Dooyeweerd intended to say about technological designs and their execution. In the first place, the intentional representational relation as he has presented it might be taken to mean the framing (read: designing) of a law for a thing to be formed subsequently. Yet, I believe it should instead be construed to mean the actual forming of a thing in conformity with a law so framed. If this construction is faithful to Dooyeweerd's intention, his definition of this relation could be reformulated without adding anything to it as follows: "The intentional object is in the *design* of the thing conceived; the intentional representational relation is the *objectification* of the intentional object in the structure of a real thing." This re-worded definition, in which the distinction between designing and execution is clear, is in harmony with what I have spoken of in the text of this study as the objectifying relation, and also with Dooyeweerd's intention as I have understood it. It is simply inconceivable to me that when he spoke of the *unfolding* or *opening* relation, he meant the actual forming or fashioning of a thing. What he must have meant instead is the *subjective* unfolding or opening of a closed objective thing structure – for example, the perception that a stone might be used as a hammer.

If I have understood Dooyeweerd correctly, the distinction he makes between the unfolding relation and the intentional representational relation is not of direct importance for an analysis of technology. In order to avoid misunderstandings about the intentional representational relation,

I shall speak of an objectifying relation when the actual forming of a thing is at issue. Furthermore, I will sharply distinguish this relation from designing. What I will take designing to mean is the conceiving, in the productive fantasy, of a new law for a thing to be formed – that is, a law which may subsequently be rendered in its correct significance in an actual design, e.g. in a blueprint (see 1.5.2).

12. F. Dessauer, *Streit um die Technik*, p. 143.
13. Dooyeweerd and Van Riessen speak of a historical function. I prefer to speak of a cultural function instead, for I regard history as the expression of the meaning-coherence of all modal aspects.
14. In contrast to Van Riessen, who speaks simply of the "scientific method," I prefer to speak of the *technological-scientific* method, thereby giving expression to the technological character of this method. The method rests on science as its foundation, but the method of science is used in a manner that is congruent with technology. Therefore it would be even better to reverse the order and speak of the *scientific-technological* method. However, the prior designation has already gained general acceptance, and therefore I will use it in this study as well. Still, it cannot be denied that behind this widely used term there can sometimes lurk the mistaken notion that technology is applied science.
15. See 1.4.5.
16. We also find mention of a complementary relation in modern physics. However, what is at issue there is not a *technological relation* such as the one between the technological operator and the energy process in the *working* of a technological fact but a *knowledge relation*: knowledge of the one, the particle, must be filled out by knowledge of the other, the wave, to account for the physical phenomenon. The operator/energy configuration in the one complementary relation is analogous to the particle/wave configuration in the other one.
17. See 1.3.6.
18. See 1.3.6, point 2.
19. The particular meaning of a technological operator is imparted to an "indifferent" or "neutral" component only as long as this component is mounted in the technological operator. Its function is particularized (or perhaps fixed or determined, to be more accurate) by the destinational function of the technological operator. Where the final product is to be a mass product, the components, too, will usually be mass-produced; their destination will be to become components of the final mass product. The parts are attuned to the whole. This is the case in the automobile industry, for example, where even the individuality or particularity of the uniform final product, the desired automobile, is leveled to the extent that this is possible (526, see 1.4.4).
20. See 1.4.2 and 1.4.4.
21. At the boundary between nature and culture as encountered, for example, in mining engineering, architectural engineering, and civil engineering, the form-giving is still elementary in character. The technological operators are still not sharply enough differentiated or delineated in their functions to allow easy identification. For example, earthmovers of various sorts may have a form-giving function as well as a transportation or conveyance function. We call such technological operators *border types* (*grenstypen*) (543).

22. Sometimes the control of *time* is at stake in a communication process: information is built into a permanent thing structure, such as a phonograph disc or a movie reel.
23. See 1.3.7.
24. See 1.3.6.
25. If the meaning of the term *information* is not clearly defined and agreed upon, innumerable misunderstandings and meaningless discussions will ensue. Bar-Hillel therefore advocates replacing *information theory* with *signal transmission theory* and reserving *information* for semantics (see "Informatietheorie onder de loep").
26. See J. Otten, "Digitale informatieverwerking," pp. 24-36.
27. See H. A. Simon, "Simulation of Human Thinking," pp. 94f.
28. See H. Van Riessen, "Denkgereedschap."
29. See C. A. Van Peursen et al., *Informatie*, pp. 157, 197. Although science and technology have become more interdependent than ever through the use of the computer, the distinction ought to be maintained. It is incorrect to say: "The information moment of information technology is the switch, which henceforth renders science and technology a unitary whole" (see *Informatie*, p. 157).
30. See 1.5.7.
31. See H. Van Riessen, "Denkgereedschap."
32. See: G. Klaus, *Kybernetik und Gesellschaft*, pp. 274-6; N. Wiener, *The Human Use of Human Beings: Cybernetics and Society*, pp. 11, 74f; Wiener, *Cybernetics; or Control and Communication in the Animal and the Machine*, p. 138; J. von Neumann, *The Computer and the Brain*, pp. 3-37; H. Röpke and J. Riemann, *Analog Computer in Chemie und Biologie*.
33. Van Riessen has spoken of technological science and technological operators as being mutually attuned (568, 569, 598). I think it would be more consistent, however, to speak of the technological-scientific method. That way we can better emphasize both its distinctness from science and the technological character of this method. I am happier with Van Riessen's later terminology (see "The Structure of Technology," pp. 303f).
34. We would do well to bear in mind that "scientific method" refers specifically to the methods of mathematics, physics, chemistry, and technological science – naturally so, since these sciences form the basis for modern technology.
35. See H. Van Riessen, "De techniek."
36. See H. Van Riessen, "The Structure of Technology," p. 304.
37. W. H. Raby, *Über Sinn und Grenzen der Technik*, p. 83.
38. See H. Van Riessen, "The Structure of Technology," p. 305.
39. The fact that perfect theoretical formulation of technological forming is impossible does not always mean that control from a distance is impossible. Aberrations of a certain magnitude, occasioned by the resistances mentioned, can be corrected with the help of a "trick." The moment a certain maximum aberration occurs, an automatic linkage based on the cybernetic principle of backcoupling may be brought into play to eliminate the aberration. When or how often such maximum aberrations will occur cannot be predicted in the theory (see the following discussion in the text).
40. See C. J. Dippel, *Verkenning en verwachting*, pp. 45-6.
41. See 1.4.5.
42. See note 19.

43. A. G. M. van Melsen, *Science and Technology*, p. 279. Radio and television can have a similar influence. Through the influence of these mass media, people can all too easily become standardized listeners and viewers. Modern technological development is accompanied by a leveled, uniform culture.

 I will discuss these trends at some length, of course. For the present, suffice it to say that these trends are present and that they can be extremely dangerous, even though they are in no manner of speaking inevitable. People, whether they are designers, entrepreneurs, reporters, journalists, ought to acknowledge their responsibilities to their neighbors and themselves in everything they do as utilizers, listeners, and viewers.

44. In the text we have already noted that given the present technological state of affairs, technological operators functioning at the boundary between nature and culture cannot be controlled perfectly. It is for this reason that the manned American lunar landings of a decade ago proved more fruitful than did the Russian robot missions. On the whole, robots cannot react to unforeseen eventualities, or can do so only with great difficulty, whereas human beings, thanks to their creativity, can react adequately in such situations. Exploration of the seabed will provide another case in point.

45. W. H. Raby, *Über Sinn und Grenzen der Technik*, p. 68.

46. See H. Van Riessen, *Roeping en probleem der techniek*, pp. 38-41, 47f, and Hannah Arendt, *Vita activa*, pp. 132f.

47. A. G. M. van Melsen, *Science and Technology*, p. 265.

48. *Ibid.*, pp. 353, 356.

49. See 1.4.2.

50. See 1.5.5.

51. I avoid using the (Dutch) term *technologie* in order to avoid confusion with special sectors of technology. Chemical technology has appropriated it. Moreover, in some sectors of modern technology, such as civil engineering, it would not gain easy acceptance for historical and emotional reasons. I use the term *technische wetenschap* instead. [English readers will appreciate that this note is addressed to a question of Dutch usage. See note 5, above. – Trans.]

52. See 1.4.3 and 1.4.5.

53. See: A. G. M. van Melsen, *Science and Technology*, pp. 84f, 182f, 244f; A. G. M. van Melsen, "De wijsgerige implicaties van de wisselwerking tussen wetenschap en techniek," pp. 172-181; A. G. M. van Melsen, "Filosofie van de techniek," pp. 245-257; W. H. Raby, *Über Sinn und Grenzen der Technik*, pp. 64f.

54. W. H. Raby, *op. cit.*, p. 66.

55. *Ibid.*, p. 83.

56. See 3.3.3.

57. F. Dessauer, *Streit um die Technik*, p. 167.

58. See, among others: M. Polanyi, *Personal Knowledge*, pp. 120f; G. Ulmann, *Kreativität*; P. Matussek, "Psychodynamische Aspekte der Kreativitätsforschung," pp. 143-51; *Wijsgerig perspectief*, Vol. 8 (1967), No. 1. This issue of the last publication is devoted entirely to human creativity; a contribution by R. F. Beerling focuses on the problematics of discovering, inventing, and creating (pp. 8-20).

59. See A. G. M. van Melsen, *Science and Technology*, p. 53.

60. See 1.5.3.
61. See J. W. Forrester, "Managerial Decision Making," p. 67; see also 3.2.6, 3.3.4 and 3.4.7.
62. See 4.10.

Chapter Two

1. Ernst Jünger, who is an older brother of Friedrich Georg Jünger (born in 1894), also wrote about technology. It is clear from his book *Der Arbeiter* (1932) that he believed technology would be the means by which the workers would achieve world dominion. The situation would be one of harmony, stability, and perfection. See also E. Jünger's "Vom Ende des geschichtlichen Zeitalters," pp. 309-42.
2. See H. Van Riessen, *Filosofie en techniek*, pp. 460-5.
3. See O. Spengler, *Der Mensch und die Technik*, p. 61.
4. *Ibid.*, p. 55.
5. F. G. Jünger, *Die Perfektion der Technik* (written in 1939; hereafter referred to as PdT).
6. See PdT 12, 13.
7. See PdT 82.
8. See PdT 17, 65-9.
9. See F. G. Jünger, *Maschine und Eigentum*, p. 258 (hereafter referred to as MuE).
10. See MuE 258.
11. See PdT 22.
12. See MuE 209.
13. See F. G. Jünger, *Die Vollkommene Schöpfung: Natur oder Naturwissenschaft*, p. 10 (hereafter referred to as DVS).
14. See MuE 241, 246, 272.
15. See MuE 286, 290f.
16. Similar ideas can be found in A. Gehlen, *Die Seele im technischen Zeitalter*, p. 72, and G. Huntemann, *Provozierte Theologie in technischer Welt*. The latter states: "Property is the appropriation of things in a personal relation. Property presupposes personal involvement. Property loses its meaning when the matter becomes one of things that have lost every 'personal connection.' Thus one who holds *deeds* cannot speak of the sacredness or security (*Heiligkeit*) of such possessions" (p. 132 – italics added).
17. See 2.2.4.
18. See 2.2.11.
19. See PdT 29, 30, 32, 78, 80, 157, 171, 194; MuE 240.
20. See PdT 195; see also W. G. Haverbeck, *Das Ziel der Technik*, p. 263: "The principle of technology is death.... Technological development means murdering nature; it means ruining the countenance of nature."
21. See 2.2.6 and DVS 260.
22. See 2.2.9.
23. See also C. J. Dippel, *Verkenning en verwachting*, p. 21.
24. See the discussion of Ellul in 2.4.

25. See PdT 33, 34.
26. See PdT 158; DVS 10, 11.
27. See MuE 209, 212, 214, 246.
28. See MuE 360.
29. See PdT 74, 76; MuE 204.
30. See PdT 40; DVS 276.
31. See Ellul (2.4 below).
32. See PdT 23, 24, 26, 51, 57, 76, 92, 93, 97, 126; MuE 203, 204, 207, 232, 233, 234; DVS 10.
33. See PdT 28.
34. See 2.2.1; see also MuE 233.
35. See 2.4.9.5.
36. See PdT 38.
37. See PdT 58, 82; DVS 260.
38. See PdT 39, 65, 74, 76; MuE 204.
39. See PdT 67.
40. See MuE 342.
41. See PdT 69, 71; MuE 305. For another interpretation of marxism, see Klaus (3.4 below).
42. See PdT 66; MuE 309f.
43. See MuE 343.
44. See PdT 72, 74; MuE 207.
45. See PdT 78; MuE 207.
46. See 1.4.5, 1.5.2, and 1.5.7.
47. Sec the following subsection.
48. See PdT 171; MuE 205.
49. See PdT 94.
50. See PdT 161.
51. See PdT 115, 116.
52. See MuE 318; DVS 63, 244.
53. "Sports presupposes the technologically organized conurbation and is inconceivable without it" (PdT 147).
54. See PdT 76, 108, 131, 136, 147, 157; MuE 302.
55. "The susceptibility of whole strata of the populace to ideologies and to the power of ideologies, which issues from them, is a sign of mass culture" (PdT 133).
56. See 2.2.9.
57. See PdT 41.
58. See PdT 27, 80.
59. See MuE 204, 228. Without adducing reasons, Jünger states that Calvinism abetted the onset of modern science more than Lutheranism did, since the Calvinists regarded God as the great clockmaker (see PdT 53).
60. See PdT 126, 127.
61. See PdT 165f.
62. See DVS 272.
63. See PdT 157, 159, 161, 192; MuE 359.
64. See PdT 95, 142.
65. See PdT 6.
66. First published as an essay in *Die Künste im technischen Zeitalter*, pp. 135ff (hereafter referred to as SuK).

67. The influences of Heidegger and Jünger on each other are discussed in 2.3.6.
68. See SuK 5, 6.
69. See DVS 63, 242, 244; see also 3.2.5, 3.2.7.2, 3.3.11.2, 3.4.3, and 3.4.12.6.
70. See SuK 13, 18.
71. See SuK 22, 26.
72. See SuK 27.
73. See PdT 39; MuE 230.
74. See MuE 229, 342; SuK 22; DVS 256f, 272.
75. See MuE 205.
76. See PdT 131.
77. See F. Dessauer, *Streit um die Technik*, pp. 42-8.
78. See PdT 41, 97, 171; DVS 272.
79. See PdT 31.
80. See PdT 143, 158; MuE 296.
81. See 1.4.2.
82. See PdT 94, 171; MuE 205.
83. See 1.5.5.
84. As far as that goes, human freedom and responsibility do not appear for the first time in invention. They are equally present in scientific practice and in the technological-scientific approach to problems. The free, responsible person is the very presupposition of such activity.
85. See MuE 219; SuK 18f.
86. See PdT 123f, 156, 185f; MuE 204, 207.
87. See MuE 352.
88. This brief consideration must suffice for now. This matter will be dealt with at greater length in Chapter 4 (see 4.5.2 and 4.5.4 below).
89. See also G. Huntemann, *Provozierte Theologie in technischer Welt*, p. 82: "Presumptuous thinking leads to darkness. *So emptiness and desolation belong to the world of technology.* Beside the unfolding of life and the increase in material prosperity there stands in an increasingly strained dialectical relation the loss of the dignity of human existence."
90. See 2.2.7.
91. Martin Heidegger (1889-1976) was one of the greatest philosophers of our time. He studied philosophy at the University of Freiburg under Husserl, among others. In 1933 (under Hitler) he was rector of that university. Later he disavowed the Nazi regime. After World War II he withdrew from all public functions and more or less lived the life of a hermit. Only a small number of students were allowed to enjoy firsthand the fruits of his philosophizing. However, access to his thought is possible through his numerous publications.
92. See 2.3.3.
93. M. Heidegger, *Sein und Zeit*, pp. 52f (hereafter referred to as SuZ).
94. See O. D. Duintjer, *De vraag naar het transcendentale*, p. 220.
95. *Ibid.*, pp. 220-2, 226.
96. See SuZ 235f, 260f.
97. For a critical analysis of SuZ, see J. van der Hoeven, *Kritische ondervraging van de fenomenologische rede*, pp. 114-60.
98. See 2.3.4.
99. M. Heidegger, *Identität und Differenz*, p. 18 (hereafter referred to as IuD).

100. M. Heidegger, *Die Technik und die Kehre*, p. 10 (hereafter referred to as TuK).
101. M. Heidegger, *Was heisst Denken?*, p. 55 (hereafter referred to as WhD).
102. M. Heidegger, *Der Satz vom Grund*, p. 99; see also p. 195 (hereafter referred to as SvG).
103. M. Heidegger, *Holzwege*, p. 86 (hereafter referred to as Hol).
104. M. Heidegger, *Vorträge und Aufsätze*, Part I, p. 79 (hereafter referred to as VuA).
105. See Hol 237.
106. M. Heidegger, *Gelassenheit*, p. 15 (hereafter referred to as Gel). Heidegger has also said: "Philosophy has reached its end in the present age. It has found its place in the practice of science.... The all-controlling mark of this is that philosophy has become cybernetic, that is to say, technological" ("La fin de la philosophie et la tâche de la Pensée," p. 178).
107. M. Heidegger, *Unterwegs zur Sprache*, p. 190 (hereafter referred to as UzS).
108. See Job 28:11, where it is said of the human being in technology: "The thing that is hid bringeth he forth to light."
109. See Hol 82, 83, 103.
110. See Hol 220, 221.
111. Heidegger speaks elsewhere of the "rendering totally dynamic" (Hol 89) and of "turmoil and frenzy" (Hol 271).
112. Elsewhere Heidegger says of this: " ... the term for the whole of the challenge which engages *man and Being* mutually is: the *Ge-stell*" (IuD 27 – italics added).
113. Heidegger also discusses the difference between earlier and modern technology in *Einführung in die Metaphysik* (hereafter referred to as EiM), where we read: "The Being that is calculable in this way and is present in calculation [i.e. objectivity] makes 'being' into the controllable in modern, mathematically based technology. This technology is *essentially* different from any hitherto familiar mechanical technology" (EiM 148).
114. See IuD 26.
115. This point is clarified elsewhere as follows: "To the extent that the *Ge-stell*, the essence of modern technicity, challenges man to command all that is in his presence as if it were purely technological fund, it [the *Ge-stell*] comes-to-presence in all happenings, but then in such a fashion that its coming-to-presence is at the same time a going-into-hiding; all commanding, i.e. technology in its totality, is consigned to calculating thought to speak therein the language of technological commanding (*Sprache des Ge-stells*)" (UzS 263).
116. See W. Heisenberg, "Das Naturbild der heutigen Physik," in *Die Künste im technischen Zeitalter*, pp. 60f. Heidegger's "Die Frage nach der Technik" is also taken up in this collection of essays. Thus Heidegger consciously opposes Heisenberg.
117. See M. Heidegger, *Hebel, der Hausfreund*, p. 30 (hereafter referred to as HdH).
118. For interpretations of the "reversal" in Heidegger's philosophy, see, among others, O. D. Duintjer, *De vraag naar het transcendentale*, pp. 228-90, and J. van der Hoeven, *Heidegger en de geschiedenis der wijsbegeerte*, pp. 7, 10, 14.

119. See Hol 285, 291, 292.
120. See S. U. Zuidema, "Cultuur als crisis," p. 18.
121. Heidegger misjudges the technological state of affairs with regard to translation machines. Only poor results have been achieved up to now. Wiener has demonstrated that in principle, translation machines cannot replace people since machines have no "feel" for the shifting meaning of language and for language development (see 3.2.6).
122. See also IuD 32.
123. See VuA 62.
124. J. Beaufret et al., *Martin Heidegger zum siebsigsten Geburtstag*.
125. See PdT 39; MuE 360; see also IuD 48; SvG 197, 198; Hol 68.
126. See PdT 144; Hol 89.
127. See PdT 100, 107; MuE 312; Hol 77f, 267.
128. See PdT 26, 39, 97; Hol 267.
129. See PdT 195, 196; TuK 37.
130. See MuE 343; Hol 279.
131. See TuK 41; HdH 28; IuD 32; UzS 262 (see 2.3.5).
132. See 2.3.1.
133. SuK 14.
134. Hereafter referred to as SuT.
135. See SuT 167f.
136. For a critical analysis of Dessauer's doctrine of ideas, see H. Van Riessen, *Filosofie en techniek*, pp. 406-22.
137. F. Dessauer, *Philosophie der Technik*, p. 6.
138. K. Tuchel, *Die Philosophie der Technik bei Friedrich Dessauer*, p. 60.
139. See TuK 17.
140. See TuK 15.
141. See SuT 50.
142. W. H. Raby, *Über Sinn und Grenzen der Technik*, p. 65.
143. *Ibid.*, p. 66.
144. See 2.3.8.8.
145. See 2.3.1 and 2.3.5.
146. I have consulted the following works, all of which are either mainly synoptic or more or less descriptive in character: R. Bakker, "Heidegger en de techniek"; R. Boehm, "Pensée et technique"; P. Dijkema, *Innen und Aussen: Die Frage nach der Integration der Künste und der Weg der Architektur*; Werner Marx, *Heidegger und die Tradition*; Simon Moser, *Metaphysik einst und jetzt*; A. T. Peperzak, *Techniek en dialoog*; R. Schaeffler, "Martin Heidegger und die Frage nach der Technik"; S. IJsseling, "De filosofie en het technische denken"; H. Berghs, "Het zakelijk karakter van Heideggers vraag naar de techniek."
147. See TuK 17.
148. VuA 14.
149. See 2.3.7.
150. See Hol 69.
151. See J. van der Hoeven, "Heidegger, Descartes en Luther," p. 71.
152. See W. H. Raby, *Über Sinn und Grenzen der Technik*, p. 77.
153. In his essay "Die Zeit des Weltbildes," Heidegger says, among other things: "Machine technology itself is such an independent transformation of praxis because of its necessarily requiring the application of mathematical physics. Machine technology is therefore until now the

most visible offshoot of the essence of modern technicity, which is identical with the essence of modern metaphysics" (Hol 69). See also *Platons Lehre von der Wahrheit; mit einem Brief über den Humanismus*, p. 88 (hereafter referred to as PH).

154. See Gel 24; WhD 53-9.
155. This term, which is introduced especially by Herbert Marcuse, apparently arises from the inspiration of Marcuse by Heidegger (see 3.3.11.6, including the notes).
156. See Hol 99.
157. See Hol 103.
158. See L. Landgrebe, *Philosophie der Gegenwart*, pp. 135f.
159. See Hol 220-2; see also S. U. Zuidema, "De cultuur als crisis," p. 11.
160. See VuA 91.
161. See Hol 99, 203, 225, 226.
162. See S. U. Zuidema, "De cultuur als crisis," p. 10.
163. See Hol 70.
164. See Hol 229f.
165. See J. van der Hoeven, "Heidegger, Descartes en Luther," pp. 105-6.
166. See 4.9 and 4.12.
167. See HdH 30, 31; VuA 65.
168. See IuD 33; Gel 27; Hol 271.
169. See O. D. Duintjer, *De vraag naar het transcendentale*, p. 307.
170. See Hol 285.
171. See L. Landgrebe, *Philosophie der Gegenwart*, p. 140.
172. See TuK 39.
173. See TuK 38, 39.
174. See TuK 32, 33.
175. In WhD 114 Heidegger states: "'Needing' (*Brauchen*) [Heidegger discerns in this word the dual sense of 'needing' and 'using'] is more than simply 'utilizing' (*Benützen*), 'wearing out' (*Abnützen*), or 'using up' (*Ausnützen*)," for "within the 'needing' there resounds the response to essence. Authentic 'needing' does not cast the 'needed' [or the 'used' – *das Gebrauchte*] aside; no, 'needing' finds its destination in leaving the needed undisturbed in its essence (*Wesen*)." Moreover, "needing" is "getting involved in essence, caring for essence."
176. See HdH 34.
177. See Hol 279.
178. J. P. A. Mekkes states: "The fact, however, that neither of these two terms [*thought* and *Being*] can have meaning without the other or outside this *two*-sided basic relation makes it clear that we are dealing here with a *dialectic* of a *fundamental* character; it is intrinsically an antinomy, for Being pretends to call forth 'remembrance' (*Andenken*) into presence, even though without it Being can have no meaning of its own in Heidegger's philosophizing" (*Radix, tijd en kennen*, pp. 45-6; see also pp. 41, 49, 53, 55, 72, 82, 110; *Scheppingsopenbaring en wijsbegeerte*, p. 46). J. van der Hoeven demonstrates at length that Heidegger's view of the "reversal" is ruled by a dialectic. He concludes that " . . . in the so-called coherence of 'thought' and 'Being,' which comes down to an irresolvable polarity between two 'basic denominators' which simultaneously presuppose and exclude one another, both are ultimately struck dumb" (*Heidegger en de geschiedenis der wijsbegeerte*, p. 14). O. D. Duintjer rejects the

notion that there is a dialectic between "thought" and "Being" in Heidegger's philosophy. According to him, "thought" is subordinate to "Being," although Heidegger grants it an exceptional position in the midst of all that is subject to "Being." Thus Duintjer states: "'Thought' is not adequate for a renewal, but it is probably an *indispensable* and *inescapable* condition" (*De vraag naar het transcendentale*, p. 320 – italics added). In this connection, H. Berghs states: "*Unremitting, courageous thought* is Heidegger's answer to the question concerning technology" ("Het zakelijk karakter van Heideggers vraag naar de techniek," p. 273 – italics added).

179. See 2.3.5 and 2.3.6.
180. See HdH 30.
181. S. U. Zuidema, "De cultuur als crisis," p. 18.
182. See Hol 69, 271.
183. Jacques Ellul (born in 1913) is a jurist and sociologist at the University of Bordeaux. He was awarded an honorary doctorate by the Free University in Amsterdam. Ellul has published two works on technology in which the relation between technology and culture is of central importance. *The Technological Society* was translated from French by John Wilkinson and then revised by the author. The original title was "La technique ou l'enjeu du siècle" (1954). The quotations are taken from the English translation (hereafter referred to as TS). The second work important for our purposes is Ellul's essay "The Technological Order" (hereafter referred to as TO).
184. See TS 128.
185. H. Schelsky has derived his idea of the technological state from Ellul (see his book *Der Mensch in der wissenschaftlichen Zivilisation*). For an analysis of Schelsky's idea of the technological state, see B. C. van Houten, *Tussen aanpassing en kritiek*. See also *Texte zur Technokratiediscussion*, ed. C. Koch and D. Senghaas, p. 6; J. Habermas, *Technik und Wissenschaft als 'Ideologie,'* pp. 81, 122; B. van Steenbergen and E. van Hengel, *Technocratie: Ideologie of werkelijkheid*, pp. 18f.
186. See TO 403.
187. See 1.4.2 and 1.5.3.
188. See TS 19; TO 402.
189. See TS 23, 24.
190. See TS 48f.
191. See TS 54, 55; TO 401.
192. See TS 72, 74.
193. See TO 394, 395.
194. See TS 85.
195. See TO 395.
196. See TO 394.
197. See TO 395.
198. See 2.4.6.
199. See TS 133, 282, 340; TO 399.
200. See TS 138.
201. See TO 463; TS 193.
202. See TO 397.
203. See TS 14.
204. See 2.4.9.4.

205. See Chapter 1, especially 1.4.
206. See TS 175, 184.
207. See TS 308f.
208. See TS 277.
209. See TO 400.
210. See TS 147.
211. See 2.4.9.5.
212. See TS 325.
213. See TS 331.
214. See TS 218, 321, 331.
215. See TS 359f.
216. See TS 363f.
217. See TO 418.
218. See TO 400.
219. See TO 397.
220. See TS 409, 410.
221. See TS 425.
222. See TS 149.
223. See TS xix.
224. See TS xxiv.
225. See TS 401-3.
226. See TS 217.
227. See 1.2.
228. See 1.4.3 and 1.4.5.
229. See 3.3.8 to 3.3.10 and 3.3.11.5.
230. See 2.4.4.
231. See 1.4.2.
232. See H. Dooyeweerd, *A New Critique of Theoretical Thought*, Vol. 2, pp. 260-2, 274.
233. See TO 463; TS 193.
234. See TS 280.
235. See TS 203f, 314.
236. See TS 423.
237. Hermann J. Meyer is more perceptive of this (see 2.5). See also H. Van Riessen, *Mondigheid en de machten*, pp. 111f, 173f, 208f.
238. See TS 421, 425.
239. See TO 460.
240. Hermann J. Meyer holds a doctorate in philosophy and teaches at the University of Mainz.
241. See 4.7.
242. See E. Kapp, *Grundlinien einer Philosophie der Technik*.
243. Hermann J. Meyer, "Technik und Kybernetik: Die Technik im Selbstverständnis des heutigen Menschen" (hereafter referred to as TiS). See also *Die Technisierung der Welt: Herkunft, Wesen und Gefahren* (hereafter referred to as TdW), p. 152.
244. See TdW 155; see also A. Gehlen, "Anthropologische Ansicht der Technik," p. 107; *Die Seele im technischen Zeitalter*, p. 7.
245. See TiS 754.
246. See TdW 160f.
247. See TdW 169.
248. See TdW 7, 69f, 98, 134.

249. See TdW 161.
250. See TdW 6, 13, 175.
251. See TdW 262.
252. See TdW 6f.
253. See TdW 2.
254. See 2.5.7.4.
255. Genesis 1:28.
256. See 2.5.7.2.
257. See TiS 757; Heidegger, SvG 99.
258. See TdW 5; TiS 757, 762, 763.
259. See TdW 35, 49, 54, 194.
260. See TdW 118, 176, 194.
261. See TdW 121; Theodor Litt, *Naturwissenschaft und Menschenbildung*, p. 38.
262. See TdW 300.
263. See TdW 46, 51, 54, 195.
264. See TdW 131.
265. See TdW 53.
266. See TdW 67, 264.
267. See TdW 67, 263.
268. See TdW 223.
269. See TdW 70, 87, 224, 235, 264.
270. See Heidegger, SvG 197.
271. See TdW 189.
272. See TdW 136.
273. Bertrand Russell, *The Scientific Outlook*, p. 210; as quoted in TdW, p. 249, from the German translation: *Das naturwissenschaftliche Zeitalter* (Stuttgart and Vienna, 1953), p. 178.
274. See TdW 133, 239, 248.
275. See TdW 287.
276. See TdW 6.
277. See TiS 772.
278. See TdW 213, 244.
279. See TdW 272.
280. See TdW 105.
281. See TdW 179; TiS 768f.
282. See TdW 151.
283. See TiS 768; TdW 230.
284. See TdW 255, 244.
285. See TdW 241, 280, 287.
286. See TiS 771; TdW 205, 215, 270.
287. See TdW 256, 270.
288. See TdW 256; TiS 771.
289. See Heidegger, VuA 71.
290. See 2.3.5, 2.3.8.9, and 2.3.8.10.
291. See 2.2.11.
292. See 2.4.9.5.
293. See A. Gehlen, *Die Seele im technischen Zeitalter*, p. 92, and TdW 288.
294. See TiS 751.
295. See TiS 752, 782.
296. See TdW 291.

297. It would have been better if Meyer had referred here to inanimate nature. Wood, oil, gas, etc. are dead organic nature, and they are therefore equally eligible for technological fashioning (see TdW 137).
298. See TdW 225.
299. See TdW 292.
300. See TdW 295.
301. See TdW 172.
302. See TiS 778.
303. See TiS 781.
304. See TdW 294.
305. This idea also occurs in N. Berdyaev, *Der Mensch und die Technik*, pp. 27, 35.
306. See TiS 772; TdW 135.
307. See K. Tuchel, "Die Technik als Gabe und Aufgabe: Die Geisteswissenschaftler müssen das Gespräch mit den Ingenieurswissenschaftlern suchen."
308. See TdW 68.
309. See TdW 6.
310. See H. Dooyeweerd, *Roots of Western Culture*, pp. 148ff.
311. See TdW 6, 262.
312. See TdW 70, 91, 264.
313. See Ortega y Gasset, *The Modern Theme*, pp. 68-70.
314. See TdW 287.
315. See TdW 198.
316. See TdW 263, 270, 285.
317. See TdW 276.
318. See H. Van Riessen, *Mondigheid en de machten*, pp. 115, 174, 203, 204, 210, 213.
319. See 2.4.9.5.
320. See TdW 277.
321. See TdW 46.
322. See A. Troost, "Personalisme en ethiek."
323. See TdW 245, 258.

Chapter Three

1. Norbert Wiener (1894-1964) was born in Columbia, Missouri. His father was a professor of Slavic languages at Harvard University. Wiener received his B.A. in mathematics at Tufts University at the age of fifteen. He completed his M.A. in 1912. In 1913 he received his Ph.D. from Harvard. He then studied in England with Bertrand Russell, and at Columbia University with John Dewey. After 1919 he was employed at the Massachusetts Institute of Technology, where he became professor of applied mathematics in 1932. He received an honorary doctorate in medicine from the University of Amsterdam in 1963. He died in Stockholm in January of 1964.
2. See 3.2.7.3.
3. See 3.2.5.

4. I have consulted the following publications by Wiener: *The Human Use of the Human Being: Cybernetics and Society* (hereafter referred to as H); *God and Golem, Inc.: A Comment on Certain Points Where Cybernetics Impinges on Religion* (hereafter referred to as GG); "Some Moral and Technical Consequences of Automation" (hereafter referred to as SM); *Selected Papers of Norbert Wiener* (hereafter referred to as SP); *Progress in Brain Research*, Vol. 2: *Nerve, Brain and Memory Models*, ed. N. Wiener and J. P. Schadé (hereafter referred to as PBR); "Scientist and Decision Making," in *Computers and the World of the Future* (hereafter referred to as CWF); Wiener's autobiography, entitled *Why I Am a Mathematician* (hereafter referred to as IAM); and *Cybernetics; or Control and Communication in the Animal and the Machine* (hereafter referred to as C). The last of these works appeared in a revised edition in 1961; all the references in the text are to the eighth printing (1950).
5. See H 2.
6. See also E. Lang, "Zur Geschichte des Wortes Kybernetik."
7. See C 19.
8. See J. C. Maxwell, *Proceedings of the Royal Society* (London), March 5, 1968.
9. See GG 88.
10. See C 116ff; see also S. T. Bok, *Cybernetica*, pp. 55f.
11. See H 12.
12. See C 133.
13. See 3.2.5.
14. At the risk of seeming redundant, I would point out again that information in the authentic sense is lingual. Wiener's use of the concept of information is directed to the analytical substratum of language as information (see 1.3.7).
15. See C 16; H 6.
16. Bok, *Cybernetica*, p. 88.
17. See H. Van Riessen, *Mededelingen van de Vereniging voor Calvinistische Wijsbegeerte* (September, 1965), p. 6.
18. See C 70f; SP 430.
19. See C 79.
20. See H 128.
21. See Bok, *Cybernetica*, pp. 184, 247; J. R. Pierce, *Symbolen en signalen*, pp. 23, 79, 88, 204f; F. J. Crosson and K. M. Sayre, *Philosophy and Cybernetics*, pp. 7, 8, 42, 105.
22. J. D. Fast and F. L. M. Stumpers close their article "Informatie, entropie en energie" as follows: "Briefly stated, the connection between information and thermodynamic energy exists because of the fact that there is no information without an energetic bearer."
23. Energy becomes structured energy – for example, signals. In 1.3.7 we have already seen that the word *information* is often misleading (see note 14 above). Information theory is really the theory of signal transmission.
24. See 3.2.6.
25. I believe that the following distinctions must be made: (1) *information* is lingually qualified, (2) a *signal* as *humanly structured energy* is technologically qualified, and (3) *energy* is physically qualified.
26. See 1.3.7.
27. See C 141f.

28. See C 13.
29. See C 54, 147; H 15.
30. See C 143, 154.
31. See 3.2.7.1.
32. See C 34, 194; GG 91f.
33. See 3.2.7.2.
34. See GG 2, 8.
35. See H 111.
36. See GG 14.
37. See C 143-4; H 201f; GG 93, 194; CWF 22f.
38. See H 71f, 205f, 212; CWF 22.
39. A. L. Samuel, "Some Studies in Machine Learning, Using the Game of Checkers," pp. 210-29.
40. See GG 25.
41. See GG 24, 77.
42. Wiener presented a mathematical-technological explanation of such propagation in Chapter 9 of his revised edition of *Cybernetics*, entitled "On learning and self-reproducing machines."
43. See GG 29f.
44. See H 168, 169.
45. Rarely is any mention made of the positive side of automation. See also Donald Brinkmann, "Der Mensch im Zeitalter der Automation."
46. See H 180.
47. See H 2, 16.
48. See H 188.
49. See "Problems of Sensory Prosthesis," in SP 431-9.
50. See GG 73f.
51. See GG 76.
52. See GG 52.
53. See H 217.
54. See CWF 26.
55. See H 212.
56. Wiener examines the problem of war games elsewhere too. He believes that machines intended for use in war games do not have the capacity to learn. "For one thing, a sufficient experience to give an adequate programming would probably see humanity already wiped out." "Moreover, the techniques of push-button war are bound to change so much that by the time an adequate experience could have been accumulated, the basis of the beginning would have been radically changed" (SM 1357).
57. See C 154.
58. A learning chess machine, too, must "play" against masters; otherwise a fiasco may result. "The best way to make a master machine would probably be to pit it against a wide variety of good chess players. On the other hand, a well-contrived machine might be more or less ruined by the injudicious choice of its opponents" (H 205).
59. See CWF 25.
60. See GG 84, 107.
61. Wiener states elsewhere that learning machines are far more dangerous than calculating machines, since it is usually not known if the conditions under which the former are operating are the correct ones. Therefore: "The importance of learning machines is not how they act as pure machines, but

how they interact with society. We thus are led to the concept of a system involving both human actions and machines. Is there any way in such a system to transfer values from the human being to the machine?" (CWF 24). On the basis of his view of translation machines, Wiener replies: "It is at least theoretically possible to transfer values from the human being to the machine in such manmachine organizations" (CWF 25).

62. See H 27.

63. See H 38f.

64. For Russell's critique of pragmatism, see D. J. O'Connor, ed., *A Critical History of Western Philosophy*, pp. 450, 451, 458, 481.

65. This notion retains currency. See 3.3 (on Steinbuch) and 3.4 (on Klaus). See also: *Philosophy and Cybernetics*, ed. F. J. Crosson and K. M. Sayre; H. Frank, *Kybernetik: Brücke zwischen den Wissenschaften; Kybernetik und Philosophie*; P. K. Schneider, *Die Begründung der Wissenschaften durch Philosophie und Kybernetik*.

66. See H 9.

67. See 1.3.7.

68. See IAM 260; GG 88.

69. See SM 1356; CWF 33; GG 89.

70. Also: "To me, logic and learning and all mental activity have always been incomprehensible as a complete and closed picture and have been understandable only as a process by which man puts himself *en rapport* with his environment" (IAM 324).

71. See GG 89.

72. See GG 92.

73. See SP 443.

74. See H 2.

75. See C 54, 147.

76. See C 168.

77. See CWF 23.

78. See CWF 33; SM 1356.

79. See GG 52, 53.

80. See CWF 32.

81. See H 217, 226.

82. The word *determined* does not imply that we are able to formulate the entire process in theory. It expresses rather the *fixed* nature of a process, irrespective of our ability to formulate it in theory. To mention two examples, then, the disturbances eliminated by a backcoupling mechanism and the elaboration of heuristic principles in modern computers are *determined*, but beyond the reach of theory.

83. See PBR 2.

84. See GG 95f; see also 3.2.5.

85. See H 102, 180, 217; see also 3.2.6.

86. See GG 89.

87. See H 23f.

88. See IAM 328.

89. See H 58.

90. See H. Van Riessen, *The Society of the Future*, pp. 157-9.

91. Karl Steinbuch (born in 1917) is professor of communications and information technology at the technical university at Karlsruhe. He has written not just on technological development but also on the future of technology,

the relation between human beings and the computer, futurology, and so forth.

92. Among the books Steinbuch has written are the following: *Automat und Mensch: Über menschliche und maschinelle Intelligenz* (hereafter referred to as AuM); *Die informierte Gesellschaft: Geschichte und Zukunft der Nachrichtentechnik* (hereafter referred to as DiG); *Falsch programmiert: Über das Versagen unserer Gesellschaft in der Gegenwart und vor der Zukunft, und was eigentlich geschehen müsste* (hereafter referred to as FP); *Programm 2000: Die konstruktive Ergänzung zu 'Falsch programmiert,' ein provokativer Aufruf zur Umorientierung* (hereafter referred to as P2); and, jointly with Simon Moser, *Philosophie und Kybernetik* (hereafter referred to as PuK).

93. See DiG 21.
94. See 2.2.
95. See DiG 31; FP 68.
96. See PuK 14.
97. See K. Steinbuch, *Kybernetik*, p. 23; see also AuM 324.
98 See K. Steinbuch in the preface to *Kybernetik: Brücke zwischen den Wissenschaften*, ed. Helmar Frank; see also K. Steinbuch and S. W. Wagner, *Neuere Ergebnisse der Kybernetik*, pp. 10-13.
99. See AuM 359.
100. See PuK 23; AuM 382.
101. See AuM 330.
102. See AuM 324, 354.
103. See 3.2.3.
104. See AuM 325.
105. See K. Steinbuch, *Kybernetik*, p. 11.
106. See AuM 293f, 340, 358; K. Steinbuch, "Technik und Gesellschaft als Zukunftsproblem," pp. 65, 66.
107. See AuM 152f.
108. See DiG 60f.
109. See DiG 86f; P2 54.
110. See DiG 57, 127, 170, 204.
111. See DiG 205f.
112. See P2 95.
113. See AuM 187f; DiG 284f; FP 110f; P2 58, 150f.
114. See DiG 282; FP 108; P2 74.
115. See DiG 259f; P2 50f; AuM 181f.
116. See DiG 266; FP 106.
117. See AuM 172, 209; P2 56f.
118. See AuM 192; DiG 274; K. Steinbuch, *Kybernetik*, p. 17.
119. See DiG 277; PuK 146f.
120. See AuM 215f, 247f.
121. See AuM 178, 225, 239; DiG 269; FP 107.
122. Only in the first printing.
123. See PuK 17; K. Steinbuch, preface to *Kybernetik: Brücke zwischen den Wissenschaften*, ed. H. Frank, pp. 169f.
124. See AuM 210. The notion that the computer will become humanity's intellectual successor also occurs in M. L. Minsky (see *Computers and the World of the Future*, ed. M. Greenberger, p. 118).
125. See AuM 247, 267.

126. See AuM 12.
127. See AuM 399f.
128. See AuM 87, 330.
129. See K. Steinbuch, *Kybernetik*, p. 18.
130. W. Wieser, *Organismen, Strukturen und Maschinen*.
131. See 1.4.2.
132. See AuM 330.
133. See AuM 10, 341, 350.
134. See AuM 247, 248.
135. See AuM 350.
136. See 3.3.4 and AuM 322.
137. See AuM 320.
138. See AuM 344.
139. See AuM 407.
140. See FP 16, 66f.
141. See FP 18, 38.
142. See FP 24, 29.
143. See FP 68; DiG 31; see also 2.2.
144. See K. Strunz, *Integrale Anthropologie und Kybernetik*, p. 47.
145. See FP 71.
146. See AuM 338; H. Reichenbach, *Der Aufstieg der wissenschaftlichen Philosophie*.
147. See FP 40, 54.
148. See FP 41, 47.
149. See FP 48.
150. See FP 43.
151. See FP 54.
152. See FP 49, 55.
153. See FP 57f.
154. See FP 55.
155. See FP 59.
156. See DiG 332; FP 126.
157. See FP 149.
158. See DiG 291, 333; FP 113.
159. See DiG 5, 294, 314, 333; FP 128.
160. See Karl Marx, ed. F. Borkenau, pp. 41-2.
161. See P2 99; DiG 294.
162. See DiG 305, 316; FP 127; P2 112.
163. See P2 100.
164. See AuM 339, 353, 399f; DiG 322; FP 137.
165. See FP 133; P2 100.
166. See FP 170.
167. See FP 138, 151.
168. See P2 76.
169. See FP 144, 170; P2 202.
170. See AuM 359; DiG 337; FP 148.
171. See FP 181.
172. See FP 165; P2 178f.
173. See P2 206.
174. See DiG 38; FP 103.
175. See, among others: C. A. Van Peursen et al., *Informatie: Een interdisci-*

plinaire studie, pp. 16-24; D. Nauta, *The Meaning of Information*, pp. 24, 36f, 52, 192f.

176. See PuK 17.
177. See FP 85.
178. See AuM 279, 350.
179. See AuM 9, 350.
180. See AuM 273.
181. See PuK 20, 139; FP 134.
182. It is for this reason that G. Schischkoff refers to Steinbuch's philosophy as cybernetical positivism ("Philosophie und Kybernetik: Zur Kritik am kybernetischen Positivismus," pp. 248-78). See also L. Heieck, *Bildung zwischen Technologie und Ideologie*. Heieck states that Steinbuch represents the ideology of cyberneticism (p. 68).
183. See AuM 6; FP 85, 94.
184. This is Steinbuch's reply to a critical review of *Automat und Mensch* by G. Schischkoff in *Zeitschrift für philosophische Forschung*, Vol. 19, No. 2, pp. 248-278.
185. See AuM 199.
186. See AuM 296.
187. See 1.4.
188. See 1.4.3.
189. *Automat und Mensch*, first edition, 1961.
190. See FP 40, 60.
191. See AuM 274, 281.
192. See K. Steinbuch, "Zur Systemanalyse des technischen Fortschritts," p. 9.
193. See DiG 9, 291, 338.
194. See DiG 5, 16, 31, 34; FP 134.
195. See H. Frank, *Kybernetik und Philosophie*, p. 103.
196. See P2 77.
197. See FP 143, 144.
198. See DiG 338.
199. K. Steinbuch, "Technik und Gesellschaft," p. 71.
200. *Ibid.*, p. 72.
201. *Ibid.*, p. 73.
202. *Ibid.*, p. 74.
203. *Ibid.*, p. 73.
204. See FP 174.
205. See P2 102.
206. See FP 174.
207. See, among others: *Katernen 2000*, 1969, Nos. 9 and 10, a publication of Werkgroep 2000, Amersfoort, The Netherlands; O. K. Flechtheim, *Futurologie: Der Kampf in die Zukunft*, pp. 19f, 218f; B. Willms, *Planungsideologie und revolutionäre Utopie: Die zweifache Flucht in die Zukunft*.
208. See 3.3.6 through 3.3.8.
209. See FP 37.
210. See also G. Zoutendijk, "Computer en macht," pp. 1-8.
211. *Ibid.*
212. See, among others: B. van Steenbergen, "Het kritische toekomstdenken van Arthur Waskow."
213. A. I. Waskow, "Looking Forward: 1999," pp. 94f; "Creating the Future in the Present," pp. 75f.

214. See 1.4.2.
215. See 1.5.5.
216. See 2.4.7.
217. See A. G. M. van Melsen, *Wetenschap en verantwoordelijkheid*, p. 166; H. Van Riessen, "Over futurologie."
218. J. Habermas, *Technik und Wissenschaft als 'Ideologie'* (hereafter referred to as TuW), quoted on p. 7.
219. See J. G. Knol, "Een dialoog met critici," pp. 11-16.
220. See TuW 55. The influence of Heidegger, his mentor, is apparent here (see 2.3.6). S. U. Zuidema also calls attention to this influence (*De revolutionaire maatschappijkritiek van Herbert Marcuse*, p. 92). Aptly enough, Marcuse has sometimes been called the "applied Heidegger." Heidegger adopted a wait-and-see attitude toward technology; Marcuse wants revolution. Heidegger's focus was "inward"; Marcuse's is "outward." See: R. Maurer, "Der angewandte Heidegger: Herbert Marcuse und das akademische Proletariat," pp. 238-59; G. Rohrmoser, "Humanität und Technologie," p. 779; R. Ahlers, "Is Technology Repressive?" p. 676. S. IJsseling also notes the points of agreement between Heidegger and Marcuse (*Tijdschrift voor Filosofie*, Vol. 32, No. 4, December 1970, p. 724). He states: "When we speak of points of agreement, we mean among other things the distinction between calculating thought and authentic thought, the problems centering on the all-controlling and all-determining power of technological thought and speech, the demand that technology no longer be considered a neutral means, the equating of Americanism and Russian communism as expressions of the same ground structure, the quest for the possibility of gaining a certain playful (*ludische*) freedom with respect to technology, the calling into question of de facto rationality and of the de facto system of inductions (the established order of thought), and finally that one-dimensionality which Heidegger so extensively discussed and designated *Eingleisigkeit*."
221. See TuW 94.
222. See TuW 81, 122, 123.
223. See TuW 80, 117.
224. See TuW 110f.
225. See TuW 91.
226. See H. Kahn and A. J. Wiener, *The Year 2000*.
227. See, among others: Ellul (2.4) and Meyer (2.5).
228. See TuW 119, 135.
229. See also J. Habermas, "Bedingungen für eine Revolutionierung spätkapitalistischer Gesellschaftssystem," pp. 212-24.
230. See TuW 91.
231. See P2 33; TuW 93.
232. See P2 206.
233. See S. U. Zuidema, *De revolutionaire maatschappijkritiek van Herbert Marcuse*, p. 136; H. Marcuse, *One-Dimensional Man: The Ideology of Industrial Society*, pp. 47, 197.
234. See FP 138f; P2 29.
235. See AuM 338.
236. See P2 135.
237. See TuW 7.
238. See P2 201.

239. This assessment is perhaps not entirely correct, for in March of 1970 Steinbuch forced the critical futurologist Robert Jungk to break with the *Gesellschaft für Zukunftsfragen*. Jungk did not want to help Steinbuch support existing industries in their attempts at development. What he wanted instead, more or less in imitation of Marcuse, was to build an ideal society in which technological development would be considered challengeable (see *Der Spiegel*, Vol. 24, 1970, No. 13, p. 195).

240. See Harmut Häusserman in a review of Steinbuch's *Program 2000* (*Der Spiegel*, Vol. 24, 1970, No. 13, pp. 196-8).

241. For a critical discussion of Habermas's view of technocracy, see B. C. van Houten, *Tussen aanpassing en kritiek*, pp. 265-84.

242. See 4.4 and 4.5.

243. Georg Klaus was professor of marxist philosophy at the University of East Berlin. He died in 1976. I refer to the following publications in my discussion of Klaus: *Kybernetik in philosophischer Sicht* (hereafter referred to as KiS); *Kybernetik und Gesellschaft* (hereafter referred to as KuG); G. Klaus and H. Schulze, *Sinn, Gesetze und Fortschritt in der Geschichte* (hereafter referred to as SGF); G. Klaus, *Spieltheorie in philosophischer Sicht* (hereafter referred to as SiS); *Kybernetik und Erkenntnistheorie* (hereafter referred to as KuE); I. A. Poletaev, *Kybernetik: Kurze Einführung in eine neue Wissenschaft* (hereafter referred to as K).

244. See K. Steinbuch, DiG 22.

245. See G. A. Wetter, *Der dialektische Materialismus: Seine Geschichte und sein System in der Sowjetunion*, p. 309. For a survey of dialectical materialism, see also I. M. J. Bochenski, *Der Sowjet-Russische Dialektisch-Materialismus*.

246. See R. T. de George, *Patterns of Soviet Thought*, pp. 151f.

247. *Ibid.*, p. 124.

248. See Wetter, *Der dialektische Materialismus*, p. 397.

249. *Ibid.*, pp. 365f; see De George, pp. 103f.

250. See Wetter, *Der dialektische Materialismus*, pp. 378, 397, 412.

251. *Ibid.*, p. 428.

252. *Ibid.*, p. 434.

253. See 3.4.6.1.

254. See Wetter, *Der dialektische Materialismus*, pp. 438f.

255. *Ibid.*, p. 448.

256. *Ibid.*, p. 460.

257. *Ibid.*, p. 301.

258. *Ibid.*, p. 465.

259. See, among others: Kostas Axelos, *Marx, penseur de la technique*; J. Hommes, *Der technische Eros: Das Wesen der materialistischen Geschichtsauffassung*; W. Hildebrandt, "Technik und Revolution: Die Funktion der Technik in der Gesellschaftslehre des Marxismus-Leninismus und Sowjetkommunismus"; Herman Ley, *Dämon Technik?*; and *Technik und Weltanschauung*; H. Klages, *Technischer Humanismus: Philosophie und Soziologie der Arbeit bei Karl Marx*; and "Marxismus und Technik"; A. A. Kusin, *Karl Marx und Probleme der Technik*; L. C. Robbins, *An Essay on the Nature and Significance of Economic Science*; F. P. A. Tellegen, *Samenleven in een technische tijd*; A. A. Zvorokine, "Technology and the Laws of Its Development."

260. W. R. Ashby has written: *An Introduction to Cybernetics*; "Design for an Intelligence Amplifier"; and *Design for a Brain: The Origin of Adaptive Behaviour.*
261. See KiS 335, 432, 371.
262. See KiS 7, 17, 95, 301, 315; KuG 30, 43, 300. G. Günther writes: "There is unquestionably a terminological affinity between cybernetics and dialectical materialism" (*Das Bewusstsein der Maschinen: Eine Metaphysik der Kybernetik*, p. 107).
263. See KuG 166, 167; see also Karl Marx, *Briefe über Das Kapital*, p. 221.
264. See KiS 219, 457; KuG x, 2, 158, 166.
265. See KiS 32.
266. See KiS 335; KuG 2.
267. See KiS 33.
268. See R. C. Kwant, "Marxistische analyse van de cybernetica."
269. See 3.4.6.2.
270. See KuG 4, 43; KiS 372, 452.
271. See KiS 79.
272. See KiS 7, 325; KuG 28; foreword to K.
273. A. M. Turing (1912-1954) was an English mathematician and logician. He designed the first modern "high-speed digital computer." Among his publications is the article "Computing Machinery and Intelligence." I shall refer to this article as reprinted in *Computers and Thought*, ed. E. A. Feigenbaum and J. Feldman (hereafter referred to as CT).
274. See CT 35.
275. See CT 22.
276. See also A. M. Turing, "On Computable Numbers, with an Application to the Entscheidungsproblem," p. 42.
277. KiS 329.
278. See KiS 35-7; KuE 45, 189, 190.
279. See KiS 42, 57.
280. See KiS 41; KuG 129.
281. See KiS 59.
282. See KuG 108.
283. See KiS 66; see also Wiener, C 155 and 3.2.7.2.
284. See KuG 16.
285. See KiS 71, 75, 77, 90, 91.
286. See KuG 111, 292; K xv; KuE 321.
287. See KiS 79.
288. See KuG 17, 18; KiS 379; KuE 205.
289. See KiS 82, 85, 95; KuE 244.
290. See KuE 192, 211.
291. See KuE 227, 281.
292. See KiS 333.
293. See KuG 21, 22.
294. See KiS 167.
295. See KiS 155-8, 375.
296. See KiS 134, 296, 305, 358; KuG 187.
297. See KiS 164f, 368.
298. See KiS 171f, 196.
299. See KuG 711, 67; KiS 197, 233; KuE 215, 216.
300. See KiS 249; KuE 215.

301. With regard to this notion, Klaus refers to N. Wiener, *Die Organisation der psychischen Funktionen*, p. 25; see also KuE 327, 328.
302. See KiS 365, 366; KuE 40; KuG 272, 276.
303. See KiS 258, 259, 270.
304. See KiS 426.
305. See KiS 123, 306, 378, 379.
306. See SiS 321; KuG 75.
307. See KuG 77; SiS 329, 331; KuE 256, 321.
308. See KuG 304.
309. See KiS 386, 439.
310. See KiS 403, 407.
311. See KiS 431, 433.
312. See KiS 399, 439.
313. See KiS 413; KuE 278.
314. See KiS 409, 411, 419.
315. See KuE 257, 269, 272, 273.
316. Sec KuE 256, 321.
317. See 3.2.7.3.
318. See KuG 123, 308f, 322; KuE 67, 68, 345f, 359.
319. See KuE 245.
320. See KuG 151; KuE 256, 322, 324, 361.
321. See KuE 130.
322. See KuG 124, 131, 135, 136, 159, 160.
323. See Karl Marx, *Grundrisse der Kritik der politischen Oekonomie*, p. 592; see also KuG 148, 158; KuE 68, 324.
324. See KuG 151.
325. See KuG 5, 136; KuE 130.
326. See KuE 389.
327. See SGF 122, 135.
328. See 284; see also Steinbuch, AuM 223.
329. See SGF 146, 229.
330. See SGF 124, 129, 152.
331. See SGF 125, 127, 156.
332. See SGF 142, 160.
333. See SGF 237.
334. For a critique of the three laws of dialectical materialism, see, among others, G. A. Wetter, *Der dialektische Materialismus*, pp. 386-9, 405-9, 419-23.
335. See also S. U. Zuidema, *Communisme in ontbinding*, p. 115.
336. *Ibid.*, p. 205.
337. R. T. de George, *Patterns of Soviet Thought*, p. 124.
338. See S. U. Zuidema, *Communisme in ontbinding*, pp. 19, 35.
339. *Ibid.*, p. 36; see also KiS 386.
340. J. Hommes, *Der technische Eros: Das Wesen der materialistischen Geschichtsauffassung*, p. 496.
341. *Ibid.*, pp. 235, 237, 238.
342. *Ibid.*, p. 247.
343. *Ibid.*, p. 373.
344. *Ibid.*, p. 3; see also p. 33.
345. *Ibid.*, p. 4; see also K. Marx and F. Engels, *MEGA: Historisch-kritische Gesamtausgabe*, I, 3, p. 155.

346. J. Hommes, *Der technische Eros*, p. 38 and p. 158; *MEGA*, I, 3, p. 83.
347. J. Hommes, *Der technische Eros*, p. 93; see also *MEGA*, I, 3, p. 89.
348. J. Hommes, *Der technische Eros*, p. 5.
349. *Ibid.*, p. 392.
350. *Ibid.*, pp. 93, 133, 151, 171, 313, 353, 387.
351. *Ibid.*, p. 179.
352. *Ibid.*, p. 185.
353. *Ibid.*, p. 247.
354. *Ibid.*, pp. 351, 440.
355. *Ibid.*, p. 353.
356. *Ibid.*, p. 500.
357. *Ibid.*, pp. 378, 379, 387, 116. From the very beginning, the notion of a "kingdom of freedom" has gone hand in hand with the idea of hierarchical planning – in the service of humanity, naturally. Marx believed that the "kingdom of freedom" would not be impaired by the planning; on the contrary, planning would make that kingdom a reality. "Marx dreamed of a utopian state in which the labor organization [as the institute of central control] would be so perfect that if someone wished to go fishing today, carpentering tomorrow, and bricklaying the day after tomorrow, he would find every opportunity to do so" (S. U. Zuidema, *Communisme in ontbinding*, p. 51; see also J. van der Hoeven et al., *Marxisme en revolutie*, pp. 33-5).
358. See 3.4.12.7.
359. See KuG 304; AuM 223.
360. See KiS 66.
361. See KiS 310.
362. See KuG 18.
363. See 3.4.5.2 under "*Re a.*"
364. See KuG 330.
365. See KiS 67, 95.
366. See KuE 12.
367. See KuE 191, 296.
368. See KuG 289, 291; AuM 190, 276; see also 3.3.5 and 3.3.11.3.
369. See AuM 200, 208, 311; KuE 137; KuG 296.
370. See KuG 305.
371. See KuE 119, 138, 241; KuG 289.
372. See KuE 217.
373. See KuG 299.
374. See KuG 25, 30, 31, 32.
375. I. B. Novik, *Some Methodological Problems of Cybernetics*, p. 47 (quoted by G. Günther, *Das Bewusstsein der Maschine: Eine Metaphysik der Kybernetik*, p. 40; see also p. 111).
376. See KuE 253.
377. See C. E. Shannon, "A Chess-playing Machine," p. 2132.
378. *Ibid.*, p. 2133; see also Shannon, "Computers and Automata," pp. 1234-41.
379. J. von Neumann, "The General and Logical Theory of Automata," p. 2078.
380. *Ibid.*, p. 2079.
381. *Ibid.*, p. 2083.
382. *Ibid.*, pp. 2084f.

383. See KuE 206.
384. See KiS 305.
385. See KuE 221, 272.
386. See KiS 142, 240.
387. See KiS 3, 5, 335, 342; KuG 24, 176, 352; SGF 6.
388. See *Erinnerungen an Karl Marx*, p. 155.
389. See KuG 21, 22, 279.
390. See KiS 457, 462, 478; KuG 149.
391. See KuE 220, 221, 235.
392. See KuE 141.
393. See R. T. de George, *Patterns of Soviet Thought*, p. 123.
394. See G. V. Plekhanov, *The Role of the Individual in History*, p. 41.
395. Marx and Engels, *MEGA*, I, 5, p. 64.
396. See KiS 262.
397. See G. A. Wetter, *Der dialektische Materialismus*, p. 448.
398. See SGF 63, 86, 142.
399. This thought originally comes from Friedrich Engels, who wrote: "The management of things and the direction of the processes of production will replace the government of men. The state will not be eliminated (*abgeschafft*), it will die out (*er stirbt ab*)" (quoted by L. Kalsbeek, *Contours of a Christian Philosophy*, p. 217).
400. See KuE 231.
401. See KuG 220; KuE 137.
402. See KiS 80, 311, 313; KuG 8, 209, 220, 279; KuE 375.
403. See note 399 above.
404. See J. Hommes, *Der technische Eros*, p. 217.
405. See KuG 93, 209, 220.

Chapter Four

1. See 4.8.3.
2. H. Beck reaches a similar conclusion when he identifies communism with technological totalitarianism (see *Philosophie der Technik: Perspektiven zur Technik, Menschheit, Zukunft*, pp. 123f).
3. See L. Heieck, *Bildung zwischen Technologie und Ideologie*, p. 36.
4. See H. Dooyeweerd, *A New Critique of Theoretical Thought*, Vol. 1, pp. 1, 4, 10f, 97, 132; Vol. 2, 3f, 30f; see also H. Van Riessen, *Wijsbegeerte*, pp. 165-77.
5. See J. P. A. Mekkes, *Radix, tijd en kennen*, p. 180. The philosophizing of Mekkes was a constant inspiration to me as I worked on the philosophical elaboration of the central themes that are indispensable to reflection on technological development. The references which follow, however, are not to be taken as a synopsis of his philosophy.
6. *Ibid.*, p. 12.
7. See H. Van Riessen, *Wijsbegeerte*, p. 146.
8. See J. P. A. Mekkes, *Teken en motief der creatuur*, p. 181.
9. See J. P. A. Mekkes, *Radix, tijd en kennen*, pp. 219, 233.
10. *Ibid.*, p. 218.

11. *Ibid.*, pp. 101, 144, 217; see also H. Dooyeweerd, *A New Critique*, Vol. 2, pp. 292f.
12. See J. P. A. Mekkes, *Radix, tijd en kennen*, pp. 84-8, 192.
13. *Ibid.*, pp. 163, 179-80, 218; *Teken en motief der creatuur*, p. 122.
14. See 1.3.2.
15. See J. P. A. Mekkes, *Radix, tijd en kennen*, pp. 47, 65.
16. *Ibid.*, pp. 32, 78; *Teken en motief der creatuur*, pp. 27, 28, 174, 200, 232.
17. See J. P. A. Mekkes, *Radix, tijd en kennen*, pp. 123, 174, 235. H. Beck, following a kindred line of questioning, also speaks of such a perverting (*Philosophie der Technik*, pp. 129-30). Beck makes the following statement, which corresponds to my own thinking on this matter: "To the extent that man responds to things as they address him with their claims and meanings, he can liberate himself from the anxiety of pretentious self-absolutization and become free by meekly submitting to a superior meaning" (p. 131).
18. See J. P. A. Mekkes, *Radix, tijd en kennen*, p. 219.
19. For a closer examination of the theme of the divine mystery in history, see M. C. Smit, "The Divine Mystery in History," especially p. 132.
20. See J. P. A. Mekkes, *Radix, tijd en kennen*, pp. 105, 154, 192, 204; *Teken en motief der creatuur*, p. 143.
21. Progress, it seems to me, should be construed not in the sense of advancement or improvement (*vooruitgang*) but in the sense of progression (*voortgang*). The former construction leads all too easily to the notion of an immanent meaning or immanent consummation of history. In addition to the tensions arising from the dialectic as sketched here, the groaning and travail of the whole creation (Romans 8:22) as a result of the fall into sin will continue to characterize history until its consummation, even if that groaning points to – and is thus a sign of – the total redemption and disclosure of the creation.
22. See J. P. A. Mekkes, *Radix, tijd en kennen*, p. 101.
23. *Ibid.*, pp. 128, 138, 141.
24. *Ibid.*, p. 161.
25. See H. Van Riessen, *Mondigheid en de machten*, p. 195.
26. See J. P. A. Mekkes, *Radix, tijd en kennen*, p. 217, and J. van der Hoeven, "Heidegger, Descartes en Luther," p. 110.
27. I will not take up the question of the existing differences of method. The natural sciences may have one method, for example, while the historical sciences use another. The differences between the methods of the various sciences are related to the differences in the nature of the aspects on which the various sciences focus as their field of inquiry.
28. The distinction between the natural sciences and the cultural sciences or human sciences is of a summary character, but it is sufficient for our purposes. Other distinctions may be found in A. G. M. van Melsen, *Wetenschap en verantwoordelijkheid*, pp. 66, 67, 72, and C. A. Van Peursen, *Wetenschap en werkelijkheid*, pp. 213, 215, 216.
29. This is also a boundary for certain sectors of natural-scientific inquiry. It is discernable in natural disasters, which are sometimes entirely unforeseeable, and sometimes partly foreseeable.
30. See H. Van Riessen, *The Society of the Future*, p. 290, and A. G. M. van Melsen, *Wetenschap en verantwoordelijkheid*, p. 166.
31. See 1.4.2.

32. For the limits of this method, see 1.4.3.
33. As I see it, there is a legitimate place for futurology as philosophy – but then as a part of the philosophy of history.
34. See 4.6.1.
35. See 1.5.3 and 1.5.4.
36. For similar ideas, see A. G. M. van Melsen, *Wetenschap en verantwoordelijkheid*, pp. 77, 78f, 83.
37. I had originally intended to discuss the problematics of futurology extensively in a thetical way (see 3.3.8, 3.3.10 and 3.3.11.5). I later decided that the material in question could more appropriately be published later in a separate work on futurology.
38. L. Heieck, *Bildung zwischen Technologie und Ideologie*, p. 77. See also F. Wagner, *Die Wissenschaft und die gefährdete Welt*, pp. 95f, and H. Beck, *Philosophie der Technik*, pp. 102f.
39. See 1.5.6.
40. See C. A. Van Peursen et al., *Informatie*, p. 142.
41. See 1.2.
42. For similar classifications, see: M. Scheler, *Die Wissensformen und die Gesellschaft*, pp. 133f; A. Gehlen, *Die Seele im technischen Zeitalter*, pp. 7f; W. van Benthem, *Das Ethos der technischen Arbeit und der Technik*, pp. 27f; K. Tuchel, *Die Philosophie der Technik bei Friedrich Dessauer*, pp. 67, 99f.
43. See C. A. Van Peursen et al., *Informatie*, pp. 135f.
44. See H. Schmidt, "Die Entwicklung der Technik als Phase der Wandlung des Menschen," pp. 118-22.
45. Quoted by K. Tuchel, *Die Philosophie der Technik bei Friedrich Dessauer*, p. 100.
46. See A. Gehlen, *Die Seele im technischen Zeitalter*, pp. 7f, 54; *Anthropologische Forschung*, p. 113.
47. See Adolf Portmann, "Zoologie und das neue Bild vom Menschen: Biologische Fragmente zu einer Lehre vom Menschen," *Rowohlts Deutsche Enzyklopädie*, Vol. 20 (Hamburg, 1959).
48. See L. Heieck, *Bildung zwischen Technologie und Ideologie*, p. 60.
49. See E. Kapp, *Grundlinien einer Philosophie der Technik*.
50. See W. G. Haverbeck, *Das Ziel der Technik*, p. 217.
51. See F. P. A. Tellegen, *Samenleven in een technische tijd*, p. 35.
52. See 1.3.7.
53. See A. G. M. van Melsen, *Science and Technology*, pp. 233, 234.
54. See F. P. A. Tellegen, *Samenleven in een technische tijd*, p. 37.
55. Nowadays people often speak of "human engineering." This term is too limited on the one hand since it does not embrace everything that should be understood by biotechnology and psychotechnology, and it is too broad on the other hand because it is used to include even the organizing of human relations. In the latter instance it would seem preferable to speak of "sociotechnology," although I would really prefer not to use the term *technology* here in any form. The words *technology* and *technique* should be reserved for form-giving below the cultural modality.
56. I am not referring here to the horrifying thought that, given the increasing knowledge of eugenics, entirely different "people" might be made in test tubes.
57. It is sometimes suggested that biotechnology will one day render the

human body immune to cosmic rays so that future space travelers will not have to undertake their voyages in heavy armor. Such notions are much easier to state than to implement, and the question of their legitimacy is seldom given serious attention. More often than not, the question is evaded through some peremptory expostulation. For example: "Nothing indicates that the altering of his natural surroundings which man has undertaken in the course of his history must halt before his own body" (G. Günther, *Das Bewusstsein der Maschinen*, p. 165). Another thinker, making a play on the famous words of Marx, says: "It is not sufficient to interpret the body; man must transform it" (G. Anders, *Die Antiquiertheit des Menschen: Über die Seele im Zeitalter der zweiten industriellen Revolution*, p. 350).

58. See the following section.

59. See O. Pedersen, *God en de techniek: Commentaar op de problemen van het atoomtijdperk*, p. 139.

60. See W. van Benthem, *Das Ethos der technischen Arbeit und der Technik*, p. 25. Van Benthem provides additional bibliography. This view is also defended by F. Klemm, *Technik: Eine Geschichte Ihrer Probleme*, pp. 89, 95, 99f; see also H. Lilje, *Das technische Zeitalter*, pp. 105f, and H. R. Müller-Schwefe, *Technik als Bestimmung und Versuchung*, p. 11.

61. See 1.4.5.

62. See P. Koessler, *Christentum und Technik*, pp. 65, 89.

63. See 1.4.4.

64. See O. Pedersen, *God en de techniek*, pp. 64, 65.

65. D. Brinkmann, *Mensch und Technik*, p. 94; see also O. Pedersen, *God en de techniek*, p. 67.

66. O. Spengler, *Der Mensch und die Technik*, p. 45.

67. See D. Brinkmann, *Mensch und Technik*, pp. 97f.

68. See W. G. Haverbeck, *Das Ziel der Technik*, pp. 223f, 235.

69. See, among others: C. J. Dippel, "Techniek en humaniteit," pp. 94-105; A. J. Toynbee, "Zijn de meesters der techniek gek geworden?" (Dutch translation of an article, published in the *NRC-Handelsblad*, April 31, 1971).

70. At this juncture I will not develop the point that human autonomy may also lead to people trying to get on without technology, which they may perceive as a threat to their freedom.

71. D. Brinkmann, *Mensch und Technik*, p. 107; see also pp. 131, 132, 140, 144; H. Lilje, *Das technische Zeitalter*, pp. 87, 113, 122.

72. See N. Berdyaev, "Man and Machine," p. 210; see also V. C. Ferkiss, *Technological Man: The Myth and the Reality*, pp. 38, 202.

73. O. Spengler, *Der Mensch und die Technik*, p. 48.

74. *Ibid.*, p. 49.

75. See O. Pedersen, *God en de techniek*, p. 78.

76. See 3.4.12.7.

77. See Ortega y Gasset, "Man the Technician," p. 151.

78. See 4.5.4 and 4.5.5.

79. See J. P. A. Mekkes, *Radix, tijd en kennen*, p. 99.

80. H. Beck, *Philosophie der Technik*, p. 98.

81. *Ibid.*, pp. 129-30.

82. See H. Van Riessen, *Wijsbegeerte*, p. 148.

83. See H. Dooyeweerd, *A New Critique*, Vol. 2, p. 285, and H. Van Riessen, *Filosofie en techniek*, p. 570.

84. See H. Dooyeweerd, *A New Critique*, Vol. 2, p. 285.
85. *Ibid.*, pp. 260-2, 274, 286.
86. *Ibid.*, p. 287, and H. Van Riessen, *Filosofie en techniek*, pp. 655, 659-60.
87. See 1.4.3.
88. See H. Van Riessen, *Filosofie en techniek*, pp. 593, 655, 661-5; "The Structure of Technology," p. 306.
89. See H. Dooyeweerd, *A New Critique*, Vol. 1, pp. 289, 290; see also H. Lilje, *Das technische Zeitalter*, pp. 120, 121, 146, 175.
90. See H. Dooyeweerd, *A New Critique*, Vol. 2, p. 291.
91. For elaboration on this theme, see 4.8.
92. See H. Dooyeweerd, *A New Critique*, Vol. 2, pp. 304, 305.
93. See 1.3.7.
94. See 3.3.10 and 3.3.11.6; see also H. D. Bahr, *Kritik der 'Politischen Technologie': Eine Auseinandersetzung mit Herbert Marcuse und Jürgen Habermas*.
95. See W. van Benthem, *Das Ethos der technischen Arbeit und der Technik*, pp. 86f; H. Beck, *Philosophie der Technik*, p. 144.
96. See 4.5.3, 4.5.4 and 4.5.5.
97. D. Brinkmann, *Mensch und Technik*, p. 143.
98. See G. Zoutendijk, "Computer en macht," pp. 1-8; see also H. Van Riessen, *The Society of the Future*, pp. 158f.
99. See G. Zoutendijk, "Computer en democratie," *Academia*, February 19, 1971, pp. 3-7; R. F. Beerling, "Technocratie en democratie," pp. 13, 21.
100. See M. Stone and W. Warner, *The Data Bank Society*, pp. 213f.
101. D. Brinkmann, "Der Mensch im Zeitalter der Automation," p. 102; see also p. 112; H. Beck, *Philosophie der Technik*, pp. 81, 86.
102. M. McLuhan, "Address at Vision '65," p. 204.
103. V. C. Ferkiss, *Technological Man*, p. 157; see also pp. 201, 205.
104. See the conclusion of 4.9.
105. See 4.4.
106. See 4.8.
107. What I mean by *nature* is not something divorced from humanity or something standing over against humanity; I mean the nature side of the creation. More specifically, I mean all nature-things and nature-facts, which have as their last subject-function the physical function. All their remaining functions are object-functions.
108. See W. H. Raby, *Über Sinn und Grenzen der Technik*, p. 97.

Bibliography*

AHLERS, ROLF. "Is Technology Repressive?" *Tijdschrift voor filosofie*. Volume 32 (1970), No. 4, pp. 651-701.

ANDERS, GÜNTER. *Die Antiquiertheit des Menschen: Über die Seele im Zeitalter der zweiten industriellen Revolution*. Munich: Beck, 1956.

ARENDT, HANNAH. *Vita activa*. Stuttgart: Kohlhammer, 1960.

ASHBY, WILLIAM ROSS. *Design for a Brain: The Origin of Adaptive Behaviour*. London: Chapman & Hall, 1960.

—— "Design for an Intelligence Amplifier." In: Shannon, Claude E., and McCarthy, John, eds., *Automata Studies*. Princeton: Princeton University Press, 1956.

—— *An Introduction to Cybernetics*. London: Chapman & Hall, 1957.

AXELOS, KOSTAS. *Marx, penseur de la technique*. Paris: Les Editions de Minuit, 1961.

BAHR, HANS-DIETER. *Kritik der 'Politischen Technologie': Eine Auseinandersetzung mit Herbert Marcuse und Jürgen Habermas*. Frankfurt: Europäische Verlagsanstalt, 1970.

BAKKER, REINOUT. "Heidegger en de techniek." *Wijsgerig perspectief*. Volume 3 (1962-63), pp. 294-306.

BAR-HILLEL, YEHOSHUA. "Informatietheorie onder de loep." *Intermediair*. Volume 6 (1970), Nos. 2-4.

BEAUFRET, JEAN, et al. *Martin Heidegger zum siebzigsten Geburtstag: Festschrift*. Pfullingen: Verlag Günther Neske, 1959.

---

* For a more extensive bibliography see: H. Beck, *Kulturphilosophie der Technik*, pp. 191-258; P. T. Durbin and C. Mitcham, eds., *Research in Philosophy and Technology*, ongoing; C. Mitcham and R. Mackey, eds., "Bibliography of the Philosophy of Technology"; H. Sachsse, ed., *Technik und Gesellschaft*, Band 1: Literaturführer; and E. Schuurman, *Techniek en toekomst*, pp. 456-533.

BECK, HEINRICH. *Kulturphilosophie der Technik*. Trier: Spee Verlag, 1979.

―――― *Philosophie der Technik: Perspektiven zu Technik, Menschheit, Zukunft*. Trier: Spee Verlag, 1969.

BEERLING, REINIER FRANCISCUS. "Technocratie en democratie." *Intermediair*. Volume 7 (1971), pp. 13-21.

BERDYAEV, NIKOLAI. "Man and Machine." In: Mitcham, Carl, and Mackey, Robert, eds., *Philosophy and Technology: Readings in the Philosophical Problems of Technology*. New York: The Free Press, 1972, pp. 203-13.

―――― *Der Mensch und die Technik*. Lucerne: Vita Nova Verlag, 1943.

BERGHS, HARRY. "Het zakelijk karakter van Heideggers vraag naar de techniek." *Tijdschrift voor filosofie*. Volume 33 (1971), No. 2, pp. 250-78.

BOCHENSKI, INNOCENT MARIE JOSEPH. *Der Sowjet-Russische Dialektisch-Materialismus*. Bern and Munich: Francke, 3rd ed., 1960.

BOEHM, RUDOLF. "Pensée et technique." *Revue Internationale de Philosophie*. Volume 14 (1960), No. 52, pp. 194-220.

BOK, SIEGFRIED THOMAS. *Cybernetica*. Utrecht: Het Spectrum, 4th ed., 1962.

BORKENAU, FRANZ, ed. *Karl Marx*. Frankfurt: Fischer Bücherei, 1956.

BRINKMANN, DONALD. *Mensch und Technik: Grundzüge einer Philosophie der Technik*. Bern: Francke, 1946.

―――― "Der Mensch im Zeitalter der Automation." In: *Festschrift H. J. de Vleeschauer*. Pretoria: Publications Committee, University of South Africa, 1960, pp. 96-113.

CROSSON, FREDERICK J., and SAYRE, KENNETH M., eds. *Philosophy and Cybernetics*. Notre Dame, Ind.: University of Notre Dame Press, 1967.

DE GEORGE, RICHARD T. *Patterns of Soviet Thought*. Ann Arbor: University of Michigan Press, 1966.

DESSAUER, FRIEDRICH. *Philosophie der Technik: Das Problem der Realisierung*. Bonn: Cohen, 1927.

―――― *Streit um die Technik*. Frankfurt: Verlag Josef Knecht, 2nd ed., 1958.

DIPPEL, CORNELIS JOHANNES. "Techniek en humaniteit." *Wijsgerig perspectief*. Volume 6 (1965), No. 3, pp. 94-105.

―――― *Verkenning en verwachting*. The Hague: Boekencentrum, 1962.

DOOYEWEERD, HERMAN. *A New Critique of Theoretical Thought*. 4 Volumes. Translated by David H. Freeman et al. Amsterdam: H. J. Paris; and Philadelphia: Presbyterian and Reformed Publishing Company, 1953-58.

―――― *Roots of Western Culture*. Translated by John Kraay. Toronto: Wedge, 1979.

―――― *De wijsbegeerte der wetsidee*. 3 Volumes. Amsterdam: H. J. Paris, 1935-36.

DUINTJER, OTTO DIRK. *De vraag naar het transcendentale (vooral in verband met Heidegger en Kant)*. Leiden: Universitaire Pers, 1966.

410

DURBIN, PAUL T., and MITCHAM, CARL, eds. *Research in Philosophy and Technology*. Volumes 1-2. Greenwich, Conn.: JAI Press, 1978-79 (ongoing).

DIJKEMA, PIETER. *Innen und Aussen: Die Frage nach der Integration der Künste und der Weg der Architektur*. Hilversum: Van Saane, 1960.

ELLUL, JACQUES. "Search for an Image." *Humanist*. Volume 33 (1973), No. 6, pp. 22-25.

—— *La technique ou l'enjeu du siècle*. Paris: Colin, 1954.

—— "The Technological Order." *Technology and Culture*. Volume 3 (1962), No. 4, pp. 394-463.

—— *The Technological Society*. Translated by John Wilkinson. London: Cape, 1965.

Erinnerungen an Karl Marx. Berlin: Dietz Verlag, 1953.

FAST, JOHAN D., and STUMPERS, FRANS L. H. M. "Informatie, entropie en energie." *Intermediair*. Volume 6 (1970), August 6.

FEIGENBAUM, EDWARD A., and FELDMAN, JULIAN, eds. *Computers and Thought*. New York: McGraw-Hill, 1963.

FERKISS, VICTOR C. *Technological Man: The Myth and the Reality*. New York: Braziller, 3rd ed., 1970.

FLECHTHEIM, OSSIP KURT. *Futurologie: Der Kampf um die Zukunft*. Cologne: Verlag Wissenschaft und Politik, 1970.

FORRESTER, JAY WRIGHT. "Managerial Decision Making." In: Greenberger, Martin, ed., *Computers and the World of the Future*. Cambridge, Mass.: MIT-Press, 1962, pp. 36-92.

FRANK, HELMAR, ed. *Kybernetik: Brücke zwischen den Wissenschaften*. Frankfurt: Umschau Verlag, 4th ed., 1964.

—— *Kybernetik und Philosophie: Materialien und Grundriss zu einer Philosophie der Kybernetik*. Berlin: Duncker & Humblot, 1966.

GEHLEN, ARNOLD. "Anthropologische Ansicht der Technik." In: Freyer, Hans, et al., eds., *Technik im technischen Zeitalter: Stellungnahmen zur geschichtlichen Situation*. Düsseldorf: Verlag Schilling, 1965, pp. 101-19.

—— *Anthropologische Forschung*. Hamburg: Rowohlt, 1961.

—— *Die Seele im technischen Zeitalter: Sozialpsychologische Probleme in der industriellen Gesellschaft*. Hamburg: Rowohlt, 1957.

GREENBERGER, MARTIN, ed. *Computers and the World of the Future*. Cambridge, Mass.: MIT-Press, 1962.

GÜNTHER, GOTTHART. *Das Bewusstsein der Maschinen: Eine Metaphysik der Kybernetik*. Krefeld and Baden-Baden: Agis Verlag, 1963.

HABERMAS, JÜRGEN. "Bedingungen für eine Revolutionierung spätkapitalistischer Gesellschaftssysteme." *Praxis*. Volume 1/2 (1969), pp. 212-24.

—— *Technik und Wissenschaft als "Ideologie."* Frankfurt: Suhrkamp Verlag, 3rd ed., 1969.

HAVERBECK, WERNER GEORG. *Das Ziel der Technik: Die Menschwerdung der Erde*. Olten and Freiburg: Walter Verlag, 1965.

411

HEIDEGGER, MARTIN. *Einführung in die Metaphysik*. Tübingen: Max Niemeyer Verlag, 1953.

——— "La fin de la philosophie et la tâche de la pensée." In: *Kierkegaard Vivant*. Paris: Gallimard, 1966.

——— *Gelassenheit*. Pfullingen: Neske, 1959.

——— *Hebel, der Hausfreund*. Pfullingen: Neske, 1957.

——— *Holzwege*. Frankfurt: Verlag Klostermann, 4th ed., 1963.

——— *Identität und Differenz*. Pfullingen: Neske, 2nd ed., 1957.

——— "Messkirch's Seventh Centennial." *Listening*. Volume 8 (1973), Nos. 1-3, pp. 40-54.

——— *Platons Lehre von der Wahrheit; mit einem Brief über den Humanismus*. Bern: Francke, 2nd ed., 1954.

——— *Der Satz vom Grund*. Pfullingen: Neske, 2nd ed., 1958.

——— *Sein und Zeit*. Tübingen: Max Niemeyer Verlag, 10th ed., 1963.

——— *Die Technik und die Kehre*. Pfullingen: Neske, 1962.

——— *Unterwegs zur Sprache*. Pfullingen: Neske, 2nd ed., 1960.

——— *Der Ursprung des Kunstwerks*. Pfullingen: Neske, 2nd ed., 1952.

——— *Vorträge und Aufsätze*, Part I. Pfullingen: Neske, 1967.

——— *Was heisst Denken?* Pfullingen: Neske, 2nd ed., 1958.

HEIECK, LUDWIG. *Bildung zwischen Technologie und Ideologie*. Heidelberg: Quelle & Meyer, 1969.

HEISENBERG, WERNER. "Das Naturbild der heutigen Physik." In: Podewils, Clemens Graf, ed., *Die Künste im technischen Zeitalter*. Munich: Oldenbourg, 1954, pp. 60ff.

HEYDE, JOHANNES ERICH. "Technischer Fortschritt – menschliche Verantwortung." *Verhandlungen der Deutschen Gesellschaft für Arbeitsschutz*. Volume 7 (1961-62), pp. 1-29.

HILDEBRANDT, W. "Technik und Revolution: Die Funktion der Technik in der Gesellschaftslehre des Marxismus-Leninismus und Sowjetkommunismus." *Studium Generale*. Volume 15 (1962), pp. 334-56.

HOMMES, JAKOB. *Der technische Eros: Das Wesen der materialistischen Geschichtsauffassung*. Freiburg: Verlag Herder, 1955.

HUNTEMANN, GEORG. *Provozierte Theologie in technischer Welt*. Wuppertal: Brockhaus Verlag, 1968.

IJSSELING, SAMUEL. "De filosofie en het technische denken." In: De Clercq, Bertrand Juliaan, et al., *Uitzicht van onze wereld*. Utrecht: Desclée de Brouwer, 1964, pp. 31-48.

JÜNGER, ERNST. *Der Arbeiter: Herrschaft und Gestalt*. Hamburg: Hanseatische Verlagsanstalt, 1932.

——— "Vom Ende des geschichtlichen Zeitalters." In: Beaufret, Jean, et al., *Martin Heidegger zum siebzigsten Geburtstag*. Pfullingen: Neske, 1959.

JÜNGER, FRIEDRICH GEORG. *The Failure of Technology*. Chicago: Regnery, 1956.

———— *Maschine und Eigentum*. Frankfurt: Verlag Klostermann, 2nd ed., 1953.

———— *Die Perfektion der Technik*. Frankfurt: Verlag Klostermann, 4th ed., 1953.

———— *Sprache und Kalkül*. Frankfurt: Verlag Klostermann, 1956.

———— *Die Vollkommene Schöpfung: Natur oder Naturwissenschaft*. Frankfurt: Verlag Klostermann, 1969.

KAHN, HERMAN, and WIENER, ANTHONY J. *The Year 2000*. New York: Macmillan, 1967.

KALSBEEK, LEENDERT. *Contours of a Christian Philosophy: An Introduction to Herman Dooyeweerd's Thought*. Edited by Bernard and Josina Zylstra. Toronto: Wedge, 1975.

KAPP, ERNST. *Grundlinien einer Philosophie der Technik: Zur Entstehungsgeschichte der Kultur aus neuen Gesichtspunkten*. Braunschweig: Westermann, 1977.

KLAGES, HELMUT. "Marxismus und Technik." In: Freyer, Hans, et al., eds., *Technik im technischen Zeitalter: Stellungnahmen zur geschichtlichen Situation*. Düsseldorf: Verlag Schilling, 1965, pp. 137-51.

———— *Technischer Humanismus: Philosophie und Soziologie der Arbeit bei Karl Marx*. Stuttgart: Ferdenand Enke Verlag, 1964.

KLAUS, GEORG. *Kybernetik und Erkenntnistheorie*. Berlin: VEB Deutscher Verlag der Wissenschaften, 3rd ed., 1969.

———— *Kybernetik und Gesellschaft*. Berlin: VEB Deutscher Verlag der Wissenschaften, 1964.

———— *Kybernetik in philosophischer Sicht*. Berlin: Dietz Verlag, 1961.

———— *Spieltheorie in philosophischer Sicht*. Berlin: VEB Deutscher Verlag der Wissenschaften, 1968.

KLAUS, GEORG, and SCHULZE, HANS. *Sinn, Gesetz und Fortschritt in der Geschichte*. Berlin: Dietz Verlag, 1967.

KLEMM, FRIEDRICH. *Technik: Eine Geschichte ihrer Probleme*. Munich and Freiburg: Verlag Alber, 1954.

KNOL, J. G. "Een dialoog met critici." *Mededelingen van de Vereniging voor Calvinistische Wijsbegeerte*. 1970, No. 3.

KOCH, CLAUS, and SENGHAAS, DIETER, eds. *Texte zur Technokratiediskussion*. Frankfurt: Europäische Verlagsanstalt, 1970.

KOESSLER, PAUL. *Christentum und Technik*. Aschaffenburg: Paul Pattloch Verlag, 1959.

KUSIN, ALEKSANDR AVRAMEVIC. *Karl Marx und Probleme der Technik*. Translated by A. Kraus. Leipzig: VEB Fachbuch Verlag, 1970.

KWANT, REMIGIUS CORNELIS. "Marxistische analyse van de cybernetica: Naar aanleiding van Georg Klaus, *Kybernetik in philosophischer Sicht.*" *Tijdschrift voor filosofie*. Volume 28 (1966), pp. 518-61.

LANDGREBE, LUDWIG. *Philosophie der Gegenwart*. Frankfurt: Ullstein, 1957.

LANG, EBERHARD. "Zur Geschichte des Wortes Kybernetik." Beiheft zu:

Grundlagenstudien aus Kybernetik und Geisteswissenschaft, Volume 9. Quickborn: Verlag Schnelle, 1968.

LEY, HERMANN. *Dämon Technik?* Berlin: VEB Deutscher Verlag der Wissenschaften, 1961.

—— *Technik und Weltanschauung*. Leipzig: Urania Verlag, 1969.

LILJE, HANS. *Das technische Zeitalter: Versuch einer biblischen Deutung*. Berlin: Furche, 1928.

LITT, THEODOR. *Naturwissenschaft und Menschenbildung*. Heidelberg: Quelle & Meyer, 2nd ed., 1954.

MCLUHAN, MARSHALL. "Address at Vision '65." *American Scholar*. Volume 35 (1966), pp. 196-205.

MARCUSE, HERBERT. *One-Dimensional Man: The Ideology of Industrial Society*. Boston: Beacon, 1964.

MARX, KARL. *Briefe über Das Kapital*. Berlin: Dietz Verlag, 1954.

—— *Grundrisse der Kritik der politischen Ökonomie*. Berlin: Dietz Verlag, 1953.

MARX, KARL, and ENGELS, FRIEDRICH. *MEGA: Historisch-kritische Gesamtausgabe*. Berlin: Marx-Engels-Verlag, 1932.

MARX, WERNER. *Heidegger und die Tradition*. Stuttgart: Kohlhammer, 1961.

MATUSSEK, P. "Psychodynamische Aspekte der Kreativitätsforschung." *Der Nervenartz*. Volume 38 (1967), pp. 143-51.

MAURER, REINHART. "Der angewandte Heidegger: Herbert Marcuse und das akademische Proletariat." *Philosophisches Jahrbuch*. Volume 77 (1970), pp. 238-59.

MEKKES, JOHAN PETER ALBERTUS. *Radix, tijd en kennen*. Amsterdam: Buijten & Schipperheijn, 1971.

—— *Scheppingsopenbaring en wijsbegeerte*. Kampen: Kok, 1961.

—— *Teken en motief der creatuur*. Amsterdam: Buijten & Schipperheijn, 1965.

MEYER, HERMANN JOSEF. "Technik und Kybernetik: Die Technik im Selbstverständnis des heutigen Menschen." In: Schwarz, Richard, ed., *Menschliche Existenz und moderne Welt: Ein internationales Symposium zum Selbstverständnis des heutigen Menschen*. Volume 1. Berlin: De Gruyter, 1967, pp. 750-83.

—— *Die Technisierung der Welt: Herkunft, Wesen und Gefahren*. Tübingen: Max Niemeyer Verlag, 1961.

—— "El tecnologismo; análisis de una actitud espiritual." *Folia Humanistica*. Volume 11 (1973), pp. 57-74.

MITCHAM, CARL, and MACKEY, ROBERT, eds. "Bibliography of the Philosophy of Technology." *Technology and Culture*. Volume 14 (1973), No. 2, Part II.

MORTON, HERBERT DONALD, ed. "Symposium: Hendrik van Riessen and Dutch Neo-Calvinist Philosophy of Technology." In: Durbin, Paul T., and Mitcham, Carl, eds., *Research in Philosophy and Technology*. Volume

2. Greenwich, Conn.: JAI Press, 1979, pp. 293-340. ["Symposium" is composed of: (1) Biography and Selected Bibliography for Van Riessen; (2) "The Structure of Technology" and "Technology and Culture" by Van Riessen; and (3) "Continuing the Neo-Calvinist Critique of Technology: Review of Egbert Schuurman's *Technology and Deliverance*," by Morton. The title *Technology and Deliverance* refers to the present title: *Technology and the Future*.]

MOSER, SIMON. *Metaphysik einst und jetzt: Kritische Untersuchungen zu Begriff und Ansatz der Ontologie*. Berlin: De Gruyter, 1958.

MÜLLER-SCHWEFE, HANS-RUDOLF. *Technik als Bestimmung und Versuchung*. Göttingen: Vandenhoeck, 1965.

NAUTA, DOEDE. *The Meaning of Information*. The Hague and Paris: Mouton, 1970.

NEUMANN, JOHN VON. *The Computer and the Brain*. New Haven Conn.: Yale University Press, 1958.

———— "The General and Logical Theory of Automata." *The World of Mathematics*. Volume 4 (1956), pp. 2078f.

NOVIK, I. B. *Some Methodological Problems of Cybernetics*. Joint Publications Research Service, 14, 592.

O'CONNOR, DANIEL JOHN, ed. *A Critical History of Western Philosophy*. London: Free Press, 1964.

ORTEGA Y GASSET, JOSÉ. "Man the Technician." In: *History as a System, and Other Essays toward a Philosophy of History*. Translated by Helene Weyl. New York: Norton, 1962, pp. 87-161.

———— *The Modern Theme*. Translated by James Cleugh. New York, Evanston and London: Harper & Row, 1961.

OTTEN, J. "Digitale informatieverwerking." *Lucerna*. Volume 8 (1969), Nos. 1, 2.

PEDERSEN, OLAF. *God en de techniek: Commentaar op de problemen van het atoomtijdperk*. Amsterdam: Het Wereldvenster, 2nd ed., 1952.

PEPERZAK, ADRIAAN THEODOOR. *Techniek en dialoog*. Delft: Waltman, 1967.

PIERCE, JOHN R. *Symbolen en signalen*. Utrecht: Het Spectrum, 1966.

PLEKHANOV, GEORGI VALENTINOVIČ. *The Role of the Individual in History*. New York: International Publishers Company, 1940.

PODEWILS, CLEMENS GRAF, ed. *Die Künste im technischen Zeitalter*. Munich: Oldenbourg, 1954.

POLANYI, MICHAEL. *Personal Knowledge: Toward a Post-critical Philosophy*. London and Henley: Routledge & Kegan Paul, 2nd ed., 1962.

POLETAEV, IGOR ANDREEVIC. *Kybernetik: Kurze Einführung in eine neue Wissenschaft*. Berlin: VEB Deutscher Verlag der Wissenschaften, 3rd ed., 1969.

POPMA, KLAAS JOHAN. *Inleiding in de wijsbegeerte*. Kampen: Kok, 1956.

PORTMANN, ADOLF. "Zoologie und das neue Bild vom Menschen: Biologische Fragmente zu einer Lehre vom Menschen." *Rowohlts Deutsche Enzyklopädie*. Volume 20. Hamburg: Rowohlt, 1959.

RABY, W. H. *Über Sinn und Grenzen der Technik.* Düsseldorf: VDI-Verlag, 1966.

REICHENBACH, HANS. *Der Aufstieg der wissenschaftlichen Philosophie.* Translated by Maria Reichenbach. Berlin: Herbig, 1951.

ROBBINS, LIONEL CHARLES. *An Essay on the Nature and Significance of Economic Science.* London: Macmillan, 1963.

ROHRMOSER, GÜNTER. "Humanität und Technologie." *Studium Generale.* Volume 22 (1969), p. 779.

RÖPKE, HORST, and RIEMANN, JÜRGEN. *Analog Computer in Chemie und Biologie.* Berlin and Heidelberg: Springer Verlag, 1969.

RUSSELL, BERTRAND. *The Scientific Outlook.* London: George Allen & Unwin, 1931.

SACHSSE, HANS, ed. *Technik und Gesellschaft.* Band 1: Literaturführer. Munich: Verlag Dokumentation, 1974.

SAMUEL, A. L. "Some Studies in Machine Learning, Using the Game of Checkers." *IBM Journal of Research and Development.* Volume 3 (1959), pp. 210-29.

SCHADEWALDT, WOLFGANG. *Natur, Technik, Kunst: Drei Beiträge zum Selbstverständnis der Technik in unserer Zeit.* Göttingen: Musterschmidt, 1960.

SCHAEFFLER, RICHARD. "Martin Heidegger und die Frage nach der Technik." *Zeitschrift für philosophische Forschung.* Volume 9 (1955), pp. 116-27.

SCHELER, MAX. *Die Wissensformen und die Gesellschaft.* Bern and Munich: Francke, 2nd ed., 1960.

SCHELSKY, HELMUT. *Der Mensch in der wissenschaftlichen Zivilisation.* Cologne and Opladen: Westdeutscher Verlag, 1961.

SCHISCHKOFF, GEORGI. "Philosophie und Kybernetik: Zur Kritik am kybernetischen Positivismus." *Zeitschrift für philosophische Forschung.* Volume 19 (1965), pp. 248-78.

SCHMIDT, HERMANN. "Die Entwicklung der Technik als Phase der Wandlung des Menschen." *Zeitschrift des Vereins Deutscher Ingenieure.* Volume 96 (1954), pp. 118-22.

SCHNEIDER, PETER KARLFRIED. *Die Begründung der Wissenschaften durch Philosophie und Kybernetik.* Stuttgart: Kohlhammer, 1966.

SCHUURMAN, EGBERT. *Reflections on the Technological Society.* Toronto: Wedge, 1977.

———— *Responsibility in the Technological Society.* Forthcoming.

———— *Techniek en toekomst: Confrontatie met wijsgerige beschouwingen.* Assen: Van Gorcum, 1972. Bibliography, pp. 456-533.

———— *Technology in Christian-Philosophical Perspective.* Potchefstroom: Potchefstroom University, 1980.

SHANNON, CLAUDE E. "A Chess-playing Machine." *The World of Mathematics.* Volume 4 (1956), pp. 2124-33.

———— "Computers and Automata." *Proceedings of the Institute of Radio Engineers.* Volume 41 (1953), pp. 1234-41.

416

SIMON, HERBERT A. "Simulation of Human Thinking." In: Greenberger, Martin, ed., *Computers and the World of the Future*. Cambridge, Mass.: MIT-Press, 1962, pp. 94-132.

SKOLIMOWSKI, HENRYK. "The Structure of Thinking in Technology." *Technology and Culture*. Volume 7 (1966), No. 3, pp. 371-83.

SMIT, MEIJER CORNELIS. "The Divine Mystery in History." *Free University Quarterly*. Volume 5 (1958), pp. 120-45.

SPENGLER, OSWALD. *Der Mensch und die Technik: Beitrag zu einer Philosophie des Lebens*. Munich: Beck, 1971.

STEINBUCH, KARL. *Automat und Mensch: Über menschliche und maschinelle Intelligenz*. Berlin, Göttingen and Heidelberg: Springer Verlag, 3rd ed., 1965.

――― *Falsch programmiert: Über das Versagen unserer Gesellschaft in der Gegenwart und vor der Zukunft, und was eigentlich geschehen müsste*. Stuttgart: Deutsche Verlagsanstalt, 1968.

――― *Die informierte Gesellschaft: Geschichte und Zukunft der Nachrichtentechnik*. Stuttgart: Deutsche Verlagsanstalt, 1966.

――― *JA zur Wirklichkeit*. Stuttgart: Seewald Verlag, 1975.

――― "Können Automaten Schrift 'lesen' und Sprache 'verstehen'?" In: Frank, Helmar, ed., *Kybernetik: Brücke zwischen den Wissenschaften*. Frankfurt: Umschau Verlag, 4th ed., 1964, pp. 169-75.

――― *Kurskorrektur*. Stuttgart: Seewald Verlag, 1973.

――― *Kybernetik*. Ludwigshafen am Rhein: Badische Anilin- und Soda-Fabrik, 1964.

――― *Programm 2000: Die konstruktive Ergänzung zu 'Falsch programmiert'; ein provokativer Aufruf zur Umorientierung*. Stuttgart: Deutsche Verlagsanstalt, 1970.

――― "Die Rolle der Technik und die Probleme unserer Gesellschaft in Gegenwart und Zukunft." *Universitas*. Volume 27 (1972), pp. 229-42.

――― "Zur Systemanalyse des technischen Fortschritts." Sonderdruck aus: *'System 69': Internationales Symposium über Zukunftsfragen*. Stuttgart: Deutsche Verlagsanstalt, 1969.

――― "Technik und Gesellschaft als Zukunftsproblem." In: Jungk, Robert, ed., *Menschen im Jahr 2000*. Frankfurt: Umschau Verlag, 1969, pp. 65-75.

――― "Technisches Handeln fordert Moral: Mässigung angesichts gewaltiger technischer Möglichkeiten." *VDI-Nachrichten*. Volume 25 (1971), No. 37, pp. 27-8.

STEINBUCH, KARL, and MOSER, SIMON, eds. *Philosophie und Kybernetik*. Munich: Nymphenburger Verlagshandlung, 1970.

STEINBUCH, KARL, and WAGNER, SIEGFRIED WILHELM, eds. *Neuere Ergebnisse der Kybernetik*. Munich: Oldenbourg, 1963.

STRUNZ, KURT. *Integrale Anthropologie und Kybernetik*. Heidelberg: Quelle & Meyer, 1965.

TELLEGEN, FRANCISCUS P. A. *Samenleven in een technische tijd*. Utrecht: Het Spectrum, 2nd ed., 1965.

TROOST, ANDREE. "Personalisme en ethiek." *Mededelingen van de Vereniging voor Calvinistische Wijsbegeerte*. March, 1966, pp. 3-5.

TUCHEL, KLAUS. *Die Philosophie der Technik bei Friedrich Dessauer: Ihre Entwicklung, Motive und Grenzen*. Frankfurt: Verlag Josef Knecht, 1964.

―――― "Die Technik als Gabe und Aufgabe: Die Geisteswissenschaftler müssen das Gespräch mit den Ingenieurswissenschaftlern suchen." *VDI-Nachrichten*. Volume 16 (1962), No. 8.

TURING, ALAN MATHISON. "On Computable Numbers, with an Application to the Entscheidungsproblem." *Proceedings of the London Mathematics Society*. Volume 42 (1936), pp. 230-65, and Volume 43 (1937), p. 544.

―――― "Computing Machinery and Intelligence." In: Feigenbaum, Edward A., and Feldman, Julian, eds., *Computers and Thought*. New York: McGraw-Hill, 1963.

ULMANN, GISELA. *Kreativität*. Weinheim and Berlin: Verlag Julius Beltz, 1968.

VAN BENTHEM, WALTER. *Das Ethos der technischen Arbeit und der Technik: Ein Beitrag zur personalen Deutung*. Essen: Ludgerus Verlag, 1966.

VAN DER HOEVEN, JOHAN. "Heidegger, Descartes en Luther." In: Bakker, Dirk Miente, et al., *Reflexies: Opstellen aangeboden aan Prof. Dr. J. P. A. Mekkes*. Amsterdam: Buijten & Schipperheijn, 1968.

―――― *Heidegger en de geschiedenis der wijsbegeerte*. Amsterdam: Buijten & Schipperheijn, 1964.

―――― *Kritische ondervraging van de fenomenologische rede*. Amsterdam: Buijten & Schipperheijn, 1963.

VAN DER HOEVEN, JOHAN, et al. *Marxisme en revolutie*. Amsterdam: Buijten & Schipperheijn, 1967.

VAN HOUTEN, BÉ CORNELIS. *Tussen aanpassing en kritiek*. Deventer: Van Loghum Slaterus, 1970.

VAN MELSEN, ANDREW G. M. "Filosofie van de techniek." *Algemeen Nederlands tijdschrift voor wijsbegeerte en psychologie*. Volume 60 (1968), Nos. 3-4, pp. 245-57.

―――― *Science and Technology*. Pittsburgh: Duquesne University Press, 1961.

―――― *Wetenschap en verantwoordelijkheid*. Utrecht: Het Spectrum, 1969.

―――― "De wijsgerige implicaties van de wisselwerking tussen wetenschap en techniek." *Algemeen Nederlands tijdschrift voor wijsbegeerte en psychologie*. Volume 57 (1965), No. 3, pp. 172-81.

VAN PEURSEN, CORNELIS ANTHONIE. *Strategie van de cultuur: Een beeld van de veranderingen in de hedendaagse denk- en leefwereld*. Amsterdam: Elsevier, 1970.

―――― *Wetenschap en werkelijkheid*. Kampen: Kok, 1969.

VAN PEURSEN, CORNELIS ANTHONIE, et al. *Informatie: Een interdisciplinaire studie*. Utrecht: Het Spectrum, 1968.

VAN RIESSEN, HENDRIK. "Denkgereedschap." *Mededelingen van de Vereniging voor Calvinistische Wijsbegeerte*. 1969, Nos. 2, 3, 4.

———— *Filosofie en techniek*. Kampen: Kok, 1949.

———— "Over futurologie." *Mededelingen van de Vereniging voor Calvinistische Wijsbegeerte*. 1970, Nos. 3 and 4.

———— *Mondigheid en de machten*. Amsterdam: Buijten & Schipperheijn, 1967.

———— *Roeping en probleem der techniek*. Kampen: Kok, 1952.

———— *The Society of the Future*. Translated by David H. Freeman. Philadelphia: Presbyterian and Reformed Publishing Company, 1957.

———— "The Structure of Technology." Translated by Herbert Donald Morton. In: Durbin, Paul T., and Mitcham, Carl, eds., *Research in Philosophy and Technology*. Volume 2. Greenwich, Conn.: JAI Press, 1979, pp. 296-313.

———— "De techniek." *Mededelingen van de Vereniging voor Calvinistische Wijsbegeerte*. 1966, No. 3.

———— "Technology and Culture." Translated by Herbert Donald Morton. In: Durbin, Paul T., and Mitcham, Carl, eds., *Research in Philosophy and Technology*. Volume 2. Greenwich, Conn.: JAI Press, 1979, pp. 313-28.

———— *Wijsbegeerte*. Kampen: Kok, 1970.

VAN STEENBERGEN, BART. "Het kritische toekomstdenken van Arthur Waskow." *Katernen 2000*. Amersfoort: Werkgroep 2000, 1969, No. 5.

VAN STEENBERGEN, BART, and VAN HENGEL, EDUARD. *Technocratie: Ideologie of werkelijkheid*. Groningen: Wolters, 1971.

WAGNER, FRIEDRICH. *Die Wissenschaft und die gefährdete Welt: Eine Wissenschaftssoziologie der Atomphysik*. Munich: Beck, 1964.

WARNER, MALCOLM, and STONE, MICHAEL. *The Data Bank Society: Organizations, Computers and Social Freedom*. London: George Allen & Unwin, 1970.

WASKOW, ARTHUR I. "Creating the Future in the Present." *The Futurist*. Volume 2 (1968), No. 4, pp. 75f.

———— "Looking Forward: 1999." In: Jungk, Robert, and Galtung, Johan, eds., *Mankind 2000*. London: George Allen & Unwin, 1969.

WETTER, GUSTAV ANDREAS. *Der dialektische Materialismus: Seine Geschichte und sein System in der Sowjetunion*. Freiburg: Verlag Herder, 5th ed., 1960.

WIENER, NORBERT. *Cybernetics; or Control and Communication in the Animal and the Machine*. New York: John Wiley & Sons, 8th ed., 1950.

———— *God and Golem, Inc.: A Comment on Certain Points Where Cybernetics Impinges on Religion*. Cambridge, Mass.: MIT-Press, 1964.

———— *The Human Use of Human Beings: Cybernetics and Society*. Cambridge, Mass.: The Riverside Press, 1950.

———— *Die Organisation der psychischen Funktionen*. Sammelband. Neuchâtel, 1951.

——— "Scientist and Decision Making." In: Greenberger, Martin, ed., *Computers and the World of the Future*. Cambridge, Mass.: MIT-Press, 1964, pp. 21-34.

——— *Selected Papers of Norbert Wiener*. Cambridge, Mass.: MIT-Press, 1964.

——— "Some Moral and Technical Consequences of Automation." *Science*. Volume 131 (1960), pp. 1355-8.

——— *Why I Am a Mathematician*. Garden City, New York: Doubleday & Company, 1956.

WIENER, NORBERT, and SCHADÉ, J. P., eds. *Progress in Brain Research*. Vol. 2: *Nerve, Brain and Memory Models*. Amsterdam, London and New York: Elsevier Publishing Company, 1963.

WIESER, WOLFGANG. *Organismen, Strukturen und Maschinen: Zu einer Lehre vom Organismus*. Frankfurt: Fischer Bücherei, 1959.

WILLMS, BERNARD. *Planungsideologie und revolutionäre Utopie: Die zweifache Flucht in die Zukunft*. Stuttgart: Kohlhammer, 1970.

Wijsgerig perspectief, "Creativiteit." Volume 8 (1967), No. 1.

ZOUTENDIJK, GUUS. "Computer en democratie." *Academia*. February 19, 1971.

——— "Computer en macht." *Wetenschap en samenleving*. 1969, No. 1, pp. 1-8.

ZUIDEMA, SYTSE ULBE. *Communisme in ontbinding*. Wageningen: Zomer & Keuning, 1957.

——— "Cultuur als crisis." *Correspondentiebladen van de Vereniging voor Calvinistische Wijsbegeerte*. Volume 17 (1953), December, pp. 8-20.

——— *De revolutionaire maatschappijkritiek van Herbert Marcuse*. Amsterdam: Buijten & Schipperheijn, 1970.

ZVORIKINE, ANATOLIJ ALEKSEEVIČ. "Technology and the Laws of Its Development." *Technology and Culture*. Volume 3 (1962), No. 4, pp. 443-57.

Index of Persons

422

424

Index of Subjects

absolutization: 79, 312, 330, 333ff.,
357, 361f., 373
– of the autonomy of technology:
172, 174
– of cybernetics: 192, 205, 211,
237, 241, 259, 307, 320
– of the economic aspect: 363
– of freedom: 325ff., 334, 373
– of the individual: 155
– of the concept of information:
202, 205, 240, 320f.
– of information theory: 240f.
– of quantification: 239, 344
– of revolution: 253
– of the social aspect: 308, 362
– of technological power: 326,
373
– of technological-scientific
method: 253, 325, 341
– of technological-scientific
thought: 75ff., 111, 164, 170,
205, 210, 249, 325f., 332, 334,
343, 355, 362
– of technology: 146, 357
absolutized
– freedom: 77, 79, 253, 259f., 328,
370, 374
– (technological) power: 260, 370,
374
– technological-scientific

thought: 75, 78, 80, 249, 328
– technology: 77, 128, 146
actualization: 9f., 12ff., 244, 272,
353
advertising: 31, 140, 356
alienation: xxii, 41, 72, 283ff., 288,
290f., 297, 305, 307, 309, 311,
316f., 322, 324, 375
automation: 7f., 18, 22, 24, 29, 34ff.,
40f., 49, 61, 142, 186, 193, 195,
197, 213, 230, 280f., 283f., 287f.,
302, 347, 349f., 352

belief, faith: 214, 227, 229f., 327f.,
330ff., 365, 375
see also: secularized
– in the autonomy of man: 359
– christian belief, faith: xxiii, 3f.,
55, 80, 115f., 145f., 156f., 161,
212, 230, 242, 248, 318, 327
– in cybernetics and information
theory: 241, 249
– and meaning-disclosure: 335f.,
361, 365, 374
– in nature: 79
– and philosophy: xxiii, 3, 55, 327
– in (technological) progress: xxi,
58, 164, 197, 295, 358
– and science: xxiii, 161, 165, 171,
173, 229ff., 318

– in technological development: 79, 298
– in technology: 249, 259, 359
biotechnology: 166, 168, 171, 318, 353

capitalism: 57f., 284f., 302, 304
– and cybernetics: 260ff., 279, 299f., 306
– and technology: 131, 133, 265, 283f., 287f., 290, 297, 299
capitalistic
– economy: 57
– production relations: 291, 296
– society: 248, 256, 283f., 290f., 322
– system: 266, 283f., 299
Christianity: 162f., 173, 178, 230, 236, 245, 321, 354
see also: culture, future
– and technology: 115f., 129ff., 144ff., 161, 172, 175f., 318, 354
collective: 57f., 65, 69, 74, 77, 152, 173, 285, 288, 308, 310
collectivism: 54, 79, 251, 255, 317, 372
communication: 181f., 188, 190, 202, 204, 210, 271, 273
– dominion-free communication: 256f., 260
– communication process: 18
– science of communication and control: 188f., 196
– system of communication: 189
– communications technology: 182, 186, 202, 215ff., 231f., 234, 237, 250, 358
– communication theory: 182, 185
– as transmission of information: 185f.
communism: 57f., 166, 288, 291, 297f., 310
– and cybernetics: 300, 305f.
– as religion: 143
– and technology: xiv, 164, 167, 283, 291, 297, 322
communist
– society: 256, 261ff., 280, 283ff.,

288, 290, 292, 297, 308, 311, 322f.
community: 139, 151f., 154, 291, 308f., 370, 373
computer: xx, xxi, 18ff., 22, 177, 202, 212, 216ff., 283, 321, 347
– and automation: 22, 49, 186
– computerocracy: 370ff.
– and culture: 19, 344, 347
– dangers of the computer: 371ff.
– and democracy: 250f., 371
– and designing: 48f., 348, 366
– and future: 231ff., 251
– learning machine: 190f., 207f., 219ff., 243f., 246, 270, 274f., 281f., 286f., 300, 308, 320, 367
– and man: 20, 22f., 186ff., 208, 214, 216, 218f., 222, 229, 241ff., 269f., 272f., 308, 319ff., 351, 367
– and quantification: 345ff.
– and science: 22, 344, 347
– structure of the computer: 20ff.
control: 12, 26ff., 40f., 46, 48, 52, 81, 93, 114, 125, 128, 133, 136, 153, 164ff., 178, 205, 209, 241, 252f., 257, 267, 338, 342, 344, 364, 374
see also: cybernetics
– and freedom: 205, 309ff.
– of the future: 52, 69, 178, 204, 214, 231ff., 253f., 309, 322, 337, 350
– of history: 52, 253, 309
– of man: 52f., 64, 128, 138, 148, 164, 166, 168, 205, 253f., 256, 310, 325, 337
– method of control: 126ff., 136ff., 146, 148
– of nature: 52, 64, 91, 164, 203, 256f., 311, 325, 337
– science of control: 241
– of society: 52f., 178, 203, 205, 253, 256f., 292, 309, 323, 325, 337
– of technology: 132, 168, 254, 289, 291, 316, 358, 367
creation: xxiii, 104, 162, 171, 191f., 210, 327ff., 351ff., 359ff., 364, 370, 374ff.

creativity: 37, 45, 49, 64, 112, 132, 170, 226, 243, 252, 285, 294, 297, 300, 304, 340, 355, 373

culture: xix, 1, 55, 60, 125, 127, 147, 157, 164, 166, 168, 173, 198, 211, 231, 251, 288, 318, 326, 328, 333f., 339, 343, 353, 374, 376
see also: computer, secularized
– and Christianity: 162f., 169, 173, 354
– cultural crisis: xxii, 2, 123, 125, 127, 158, 161, 166, 170, 174
– cultural progress: xxi, 177, 231, 314
– and nature: 28, 185, 321
– problems of our culture: xx, 125, 314
– technological culture: xv, 123, 125, 129, 132, 317, 324, 326
– and technology: xix, xx, xxi, 1ff., 8, 19, 42, 55, 60, 74, 111, 124, 126f., 129, 132, 136f., 146, 150, 161, 170, 173, 177, 198, 231, 237, 267, 306, 317f., 324, 351ff., 358, 363
– world culture: 153, 210, 324, 367

cybernetics: 71, 140, 181f., 213ff., 247, 267f., 290, 306, 314, 319ff., 364
see also: absolutization, belief, capitalism, communism
– and control: 12, 267
– and control of man: 205
– and control of the future: 203, 231, 310, 322
– and control of society: 205, 267, 309ff., 323
– and dialectical materialism: 261f., 264, 266f., 275f., 278, 290, 294, 298, 305ff.
– and freedom: 206, 209, 246, 298, 305, 309ff., 322
– and the future: xxi, 178, 180, 193ff., 209ff., 247, 298
– and man: 187f., 209, 216, 223, 242, 246, 267
– method of cybernetics: 180, 188, 206, 247, 276ff., 314, 325
– and philosophy: 180, 201, 237, 261, 266f., 271, 298f.
– as philosophy: 188ff., 201ff., 228, 241, 319
– principles of cybernetics: xxii, 30, 71, 178, 180, 182ff., 188, 200, 214, 216, 244, 290, 321
– and sciences: 71, 178, 188, 201f., 204f., 215f., 241, 247, 266ff., 294, 306, 321
– as science of guidance, communications and control: xxi, 182, 189, 196, 216, 237, 241
– and society: 298ff., 306f., 312
– cybernetic state: 234, 259, 321f.
– and technology: 34, 182, 188, 201, 215f., 237, 267, 305, 319
– and technological science: 182, 188, 201

deliverance: xvii, xxiii, 53, 55, 57, 68ff., 78, 94ff., 102, 108ff., 117f., 121ff., 167, 170, 259, 302, 317, 368

democracy: 137, 141, 143, 154, 233ff., 237, 249ff., 256, 258, 292, 310f., 321ff., 369ff.

design: 11f., 24f., 27f., 37ff., 43ff., 84ff., 103, 117, 252, 362

designing: 7, 11f., 14, 24f., 37ff., 45, 48f., 64, 113, 128, 148ff., 217, 340f., 344f., 347f., 362, 366

destination: 9f., 13, 15f., 18, 22, 26f., 33f., 37

destinational function: 10f., 13ff., 17, 24, 32, 332

dialectic: 121ff., 263, 276, 282, 295, 333ff., 337, 339
– of behavior and structure: 264, 278
– of Being and man: 109, 122, 124
– of freedom and (technological) power: 79, 157f., 174, 176, 260, 326, 328
– of freedom and necessity: 157, 265, 305, 309
– of man and machine: 287f.

religion: 3, 65, 79f., 107, 126, 129, 135, 141, 143, 145ff., 155ff., 164f., 168, 176, 178, 189f., 209, 295, 311, 320, 324, 327, 330ff., 339, 359, 368, 370, 374

Renaissance: 130, 162, 169, 172f.

responsibility: 12, 24, 35f., 41, 49, 59f., 62, 77, 80, 96, 119, 124, 129, 137, 140, 149, 151ff., 158, 168f., 195, 199, 206ff., 245f., 253f., 259, 301, 308, 311, 315, 318, 320f., 323, 327, 329, 331ff., 335ff., 343, 350f., 357, 359, 361f., 365f., 368ff., 372f., 375

revolution: 214, 235ff., 250ff., 257ff., 263, 291, 296, 352, 368f.

science: 25ff., 53, 62, 65f., 70f., 79, 83, 86, 89, 92f., 96, 101, 108, 116, 118, 125, 148ff., 155, 160ff., 172, 178, 180, 188f., 200ff., 205f., 214, 227ff., 235f., 238, 241, 243, 245f., 248f., 251f., 254ff., 274, 294, 298, 301, 303f., 320f., 325, 327, 336ff., 347f.

see also: belief, communication, computer, control, cybernetics, future, technological
- application of science: 76, 92, 118f., 128, 237, 315, 338, 340ff., 344
- autonomous science: 76, 150f., 153, 155, 161, 200, 251, 257
- and freedom: 205, 208, 245, 255, 304f., 338f.
- and technology: xx, xxii, 1ff., 7f., 11f., 22, 26f., 34, 37, 42, 49, 53, 64, 66, 73f., 76, 91f., 101, 103, 106, 108f., 111ff., 118f., 126, 128, 130, 132, 135, 146, 148ff., 154, 160, 165, 171, 173f., 237, 317, 325, 342, 352f., 355

scientific method: 7, 25ff., 31, 37, 40, 45, 106, 132, 148ff., 157f., 162f., 168, 170, 201, 253, 315, 318, 337f., 340ff., 344

secularization: xxii, 146, 152, 156, 173, 178, 319, 353

secularized
- belief in salvation: 360
- culture: 318, 326, 361
- motives: xiv, xxiii, 328, 354, 359f., 375
- worldview: 311, 329, 354

skill: 5f., 24, 34f., 76, 263, 286, 349, 351

social engineering: 129, 133, 318

society: 129, 137, 142, 154, 164, 173, 188, 204f., 215, 227, 233ff., 247ff., 253, 256f., 261, 263ff., 267, 283ff., 288ff., 296ff., 300ff., 306ff., 322f., 358

see also: absolutization, capitalistic, communist, control, cybernetics, dialectic, future, technological
- and technology: 131f., 166, 178, 204f., 231, 247, 261, 279f., 317, 321, 323, 352, 368, 372

specialization: 35, 39, 62, 107

standardization: 33, 35f., 64, 128, 306

state: 64, 101, 126f., 131, 136ff., 140, 153, 164, 255, 282, 302, 306, 309f., 314, 323, 367

see also: cybernetics, technological
- world state: 153, 234, 249, 259, 321, 367

subject function: 9ff., 23, 46, 244, 272

synthetics: 6, 28, 112, 348, 351

technicism: 53, 178, 180, 189, 205f., 325, 342, 358

technocracy: 53ff., 64f., 77, 126, 137, 146, 164, 174, 178f., 236, 253, 256ff., 294, 297f., 302, 307ff., 314, 316, 318, 323ff., 344, 354, 359, 367f., 370, 373

technological
- development: xiv, xix, xx, xxii, xxiii, 1ff., 5, 8ff., 19, 22, 25, 36f., 39, 41f., 44, 48ff., 52ff., 62f., 65ff., 72ff., 96, 101, 106ff., 119, 121ff., 127ff., 136, 138, 141ff.,

Illinois, Indiana & Ohio

Published by:
AAA Publishing
1000 AAA Drive
Heathrow, FL 32746-5063
Copyright AAA 2001

Send Written Comments to:
AAA Member Comments
1000 AAA Drive, Box 61
Heathrow, FL 32746-5063

**Advertising Rate and Circulation
Information**
Call: (407) 444-8280

Printed in the USA by Quebecor
World, Buffalo, NY

◆ ◆

Illinois, Indiana & Ohio

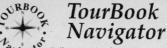

TourBook Navigator
Follow our simple guide to
make the most of this member benefit 7-23

Comprehensive City Index
Alphabetical list for the entire book 822

■ *Ohio*

Featured Information

The sign it's time to pull off the road.